Ultrasonic Waves in Solid Media

Ultrasonic wave techniques are used increasingly in areas ranging from nondestructive inspection of materials to medical diagnosis. These modern techniques evolved from basic physical principles of wave mechanics.

This book brings together basic physics and modern applications. It explains the physical principles of wave propagation and then relates them to ultrasonic wave mechanics and the more recent guided wave techniques used to inspect and evaluate aircraft, power plants, and pipelines in chemical processing plants.

Mechanics, mathematics, and modeling are stressed throughout the book, establishing the framework for practical applications. Among topics covered are wave propagation in plates, rods, hollow cylinders, and multiple layers in solid and composite materials; reflection and refraction; surface and subsurface waves; and horizontal shear wave propagation. Appendices provide background information on ultrasonic nondestructive testing, elasticity theory, complex variables, and key wave propagation experiments. The text is amplified with numerous examples, laboratory experiments, and homework exercises.

Graduate students, researchers, and practicing engineers will find *Ultrasonic Waves in Solid Media* an invaluable reference to this active field.

Joseph L. Rose is the Paul Morrow Professor in Design and Manufacturing in the Engineering Science and Mechanics Department at The Pennsylvania State University.

Ultrasonic Waves in Solid Media

JOSEPH L. ROSE

The Pennsylvania State University

CAMBRIDGE
UNIVERSITY PRESS

PUBLISHED BY THE PRESS SYNDICATE OF THE UNIVERSITY OF CAMBRIDGE
The Pitt Building, Trumpington Street, Cambridge, United Kingdom

CAMBRIDGE UNIVERSITY PRESS
The Edinburgh Building, Cambridge CB2 2RU, UK
40 West 20th Street, New York NY 10011–4211, USA
477 Williamstown Road, Port Melbourne, VIC 3207, Australia
Ruiz de Alarcón 13, 28014 Madrid, Spain
Dock House, The Waterfront, Cape Town 8001, South Africa

http://www.cambridge.org

First published 1999
First paperback edition 2004

Typeface Times 10.5/13 pt.   *System* AMS-T$_{\text{E}}$X   [FH]

A catalogue record for this book is available from the British Library

ISBN 0 521 64043 1 hardback
ISBN 0 521 54889 6 paperback

To Olek

Contents

Preface

As a student of wave mechanics for over 30 years, I have been fascinated by the various attributes of dynamics, vibration, and especially wave propagation. An example of my fascination is how the striking or rubbing of a taut string between two supports (as with a piano, guitar, or violin) can reveal the change in wave velocity as the tension increases and so leads to changes in pitch or frequency. Higher pitch is a consequence of increased tension for a given length of string. This observation demonstrates pulse propagation changes and resulting standing waves from one end to the other. Similar wave propagation phenomena occur in structures constructed from metals, plastics, ceramics, and composite materials; here, however, the wave velocity values are typically much higher and the motions are more or less invisible to the naked eye.

I have also been interested in safety and improved diagnostics utilizing wave propagation concepts. Wave phenomena can be used to evaluate material properties nondestructively as well as to locate and measure defects in critical structures. This work has led to devices that have become valuable quality control tools and/or in-service inspection procedures for structures whose integrity is vital to the public safety, such as critical aircraft and nuclear power components.

Early work (1970 to 1985) in ultrasonic nondestructive evaluation – beyond basic pulse-echo and through-transmission testing – focused on signal processing and pattern recognition. New tomographic ultrasonic imaging procedures were developed that employed special features to assist in defect classification; these procedures supplemented or replaced the standard amplitude-based "C" scanning. In the late 1970s, ultrasonic research was extended to medical applications. I explored linear phased array transducer systems used in real-time medical imaging, and I became involved with tissue classification: differentiating malignant from benign growth.

Around 1985, a newer vision was conceived of faster and more sensitive ultrasonic examination. Some pioneering work on oblique incidence for adhesive bond inspection was responsible for this new direction. Further research revealed that guided waves – waves that travel along a surface or along a rod, tube, or platelike structure – could not only produce the same kind of two-dimensional particle velocity as that in oblique incidence but could also be much more efficient than the traditional technique of point-by-point examination.

These guided wave efforts continue today. Guided wave concepts have been extended to help examine the tubing in power plants, the pipelines in chemical processing facilities, and the safety aspects of large petroleum and gas pipelines. Because of special test modes, guided wave techniques can be used to find tiny defects – over large distances, under adverse conditions, in structures with insulation and coatings, and in harsh environments.

For example, this concept is incorporated in a wing ice detection program for commercial aircraft. Beyond the development and current use of guided wave techniques for aircraft and power-plant inspection, future directions point to research and applications of guided waves in America's aging infrastructure and in many other new commercial efforts.

Engineers, technicians, and students involved in ultrasonic nondestructive evaluation (NDE) will appreciate the usefulness of this textbook. Managers of such evaluation and testing will be able to develop an understanding of what the engineers, technicians, and students are talking about. Even though the mathematics is sometimes detailed and sophisticated, the treatment can often be considered from a "black box" point of view. An understanding of the appropriate input parameters for data acquisition can provide reasonable output to evaluate phase velocity, dispersion curves, group velocity curves, and wave structure analysis.

Overall, the material presented here in wave mechanics – and, in particular, guided wave mechanics – establishes a framework for the creative data collection and signal processing needed to solve many problems in ultrasonic nondestructive evaluation. I therefore hope that this book will be used as a reference in ultrasonic NDE by individuals at any level and as a textbook for seniors and graduate students. It is also hoped that this book will help expand and promote guided wave technology on both a national and international level.

Acknowledgments

Thanks are given to many individuals over the past ten years for their work efforts, discussions, and contributions in wave mechanics. A special tribute is made to Dr. Aleksander Pilarski, who passed away on January 6, 1994. "Olek" worked with me as visiting Professor at Drexel University and at Penn State University from 1986 to 1988 and from 1992 to 1994. His energetic and enthusiastic style, as well as his technological contributions, had a strong influence on many of us. He was a dear friend whose memory will remain forever.

Special thanks are given to John Ditri, a superior doctoral student, for many contributions and in particular to Chapters 13, 14, 15, and 19. Thanks are also given to Sam Pelts for many contributions, especially to Chapters 6, 9, 11, 14, and 18. Appreciation is also expressed to Mike Avioli for work on complex variables (Appendix C) and the graphics. Thanks are also given to Dale Jiao for contributions in Chapter 5 and to Younho Cho for work on Chapters 12 and 20. I thank Hyeon Jae Shin, Kris Rajana, Derrick Hongerholt, Luis Soley, Jian Li, Ken Lohr, Mike Quarry, and Jim Barshinger for their support.

I also thank Jaimie Rose for her superb editing efforts, and Ann Hibbert for her patience and excellent organizational and typing skills.

Thanks are also given to all of my Ph.D. students over the past 28 years for their work efforts and support in the field of ultrasonics, which have played an important role in our journey and in the integration of signal processing and pattern recognition with wave mechanics. Thanks also to Penn State University and for all who have funded my research efforts over the years, which made development of this material possible. Final thanks go to my Mom and Dad, son Joe, daughters Debbie, Terry, Jaimie, and Kristina, and especially to my wife, Carole, for their support of my endeavors for decades.

1

Introduction

1.1 Background

Ultrasonic waves in solid media have become a critically important subject in nondestructive evaluation (NDE). New ways of looking at materials and structures have become possible that are faster, more sensitive, and more economical than previous ultrasonic techniques. For example, rather than inspecting an insulated pipe by removing insulation and using a single probe to check along the pipe with thousands of waveforms, one can now use a guided wave probe at a single location, leave the insulation intact, and perhaps inspect the entire pipe by examining just one waveform. An understanding of ultrasonic waves in solid media is necessary to make use of this technique (an example of mode and frequency choice in guided wave analysis of multilayer cylinders).

Wave propagation studies are not limited to ultrasonic nondestructive evaluation. Three major areas of study in elastic-wave analysis are:

(1) transient response problems, including dynamic impact loading;
(2) stress waves as a tool for studying mechanical properties, such as the modulus of elasticity and other constants in constitutive equations (the formulas relating stress with strain and/or strain rate can be computed from the values obtained in various wave propagation experiments); and
(3) industrial and medical ultrasonics or acoustic-emission nondestructive testing analysis.

A study of wave propagation can also aid in the understanding of vibration analysis. Problems in wave propagation – in a limit process as time values become large – approach the various solution techniques available in vibration analysis. Elastic-wave propagation theory handles both transient response and the steady-state character of vibration problems. Vibration solution techniques, on the other hand, cannot generally deal effectively with problems in wave propagation.

Historically, the study of wave propagation has been of tremendous interest to investigators in the area of mechanics. Early work was carried out by such famous individuals as Stokes, Poisson, Rayleigh, Navier, Hopkinson, Pochhammer, Lamb, Love, Davies, Mindlin, Viktorov, Graff, Miklowitz, Auld, and Achenbach. Graff (1991) presents an interesting history; for some history and basics of wave propagation, see also Achenbach (1984), Auld (1990), Beranek (1990), Eringen and Suhubi (1975), Ewing, Jardetsky, and Press (1957), Fedorov (1968), Kino (1987), Kinsler et al. (1982), Kolsky (1963), Miklowitz (1978), Musgrave (1970), Pollard (1977), Rayleigh (1945), Redwood (1960), and Viktorov

(1967). Many other interesting and useful references are cited throughout this text; see the Bibliography for a detailed listing.

Recent technology transfer from a number of areas has led to a re-examination of wave propagation literature and to many modifications and new developments for practical use. Computer programs are now available for solving all kinds of problems in engineering and science. The field of ultrasonic nondestructive testing, as an example, has grown rapidly over the last three decades, helped by advances in finite element and boundary element techniques. These sophisticated mathematical treatments of boundary value problems – together with improved computational power and efficiency – can assist in solving bulk and guided wave propagation problems. Yet even though countless computer packages are available, it is essential to interpret the results intelligently and to use them in a practical engineering environment. Scientists and engineers are needed to write the programs, develop physical models of new situations, and use a variety of techniques to handle new boundary value problems in wave mechanics and material characterizations.

1.2 Text Preview

Although there are numerous applications of ultrasonic waves in solid media, the focus throughout this text will be on ultrasonic NDE, with emphasis on material characterization analysis. We begin with a discussion in Chapter 2 on dispersion principles. Basic dispersion concepts are encountered whenever wave velocity becomes a function of frequency or angle of propagation. The more traditional work in ultrasonic nondestructive testing employs *non*dispersive wave propagation concepts, but dispersion phenomena become more pronounced with anisotropic media and guided wave technology.

Chapter 3 describes wave propagation in unbounded isotropic and anisotropic media. The Christoffel equations are reviewed in detail to show the steps required to study waves in anisotropic media. The wave velocity is no longer independent of angle and in fact often changes strongly with angle. As a result, interference phenomena change drastically, affecting the group velocity of waves in different directions as well as producing skew angle effects that occur as the waves propagate through the material. Detailed mathematical treatment and sample problems are presented.

Reflection and refraction are examined in Chapter 4. The initial emphasis is on isotropic media, followed by Snell's law and mode conversion. A variety of different models and boundary conditions are used to tackle wave propagation problems in various structures and anisotropic media, which will be discussed in later chapters.

Chapter 5 treats the more general problem of reflection and refraction for oblique incidence, including study of slowness profiles and critical angle analysis. A brief treatment of wave scattering principles in solid media is given in Chapter 6. Chapter 7 covers surface and subsurface waves. The famous Rayleigh surface wave problem is studied in detail, along with subsurface waves. These bulk waves have unusual propagation characteristics that can be of great value in ultrasonic nondestructive evaluation. A mathematical and practical treatment is presented.

Chapter 9 deals with interface waves, including Love, Stoneley, and Scholte waves. As waves propagate through a material to encounter an interface between two materials, certain physical conditions and properties can allow these waves to travel directly along the interface. Wave propagation characteristics and leaky wave phenomena are discussed theoretically and with respect to using this information for nondestructive evaluation.

Chapter 10 looks at a layer on a half-space, which can be used to handle a number of density gradient, property gradient, and anisotropic media composite material problems. One can use the dispersive character of surface waves to estimate property gradients with depth.

Chapters 8, 11, and 12 deal with waves in plates, rods, and hollow cylinders, respectively. The results are dispersive in nature, and the assumed displacement solutions lead to determinants whose solutions lead to the dispersion curves of phase, group velocity, and attenuation versus frequency. Much time is spent discussing interpretation techniques, mode choice, and how the information might be used in nondestructive evaluation.

Guided waves in multiple layers is the subject of Chapter 13. This is a very useful modeling technique in ultrasonic NDE, where one can model adhesive bonding, diffusion bonding, or composite material situations. Again, the higher-order determinants are found and the root extraction techniques are discussed in detail. Many sample results are included to illustrate how these techniques can be applied to the material characterization aspects of nondestructive evaluation.

Chapter 14 looks at a source influence phenomena associated with guided waves. The source influence accompanies transducer loading and can affect control of phase velocity and frequency spectrum when generating guided waves. This is a critical and often overlooked circumstance that allows us to generate experimentally the modes of interest with a significant improvement in signal-to-noise ratio.

Horizontal shear wave propagation is discussed in Chapter 15. This mode of wave propagation – where the particle velocity is perpendicular to the wave velocity vector – achieves some very interesting effects with respect to waves traveling over rough surfaces (across welds, etc.). Again, the dispersion curves are generated by solving the determinants associated with this problem; physical interpretations are studied as well.

In Chapter 16 we examine wave propagation in an anisotropic layer and find that the dispersion curve changes quite drastically with direction in that layer. It is shown, however, that this complexity can become a benefit with respect to material characterization in NDE. Once again, the determinant is set up, the roots are extracted, and sample problems are presented.

Chapter 17 deals with elastic constant determination based on a series of wave velocity measurements. Using different bulk and guided wave measurements, the inverse problem is studied in order to calculate the anisotropic elastic constants of a material. This work is of great value to both the design engineer and the nondestructive evaluation engineer.

Wave propagation in viscoelastic media is studied in Chapter 18. The basic principles are presented along with a sample problem of wave propagation in a viscoelastic layer. The phase, group, and attenuation-versus-frequency dispersion curves are generated by extracting the complex constants from the determinant. Physical interpretations are presented. In this case, the phase velocity, group velocity, and attenuation dispersion curves effectively show the viscoelastic influence. Chapter 19 deals with stress influence on wave propagation. We discuss various nonlinear influences with respect to wave propagation in a structure as well as the possibility of determining the stress state of a material.

Chapter 20 is on boundary element methods – in particular, dealing with guided wave propagation. It is found that the boundary element technique is computationally efficient and can treat very nicely the reflection and transmission factor of guided wave modes impinging on various shapes and defects inside a structure. This provides valuable assistance in ultrasonic nondestructive evaluation with respect to data acquisition and analysis required for defect detection, classification, and sizing.

The book concludes with five appendices. Appendix A briefly considers some background and fundamental concepts to assist and motivate us in our journey through the world of the physics and mechanics of wave propagation. Basic material is presented on ultrasonic nondestructive evaluation, state-of-the-art practice, analysis, and instrumentation. Also treated are such items as interference phenomena, ultrasonic field analysis, near field, beam divergence, axial and lateral resolution, and display technology. Appendix B includes basic formulas in the theory of elasticity. It is recommended that the student have some knowledge of elasticity before tackling the subject of wave propagation, but a reasonable background in strength of materials enables one to use the formulas of Appendix B in a reasonably efficient way to move forward with the study of wave propagation in solids. Appendix C includes some basic formulas in complex variables, an absolutely essential topic for understanding wave propagation in solid media. This appendix is not included to replace a textbook analysis of complex variables but rather to point out basic highlights – so students can review the material quickly or learn the basic elements if necessary.

Appendix D, on Schlieren imaging and dynamic photoelasticity, is included to show how these waves might be visualized in various materials in order to demonstrate propagation principles discussed throughout the textbook. Appendix E reviews key wave propagation experiments. This is probably one of the highlights of the textbook for students, since many essential principles are illustrated here with some sample wave propagation experiments and recommendations for further study. Such topics include basic bulk and guided wave velocity measurements, refraction and the use of angle beam transducers, skew angle measurement in composite materials, and wave propagation characteristics in plates, rods, and tubes. An experiment on horizontal shear wave propagation, generation, and reception is also included.

Exercises are included at the end of each chapter and appendix in order to highlight and summarize some key elements of the text. They are included to assist both instructor and student.

2

Dispersion Principles

2.1 Introduction

Before studying stress wave propagation in such structures as solid rods, bars, plates, hollow cylinders, or multiple layers, it is useful and interesting to review some applicable concepts taken from studies of waves and vibrations in strings. Wave propagation essentials and dispersion principles can be introduced for strings and then carried over to the wave propagation studies in many structures. Even though wave dispersion can be considered for anisotropic media (where wave velocity is a function of direction), the emphasis in this chapter is on dispersion due to structural geometry. Some basic terms are introduced, including wave velocity, wavenumber, wavelength, material and geometrical dispersion, phase velocity, group velocity, attenuation, cutoff frequency, frequency spectrum, and energy transmission, all of which will be useful in further studies. Graphical interpretations and analysis of phase and group velocity are also covered in this chapter. See Graff (1991) for more details.

2.2 Waves in a Taut String

2.2.1 Governing Wave Equation

In order to derive a governing wave equation for wave propagation in a string, consider a differential string element as shown in Figure 2-1. Let's initially consider infinite length; later, we'll consider specific boundary conditions. Displacement due to tension F will also be assumed as negligible. Recall that

$$u_{,xx} = \frac{\partial^2 u}{\partial x^2}, \quad \ddot{u} = \frac{\partial^2 u}{\partial t^2}, \quad \text{and} \quad u' = \frac{\partial u}{\partial x}.$$

The equation of motion in the u direction, following Newton's second law $\left(\sum F = ma\right)$, is as follows:

$$-\text{F}\sin\theta + \text{F}\sin\left(\theta + \frac{\partial\theta}{\partial x}\,dx\right) + q\,ds = \rho\,ds\,\ddot{u}. \tag{2.1}$$

We assume small deflections ($ds \approx dx$, $\sin\theta \approx \theta$, and $\theta \cong \partial u/\partial x$) and so obtain

$$-\text{F}\theta + \text{F}\left(\theta + \frac{\partial\theta}{\partial x}\,dx\right) + q\,dx = \rho\,dx\,\ddot{u};$$

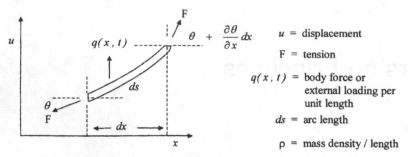

u = displacement

F = tension

$q(x, t)$ = body force or external loading per unit length

ds = arc length

ρ = mass density / length

Figure 2-1. Differential element of a taut string.

as a result,

$$F\frac{\partial^2 u}{dx^2} + q = \rho\ddot{u}. \tag{2.2}$$

This is a second-order hyperbolic partial differential equation that is homogeneous if $q = 0$. Without external forcing we have

$$u_{,xx} = \frac{1}{c_0^2}\ddot{u}, \quad \text{where} \quad c_0 = \sqrt{F/\rho}. \tag{2.3}$$

This is a one-dimensional, homogeneous, simple wave equation. Here c_0, which arises as a natural consequence of the mathematical solution, denotes the velocity of wave propagation.

2.2.2 Solution by Separation of Variables

Let $u(x, t) = X(x)T(t)$. Substituting this into (2.3) yields

$$\frac{\partial^2 u}{\partial x^2} = X''T, \quad \frac{\partial^2 u}{\partial T^2} = XT'', \quad \text{and} \quad TX'' = \frac{1}{c_0^2}XT'',$$

where $X = X(x)$ and $T = T(t)$. This simplifies as

$$\frac{X''}{X} = \frac{T''}{c_0^2 T} = \text{const} = -k^2. \tag{2.4}$$

Note that the constant in (2.4) must equal $-k^2$ in order to guarantee a solution. Consequently,

$$X'' + k^2 X = 0 \quad \text{and} \quad T'' + k^2 c_0^2 T = 0.$$

These are two simple equations of harmonic motion, about which a great deal is known. For example,

$$X(x) = A_1 \sin kx + A_2 \cos kx, \qquad T(t) = A_3 \sin kc_0 t + A_4 \cos kc_0 t.$$

In general, choosing $+k^2$ would produce sinh and cosh (hyperbolic) solutions, which are often used for guided waves.

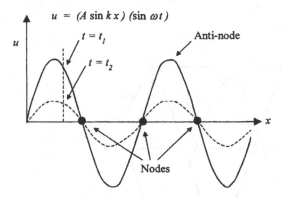

Figure 2-2. Standing or stationary wave example.

These results may be interpreted as follows. We first present the relationships between wave number k, phase velocity c_p, circular frequency ω, and wavelength λ:

$$k = \frac{2\pi}{\lambda} = \frac{\omega}{c_p}, \quad \text{where} \quad \omega = 2\pi f \text{ and } c_p = f\lambda. \tag{2.5}$$

Since $\omega = kc_0$, for the special case of $c_p = c_0$ we may write the the general solution $u(x, t) = X(x)T(t)$ as

$$u(x, t) = (A_1 \sin kx + A_2 \cos kx)(A_3 \sin \omega t + A_4 \cos \omega t).$$

Regrouping and multiplying, we obtain

$$\begin{aligned} u(x, t) = {} & A_1 A_4 \sin kx \cos \omega t + A_2 A_3 \cos kx \sin \omega t \\ & + A_2 A_4 \cos kx \cos \omega t + A_1 A_3 \sin kx \sin \omega t. \end{aligned} \tag{2.6}$$

Consider any of the four terms in (2.6) and examine the illustration of Figure 2-2. As shown in the figure, there is no shifting of the waveform and hence standing or stationary waves result.

By using such trigonometric identities as $\sin(\alpha + \beta) = \sin \alpha \cos \beta + \cos \alpha \sin \beta$, equation (2.6) becomes

$$\begin{aligned} u(x, t) = {} & B_1 \sin(kx + \omega t) + B_2 \sin(kx - \omega t) \\ & + B_3 \cos(kx + \omega t) - B_4 \cos(kx - \omega t). \end{aligned} \tag{2.7}$$

Consider a typical version of the four terms in (2.7): $u(x, t) = A \cos(kx - \omega t)$. Because $\omega = kc_0$,

$$u(x, t) = A \cos k(x - c_0 t). \tag{2.8}$$

This is a typical term showing wave propagation in the positive direction of x for a particular wavelength λ. Since $k = 2\pi/\lambda$, we could then plot $u(x)$ for various t values. The phase angle is $\phi = kx - \omega t$ or $k(x - c_0 t)$, the argument of the cosine function. Note that, as time increases, x also must increase by $c_0 * \Delta t$ in order to maintain a constant ϕ. Since $x = c_0 * \Delta t$, we have:

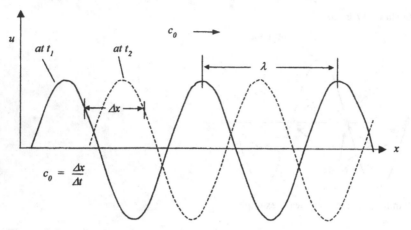

Figure 2-3. A disturbance propagation at constant phase velocity.

(a) $f(x - c_0 t)$ is a right-traveling wave, since x must be positive;
(b) $f(x + c_0 t)$ is a left-traveling wave, since x must be negative.

In order to introduce c_p, the phase velocity, we now consider waves traveling with constant phase. The physical explanation is as follows. For (2.8), note that t increases as x increases for constant phase ϕ, and that the propagation velocity of constant phase is c_0. See Figure 2-3, where $\Delta t = t_2 - t_1$.

The exponential representation of the solution equivalent to (2.6) is

$$u = A_1 e^{i(kx+\omega t)} + B_1 e^{-i(kx-\omega t)},$$

which may be simplified as $u = (A_1 e^{ikx} + B_1 e^{-ikx})e^{i\omega t}$. This can be shown using Euler's formula $e^{i\theta} = \cos\theta + i\sin\theta$ (see Appendix C).

2.2.3 D'Alembert's Solution

We will now discuss an alternative approach to equation (2.3). The D'Alembert solution to the simple wave equation, which was considered as early as 1750, is

$$u(x, t) = f(x - ct) + g(x + ct). \tag{2.9}$$

Equation (2.9) satisfies (2.3) for any arbitrary function f and g, as long as the initial and boundary conditions can eventually be satisfied. Alternatively, let $\xi = x - ct$ and $\eta = x + ct$. Equation (2.3) then becomes $\partial^2 u/\partial\xi \partial n = 0$ and hence has the same solution. These waves propagate without distortion, and f and g are arbitrary functions; see Figure 2-4.

2.2.4 Initial Value Considerations

Consider now the following initial value problem:

$$u(x, 0) = U(x), \qquad \dot{u}(x, 0) = V(x).$$

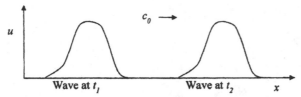

a. Example of wave traveling without distortion

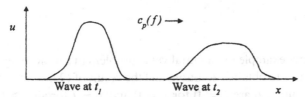

b. Example of wave envelope traveling with distortion

Figure 2-4. Wave motion possibilities.

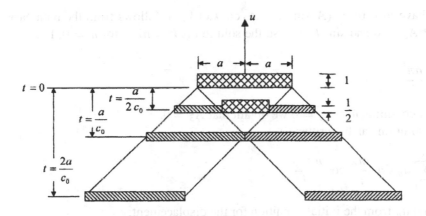

Since $x = c_0 t$, the arrival time at $x = a/2$ would be $t = a/2 c_0$

Figure 2-5. Propagation of an initial condition displacement in a string.

We can substitute these expressions into D'Alembert's solution, (2.9), to obtain

$$f(x) + g(x) = U(x) \quad \text{(since } t = 0\text{)};$$
$$-c_0 f'(x) + c_0 g'(x) = V(x).$$

Because $u(x, t) = f(x - c_0 t) + g(x + c_0 t)$, we could proceed to find f and g.

We will now discuss the propagation of an initial condition displacement in a string, where $u(x, 0) = U(x) = +1$ over the region $-a < x < a$ and $U(x) = 0$ for $x > a$. The solution,

$$u(x, t) = \tfrac{1}{2} U(x - c_0 t) + \tfrac{1}{2} U(x + c_0 t),$$

is shown in Figure 2-5.

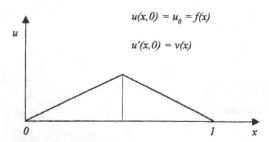

Figure 2-6. An initial value problem.

Figure 2-6 depicts a more general example of an initial value problem in a finite-length string, one that utilizes Fourier series to obtain a solution, and the result of (2.6). Boundary conditions for the string in Figure 2-6 are $u = 0$ for $x = 0$ and $x = l$ for all $t \geq 0$. The string is fastened at $x = 0$ and l. Consider the general solution given $\omega = kc_0$:

$$u(x, t) = (A_1 \sin kx + A_2 \cos kx)(A_3 \sin \omega t + A_4 \cos \omega t).$$

At $t = 0$, we have $u(x, 0) = (A_1 \sin kx + A_2 \cos kx)A_4$. It follows from the boundary conditions that $A_2 = 0$ and $\sin kl = 0$, so the solution is $k = n\pi/l$ for $n = 0, 1, 2, \ldots$ and hence

$$\omega = kc_0 = \frac{n\pi}{l}c_0.$$

From the initial condition $\dot{u}(x, 0) = 0$ we obtain that $A_3 = 0$.

Finally, the solution can be written as

$$u(x, t) = \sum_{n-1}^{\infty} a_n \sin \frac{n\pi x}{l} \cos \frac{n\pi c_0 t}{l}.$$

We can now find a_n from the initial condition for the displacement:

$$\sum_{n=1}^{\infty} a_n \sin \frac{n\pi x}{l} = f(x).$$

From the Fourier series analysis, the coefficients can be obtained as follows:

$$a_n = \frac{2}{l} \int_0^l f(x) \cdot \sin \frac{n\pi x}{l} \, dx \quad (n = 1, 2, \ldots).$$

This completes the solution.

2.2.5 Characteristic Line Concepts

Recall that $\xi = x - c_0 t$ is a straight line in the (x, t)-plane and that $f(x - c_0 t) = f(\xi) =$ constant along this line. Similarly, $\eta = x + c_0 t$ is a straight line but opposite in slope; $g(x + c_0 t)$ is also constant. These lines are called *characteristic* lines. Characteristic lines are useful in many areas of study of wave propagation in solids but are introduced here only to obtain a physical feeling of the wave propagation process in a structure.

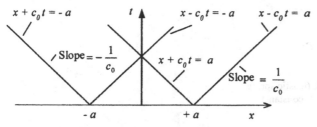

Figure 2-7. Lines showing position of wavefronts in (x, t)-domain (for problem illustrated in Figure 2-5).

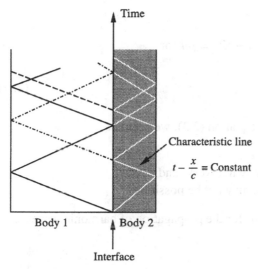

Figure 2-8. Characteristic lines when one body impacts another (considering just one wave type; no mode conversion).

Characteristic lines are featured in the general theory of hyperbolic partial differential equations, where they are in general curved lines. Straight lines only are considered in Figure 2-7, where slopes and y-intercept points are used to derive equations of the straight lines.

Let's now move our attention to a different kind of problem, still utilizing characteristic lines. Visualize one body impacting another and consider the wave propagation back and forth inside each structure, as illustrated in Figure 2-7. Each impingement on each interface produces new reflected and refracted waves. Wavefront travel is illustrated in the x-versus-t profile in Figure 2-8.

2.3 String on an Elastic Base

Consider now a different problem that might involve wave distortions: a string on an elastic base (see Figure 2-9). The relationship between force q and displacement is

$$q(x, t) = -Ku(x, t). \tag{2.10}$$

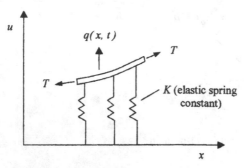

Figure 2-9. Differential element of a string on an elastic base.

Substituting (2.10) into (2.2), we obtain $u,_{xx} - Ku = \rho\ddot{u}$, or

$$u,_{xx} - \frac{K}{F}u = \frac{1}{c_0^2}\ddot{u}, \quad \text{where} \quad c_0 = \sqrt{\frac{F}{\rho}}. \tag{2.11}$$

Comparing (2.11) with the simple wave equation (2.3), we see that

(1) there is an extra term;
(2) a solution of the form $f(x - c_0 t)$ may not work; and
(3) wave propagation with no distortion may not be possible.

We may now examine possible conditions for the propagation of harmonic waves. Assume that

$$u(x, t) = Ae^{i(kx - \omega t)}. \tag{2.12}$$

Substituting (2.12) into (2.11), we have

$$\frac{\partial u}{\partial x} = Ae^{i(kx - \omega t)}(ik),$$

$$\frac{\partial^2 u}{\partial x^2} = Ae^{i(kx - \omega t)}(ik)(ik) = -Ae^{i(kx - \omega t)}k^2,$$

$$\frac{\partial u}{\partial t} = Ae^{i(kx - \omega t)}(-i\omega),$$

$$\frac{\partial^2 u}{\partial t^2} = Ae^{i(kx - \omega t)}(-i\omega)(-i\omega) = -Ae^{i(kx - \omega t)}\omega^2.$$

Now, substitute the foregoing terms into (2.11):

$$-Ae^{i(kx - \omega t)}k^2 - \frac{K}{F}Ae^{i(kx - \omega t)} = \frac{1}{c_0^2}(-Ae^{i(kx - \omega t)}\omega^2),$$

$$-k^2 e^{i(kx - \omega t)} - \frac{K}{F}e^{i(kx - \omega t)} = \frac{-\omega^2}{c_0^2}e^{i(kx - \omega t)},$$

$$\left(-k^2 - \frac{K}{F} + \frac{\omega^2}{c_0^2}\right)e^{i(kx - \omega t)} = 0. \tag{2.13}$$

This expression is called a *characteristic equation* (sometimes called a *dispersive, frequency,* or *secular* equation). For a general nontrivial solution,

$$\left(-k^2 - \frac{K}{F} + \frac{\omega^2}{c_0^2}\right) = 0.$$

There are many ways to tackle this expression. For example, we may write

$$\omega^2 = c_0^2\left(k^2 + \frac{K}{F}\right) \quad \text{where} \quad \omega = \omega(k) \tag{2.14}$$

or, alternatively,

$$k^2 = \frac{\omega^2}{c_0^2} - \frac{K}{F} \quad \text{where} \quad k = k(\omega); \tag{2.15}$$

note that $c_p \neq c_0$ as in the earlier taut string example. Hence, given $\omega = kc_p$, by equation (2.14) we have

$$c_p^2 = c_0^2\left(1 + \frac{K}{Fk^2}\right). \tag{2.16}$$

If $k \to \infty$ then

$$\omega = c_0 k\sqrt{1 + \frac{K}{Fk^2}} \to c_0 k. \tag{2.17}$$

The same result is obtained when $K = 0$ in (2.14), and this gives $c = c_0$ (see (2.16)). Hence, $c = c(k)$ or $k = k(c)$, and thus

$$k^2 = \frac{K/F}{(c^2/c_0^2) - 1}.$$

This yields three important results.

(1) Wave velocity is a function of frequency. Therefore, from a consideration of Fourier harmonic analysis, pulse distortion must occur.
(2) There are two real roots of (2.15):

$$k = \pm\sqrt{\frac{\omega^2}{c_0^2} - \frac{K}{F}}.$$

If $\omega^2/c_0^2 > K/F$ then propagation is possible to the right or left and so $u = Ae^{-i(\pm kx + \omega t)}$. On the other hand, if $\omega^2/c_0^2 < K/F$ then k is imaginary. We do not deal with this case, which represents a nonpropagating disturbance. If we write $\bar{k}^2 = -k^2$ then $u(x, t) = A(e^{\pm \bar{k}x})(e^{-i\omega t})$ as before, thus producing standing waves.
(3) For the special case of $k = 0$, by using $\omega^2/c_0^2 = K/F$ from (2.15) and recalling that $k = 2\pi/\lambda$, we can see that, as $\lambda \to \infty$ (the long wavelength limit), $u = Ae^{-i\omega_0 t}$. By (2.14) this represents uniform vibration, and the cutoff frequency $\omega_0 = c_0\sqrt{K/F}$.

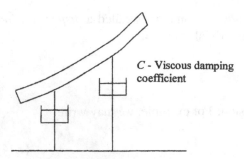

C - Viscous damping coefficient

Figure 2-10. String on a viscous foundation.

Because $c_p = \omega_0/k$, it follows that c_p increases if k decreases; hence, c_p approaches infinity as k approaches zero.

2.4 String on a Viscous Foundation

Figure 2-10 shows a string on a viscous foundation, an example of a highly dispersive and attenuative system. Let $q(x,t) = -C\dot{u}(x,t)$. If we consider the governing wave equation and a possible solution, we obtain

$$Tu,_{xx} - C\dot{u} = \rho\ddot{u}. \tag{2.18}$$

Assume that

$$u = Ae^{i(kx-\omega t)}. \tag{2.19}$$

Substituting (2.19) into (2.18) yields

$$k^2 = i\frac{C\omega}{F} + \frac{\omega^2}{c_0^2}. \tag{}$$

Therefore, k is complex and so free propagation of waves is not possible.

In a more conventional notation, we can consider the following expression for wavenumber with real k_r and imaginary components:

$$\bar{k} = k_r + i\alpha.$$

Substituting this into (2.19) and (2.18), we obtain $u = Ae^{(i(k_r+i\alpha)x-\omega t)}$, which can be simplified as follows:

$$u = \underbrace{Ae^{(-\alpha x)}}_{\substack{\text{new element} \\ \text{responsible for} \\ \text{attenuation}}} * \underbrace{e^{i(k_r x - \omega t)}}_{\substack{\text{old form of} \\ \text{wave motion}}}.$$

2.5 String on a Viscoelastic Foundation

Now imagine a string on a viscoelastic foundation. It is useful to introduce some basic aspects of viscoelastic model analysis in considerations of material dispersion. In dispersive systems, the phase velocity c_p is a function of frequency. Parts (a) and (b) of

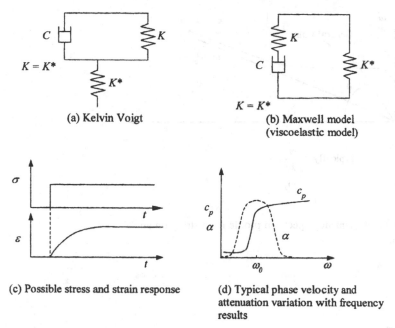

(a) Kelvin Voigt

(b) Maxwell model
(viscoelastic model)

(c) Possible stress and strain response

(d) Typical phase velocity and
attenuation variation with frequency
results

Figure 2-11. Possible viscoelastic models and response functions.

Figure 2-11 show electrical analogs of the material system. For the classical Kelvin and Maxwell models, attenuation occurs as a function of frequency; this is difficult to deal with both theoretically and experimentally. Parts (c) and (d) of Figure 2-11 show sample stress and strain response and corresponding sample phase velocity and attenuation curves as a function of frequency. These topics will be discussed more thoroughly in Chapter 18.

2.6 Graphical Representations of a Dispersive System

In order to study some basic aspects of dispersive systems, we now consider some graphical representations of the results for the string on an elastic foundation. Two major representations are used to graph dispersive character: a frequency spectrum and a dispersion curve. We will first consider the frequency spectrum.

Visualize a spring on an elastic foundation and recall (2.14):

$$\omega^2 = c_0^2 \left(k^2 + \frac{K}{F} \right).$$

If $k = 0$ then $\omega = \omega_0 = c_0\sqrt{K/F}$ as a cutoff frequency and, as $k \to \infty$, $\omega = c_0 k$ is a straight line (an asymptote). Now consider k_{Re} and k_{Im} and the relationship with ω_0 from (2.15):

$$k^2 = \frac{\omega^2}{c_0^2} - \frac{K}{F}.$$

Since $\omega_0 = c_0\sqrt{K/F}$, it follows that $\omega_0^2 = c_0^2(K/F)$ and $K/F = \omega_0^2/c_0^2$. As a result,

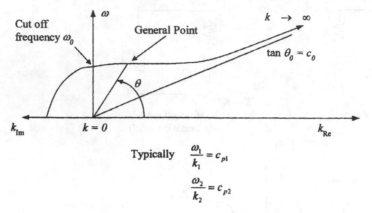

Typically $\dfrac{\omega_1}{k_1} = c_{p1}$

$\dfrac{\omega_2}{k_2} = c_{p2}$

Figure 2-12. A typical frequency spectrum profile for a string on an elastic foundation.

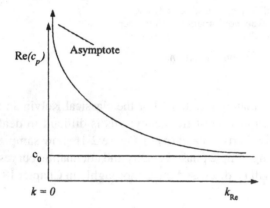

Figure 2-13. Typical dispersion curve of phase velocity versus wavenumber.

$$k^2 = \frac{\omega^2}{c_0^2} - \frac{\omega_0^2}{c_0^2} \quad \text{and} \quad k = \sqrt{\frac{\omega^2}{c_0^2} - \frac{\omega_0^2}{c_0^2}}.$$

The term k_{Im} exists when $\omega < \omega_0$, and k_{Re} exists when $\omega > \omega_0$. In a frequency spectrum, we consider frequency versus wavenumber; in a dispersion curve, phase velocity versus wavenumber.

A graphical form of the frequency spectrum is illustrated in Figure 2-12. Phase velocity information can be extracted from the frequency spectrum in order to produce a dispersion curve as follows. Use $\tan\theta = \omega/k = c_p$ to derive phase velocity from the frequency spectrum. Figure 2-13 shows that, as $k \to 0$, $c_p \to \infty$.

Note that, for small k and long λ, the cutoff frequency is ω_0 (here, $c_p \to \infty$), giving rise to uniform vibration. For $k \gg 1$ and $\lambda \ll 1$, the foundation effect is minimized. Note that we can have waves (e.g. Lamb waves) with multimode characteristics. Figure 2-14 shows

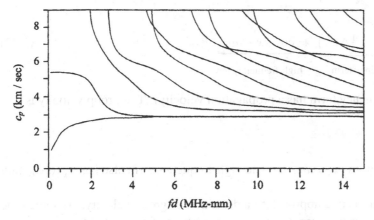

Figure 2-14. Phase velocity dispersion curves for an aluminum plate with $c_L = 6.3$ km/s and $c_T = 3.1$ km/s.

a multimode dispersion curve used in ultrasonic evaluation. Such curves will be developed later in this textbook for applications to plates, rods, multiple layers, and other structures.

2.7   Group Velocity Concepts

Group velocity is associated with the propagation velocity of a group of waves of similar frequency. From Lord Rayleigh: "It has often been remarked that when a group of waves advances into still water, the velocity of the group is less than that of the individual waves of which it is composed; the waves appear to advance through the group, dying away as they approach its interior limit" (1945, vol. I, p. 475). Also note that, in wave mechanics, Heisenberg used the term "velocity of wave packets" (see Serway 1990).

The simplest analytical explanation (Stokes 1876) is to consider two propagating harmonic waves of equal amplitude but of slightly different frequency, ω_1 and ω_2. Then

$$u = A \cos(k_1 x - \omega_1 t) + A \cos(k_2 x - \omega_2 t), \tag{2.20}$$

where $k_1 = \omega_1/c_1$ and $k_2 = \omega_2/c_2$. Using trigonometric identities,

$$A(\cos \alpha + \cos B) = 2A\left[\cos\left(\frac{\alpha - B}{2}\right) * \cos\left(\frac{\alpha + B}{2}\right)\right].$$

We can rewrite (2.20) as

$$u = 2A \cos\left\{\tfrac{1}{2}(k_2 - k_1)x - \tfrac{1}{2}(\omega_2 - \omega_1)t\right\} * \cos\left\{\tfrac{1}{2}(k_2 + k_1)x - \tfrac{1}{2}(\omega_2 + \omega_1)t\right\}, \tag{2.21}$$

noting that the cosine is an even function. We now make the following substitutions:

$$\Delta\omega = \omega_2 - \omega_1, \quad \Delta k = k_2 - k_1;$$
$$\tfrac{1}{2}(\omega_2 + \omega_1) = \omega_{AV}, \quad \tfrac{1}{2}(k_2 + k_1) = k_{AV};$$
$$c_{AV} = \frac{\omega_{AV}}{k_{AV}}.$$

Hence,

$$u = 2A \underbrace{\cos\{\tfrac{1}{2}\Delta k x - \tfrac{1}{2}\Delta \omega t\}}_{\text{low-frequency term}} * \underbrace{\cos(k x - \omega t)}_{\text{high-frequency term}} \qquad (2.22)$$

(note that the low-frequency term has a propagation velocity). The group velocity is

$$c_g = \frac{\Delta \omega}{\Delta k},$$

which in the limit becomes $c_g = d\omega/dk$. The high-frequency term also has a propagation velocity, $c_p = \omega/k$.

Consider now an alternative approach to a definition of group velocity. At some time increment $t = t_0 + dt$, we may represent changes in phase of any individual component as follows:

$$dP_i = \{k_i(x_0 + dx) - \omega_i(t_0 + dt)\} - \{k_i x_0 - \omega_0 t\} = k_i \, dx - \omega_i \, dt.$$

In order for the wave group to be maintained, the changes in phase for all components should be the same: $dP_j - dP_i = 0$.

With regard to the phase angle $kx - \omega t$, we have

$$\underbrace{(k_j - k_i)\, dx}_{dk} - \underbrace{(\omega_j - \omega_i)\, dt}_{d\omega} = 0 \quad \text{and}$$

$$\frac{dx}{dt} = \frac{d\omega}{dk} = c_g \Rightarrow \frac{\Delta\omega}{\Delta k}. \qquad (2.23)$$

This represents a classical definition of group velocity.

We may further compare $c_p = \omega/k$ and $c_g = d\omega/dk$ as follows:

$$c_g = \frac{d(k c_p)}{dk} = c_p + k\frac{dc_p}{dk}, \qquad (2.24)$$

where $c_p = c_p(k)$. Three cases ($c_g > 0$, $c_g = 0$, or $c_g < 0$) can be graphed on a frequency spectrum of ω versus k.

Superposition of a group of waves of similar frequency leads to the typical result shown in Figure 2-15. The individual harmonics travel with different phase velocities c_p, but the superimposed packet travels with the group velocity c_g. Realistically, we should therefore consider a superposition of a number of waves, rather than just two as used in the earlier example. Thus,

$$u = \sum_{i=1}^{n} A_i \cos(k_i x - \omega_i t), \qquad (2.25)$$

where k_i and ω_i differ only slightly.

Now we consider phase and group velocity from the perspective of a frequency spectrum. Group velocity is indicated as a local slope to the curve at the point, as illustrated in Figure 2-16. Physical examples of the three cases can now be supplied:

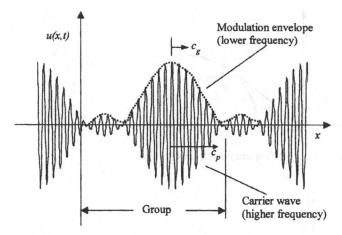

Figure 2-15. Group velocity example.

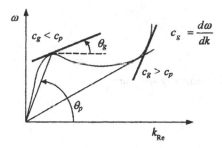

Figure 2-16. Group velocity variation with phase velocity.

(1) $c_p > c_g$ – *classical* (normal) dispersion, as in a Rayleigh example, that appears to originate behind the group, travel to the front, and disappear;

(2) $c_p = c_g$ – no dispersion;

(3) $c_p < c_g$ – *anomalous* dispersion, which appears to originate at the front, travel to the rear, and then disappear.

Alternative forms of the expressions for c_g should also be addressed. Starting with (2.24),

$$c_g = \frac{d(kc_p)}{dk} = c_p + k\frac{dc_p}{dk},$$

we note that $\lambda = 2\pi/k = 2\pi k^{-1}$. Thus

$$c_g = c_p + k\frac{dc_p}{d\lambda}\frac{d\lambda}{dk} = c_p + \frac{2\pi}{\lambda}\frac{dc_p}{d\lambda}(-2\pi k^{-2})$$

$$= c_p + \frac{2\pi}{\lambda}\frac{-2\pi}{(2\pi)^2}\lambda^2\frac{dc_p}{d\lambda} = c_p - \lambda\frac{dc_p}{d\lambda}$$

and so

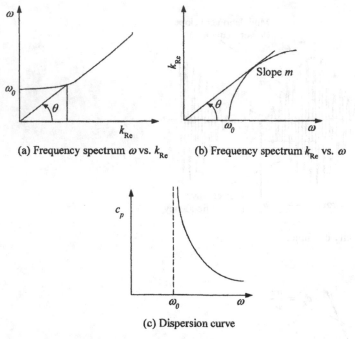

(a) Frequency spectrum ω vs. k_{Re} (b) Frequency spectrum k_{Re} vs. ω

(c) Dispersion curve

Figure 2-17. Construction of dispersion curve from the frequency spectrum.

$$c_g = c_p - \lambda \frac{dc_p}{d\lambda}. \tag{2.26}$$

Next, we look at group velocity and energy transmission. Consider the simple group

$$u = 2A \cos\left(\tfrac{1}{2}\Delta kx - \tfrac{1}{2}\Delta\omega t\right) \cos(kx - \omega t).$$

Define the energy density as

$$\hat{E} = \rho \dot{u}^2$$
$$= 4\rho\omega^2 A^2 \cos^2\left(\tfrac{1}{2}\Delta kx - \tfrac{1}{2}\Delta\omega t\right)\sin^2(kx - \omega t) + \cdots.$$

Taking the time average of this expression over several periods T (during which the modulation does not change much), from Miklowitz (1978) we find for lossless media

$$\langle\hat{E}\rangle \cong 2\rho\omega^2 A^2 \cos^2\left(\tfrac{1}{2}\Delta kx - \tfrac{1}{2}\Delta\omega t\right).$$

This suggests that the time-averaged energy density propagation has the velocity

$$c_E = c_g = \frac{d\omega}{dk}. \tag{2.27}$$

Hence, group velocity is the velocity of energy transportation.

Now imagine the possibility of going from a frequency spectrum to the more popular engineering dispersion curve of $c_p(\omega)$ (see Figure 2-17 for graphical details). In order to plot c_p versus f, it might be easier to change the frequency spectrum from ω versus k to

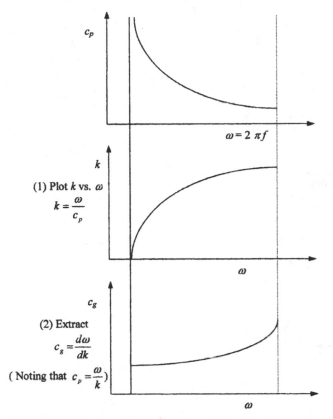

Figure 2-18. Construction of the group velocity dispersion curve from the phase velocity dispersion curve.

k versus ω, as shown in parts (a) and (b) of Figure 2-17. The curve c is then plotted from (b) on a point-by-point basis, using $c_p = \omega/k$ from this curve. A graphical procedure for plotting a group velocity dispersion curve from the phase velocity dispersion curve is illustrated in Figure 2-18.

2.8 Exercises

1. (a) What percent tension increase is necessary to double the frequency of a guitar string of length L?

   (b) What change in cross-sectional diameter of the string would also give the same doubled frequency?

2. (a) What is the wave velocity on a string if tension doubles?

   (b) Determine the maximum wave velocity on a string.

3. Derive the wave equation for a string on a viscoelastic foundation.

4. Select a specific frequency spectrum. Derive the dispersion curve from the frequency spectrum. Also, determine the group velocity curve.

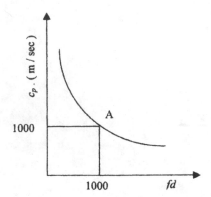

Figure 2-19. Exercise 7.

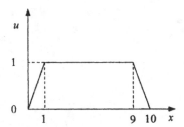

Figure 2-20. Exercise 8.

5. Solve the wave propagation problem in a string, given boundary conditions $u(0, t) = u(l, t) = 0$ and initial conditions $u(x, 0) = f(x)$ and $(du/dt)(x, 0) = 0$.

6. Derive an expression for group velocity as a function of phase velocity and frequency.

7. For the specific mode shown in Figure 2-19, estimate the numerical value of the group velocity at point A. Also estimate the value of the cutoff frequency for the curve shown.

8. Solve the initial value problem shown in Figure 2-20. The string is fixed at $x = 0$ and $x = 10$; displacement function is shown at $t = 0$.

9. On Figure 2-7, confirm equations of straight lines.

10. For the characteristics in Figure 2-8, plot a diagram of steel impacting Plexiglas of specific thickness values.

11. Derive the characteristic equation for a string on a viscoelastic foundation. Provide a graphical presentation.

12. Given a sample frequency spectrum of $\omega(k)$, plot $c_p(k)$.

13. Given a sample frequency spectrum of $\omega(k)$, plot $c_g(k)$.

14. Given a sample dispersion curve $c_p(\omega)$, plot the group velocity dispersion curve $c_g(\omega)$.

15. What are possible conditions in $k(\omega)$ or $\omega(k)$ curves for a negative group velocity?

16. What are possible conditions in $c_p(f)$ curves for a negative group velocity?

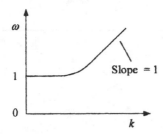

Figure 2-21. Exercise 18.

17. Plot a frequency spectrum similar to that shown in Figure 2-12 for sample realistic values of tension, density, and spring constant.

18. Given the sample frequency spectrum curve of ω versus k shown in Figure 2-21, graphically estimate the phase and group velocity curves as a function of the wavenumber k. (Concentrate on values when $k = 0$, k large, and then a general trend in between.)

3

Unbounded Isotropic and Anisotropic Media

3.1 Introduction

Bulk wave propagation refers to wave propagation in infinite media; guided waves are those that require a boundary for their existence, such as surface waves, Lamb waves, and interface waves. This chapter will focus on bulk wave propagation in infinite (or semi-infinite) media. Keep in mind that a thin structure can, for all practical purposes, still be considered a half-space or semi-infinite media if the wavelength of excitation is small with respect to the thickness of the test object.

We shall explore some interesting phenomena of phase velocity variation with angle of propagation into solid media. This leads to a dispersive influence as a result of differences in phase velocity and energy velocity. For isotropic materials, phase velocity is independent of entry angle. For lossless media, the energy velocity is equal to the group velocity. However, because of the wave velocity variations with angle, interference phenomena will lead to a skew angle. Trying to send waves or ultrasonic energy in a specific direction may be more difficult than you think!

3.2 Isotropic Media

3.2.1 Equations of Motion

The development of the equation of motion for an elastic isotropic solid is a classical topic covered in many elasticity textbooks (e.g., Kolsky 1963 and Pollard 1977; see also Appendix B). The Navier governing equations are:

$$(\lambda + \mu)u_{j,ij} + \mu u_{i,jj} + \rho f_i = \rho \ddot{u}_i \quad (i, j = 1, 2, 3). \tag{3.1}$$

In scalar Cartesian notation, this represents three equations:

$$(\lambda + \mu)\frac{\partial}{\partial x_1}\left(\frac{\partial u_1}{\partial x_1} + \frac{\partial u_2}{\partial x_2} + \frac{\partial u_3}{\partial x_3}\right) + \mu\nabla^2 u_1 + \rho f_x = \rho\frac{\partial^2 u_1}{\partial t^2},$$

$$(\lambda + \mu)\frac{\partial}{\partial x_2}\left(\frac{\partial u_1}{\partial x_1} + \frac{\partial u_2}{\partial x_2} + \frac{\partial u_3}{\partial x_3}\right) + \mu\nabla^2 u_2 + \rho f_y = \rho\frac{\partial^2 u_2}{\partial t^2}, \tag{3.2}$$

$$(\lambda + \mu)\frac{\partial}{\partial x_3}\left(\frac{\partial u_1}{\partial x_1} + \frac{\partial u_2}{\partial x_2} + \frac{\partial u_3}{\partial x_3}\right) + \mu\nabla^2 u_3 + \rho f_z = \rho\frac{\partial^2 u_3}{\partial t^2},$$

where

$$\nabla^2 = \frac{\partial^2}{\partial x_1^2} + \frac{\partial^2}{\partial x_2^2} + \frac{\partial^2}{\partial x_3^2}. \tag{3.3}$$

We introduce the following notation:

$$\nabla = \bar{i}_1 \frac{\partial}{\partial x_1} + \bar{i}_2 \frac{\partial}{\partial x_2} + \bar{i}_3 \frac{\partial}{\partial x_3}; \tag{3.4}$$
$$\bar{u} = (u_1, u_2, u_3).$$

In the absence of body forces, the system of equations (3.2) can be expressed in vector form as

$$(\lambda + \mu)\nabla\nabla \bullet \bar{u} + \mu\nabla^2\bar{u} = \rho\frac{\partial^2\bar{u}}{\partial t^2}. \tag{3.5}$$

The solution for the body force problem is discussed for example in Graff (1991). One can use the vector identity

$$\nabla^2\bar{u} = \nabla\nabla \bullet \bar{u} - \nabla \times \nabla \times \bar{u}, \tag{3.6}$$

where

$$\nabla \times \bar{u} = \begin{vmatrix} \bar{i}_1 & \bar{i}_2 & \bar{i}_3 \\ \dfrac{\partial}{\partial x_1} & \dfrac{\partial}{\partial x_2} & \dfrac{\partial}{\partial x_3} \\ u_1 & u_2 & u_3 \end{vmatrix}.$$

By substituting (3.6) into (3.5), the equation of motion can alternatively be expressed as

$$(\lambda + 2\mu)\nabla\nabla \bullet \bar{u} - \mu\nabla \times \nabla \times \bar{u} = \rho\frac{\partial^2\bar{u}}{\partial t^2}. \tag{3.7}$$

The equation of motion can also be expressed in a more simplified form. First, the vector displacement $\bar{u}$ can be expressed via Helmholtz decomposition as the gradient of a scalar and the curl of the zero divergence vector (see Morse and Feshbach 1953):

$$\bar{u} = \nabla\Phi + \nabla \times \bar{H}, \quad \nabla \bullet \bar{H} = 0, \tag{3.8}$$

where Φ and $\bar{H}$ are scalar and vector potentials, respectively. Then, substituting (3.8) into Navier's equation of motion (3.5), we obtain

$$(\lambda + \mu)\nabla\nabla \bullet (\nabla\Phi + \nabla \times \bar{H}) + \mu\nabla^2(\nabla\Phi + \nabla \times \bar{H})$$
$$= \rho\left(\nabla\frac{\partial^2\Phi}{\partial t^2} + \nabla \times \frac{\partial^2\bar{H}}{\partial t^2}\right). \tag{3.9}$$

Next, using (3.6), equation (3.9) can be regrouped as

$$\left[(\lambda + 2\mu)\nabla\nabla \bullet (\nabla\Phi) - \rho\nabla\frac{\partial^2\Phi}{\partial t^2}\right] - \mu\nabla \times \nabla \times \nabla\Phi$$
$$+ (\lambda + \mu)\nabla\nabla \bullet \nabla \times \bar{H} + \left[\mu\nabla^2\nabla \times \bar{H} - \nabla \times \frac{\partial^2\bar{H}}{\partial t^2}\right] = 0. \tag{3.10}$$

The following identities can be used:

$$\nabla \bullet \nabla \Phi = \nabla^2 \Phi; \quad \nabla \times \nabla \times \nabla \Phi = 0; \quad \nabla \bullet \nabla \times \bar{H} = 0. \tag{3.11}$$

Finally, by using (3.11), from (3.10) we have

$$\nabla \left[(\lambda + 2\mu) \nabla^2 \Phi - \rho \frac{\partial^2 \Phi}{\partial t^2} \right] + \nabla \times \left[\mu \nabla^2 \bar{H} - \rho \frac{\partial^2 \bar{H}}{\partial t^2} \right] = 0, \tag{3.12}$$

which is satisfied if both terms vanish. This leads to the equations

$$\nabla^2 \Phi = \frac{1}{c_L^2} \frac{\partial^2 \Phi}{\partial t^2} \quad \text{and} \tag{3.13}$$

$$\nabla^2 \bar{H} = \frac{1}{c_T^2} \frac{\partial^2 \bar{H}}{\partial t^2}, \tag{3.14}$$

where

$$c_L^2 = \frac{\lambda + 2\mu}{\rho} \quad \text{and} \quad c_T^2 = \frac{\mu}{\rho}. \tag{3.15}$$

As a result, the equation of motion (3.5) is decomposed as two simplified wave equations, (3.13) and (3.14).

3.2.2 Dilatational and Distortional Waves

Now suppose that the rotational part $\nabla \times \bar{H}$ in (3.8) is zero and that

$$\bar{u} = \nabla \Phi. \tag{3.16}$$

In this case, (3.12) and (3.13) give:

$$\nabla^2 \bar{u} = \frac{1}{c_L^2} \frac{\partial^2 \bar{u}}{\partial t^2}. \tag{3.17}$$

This means that dilatational disturbance propagates with the velocity c_L. In a similar way, suppose that the displacement in (3.8) has *only* a rotational part:

$$\bar{u} = \nabla \times \bar{H}, \quad \nabla \bullet \bar{H} = 0. \tag{3.18}$$

In this case, from (3.12) and (3.14) it follows that:

$$\nabla^2 \bar{u} = \frac{1}{c_T^2} \frac{\partial^2 \bar{u}}{\partial t^2}. \tag{3.19}$$

Equation (3.19) shows that rotational waves propagate with velocity c_T. Equations (3.17) and (3.19) are independent of each other, which means that the longitudinal and shear (or torsional) waves propagate without interaction in unbounded media. These two types of waves are coupled only on the boundary of the elastic body, an obvious consequence of satisfying the boundary conditions.

3.3   The Christoffel Equation for Anisotropic Media

Imagine waves in an infinite elastic anisotropic solid. We shall consider waves in pure crystals, where homogeneity and pure anisotropy is assumed. There are many different ways to approach this problem, but for now we consider the following.

Using indicial (or tensor) notation and Newton's law, set

$$\frac{\partial \sigma_{ik}}{\partial x_k} = \rho \ddot{u}_i \quad \text{or} \quad \rho \ddot{u}_i = \sigma_{ik,k} \tag{3.20}$$

as a governing wave equation, where σ denotes stress. Using Hooke's law, we have

$$\sigma_{ik} = C_{iklm} \varepsilon_{lm} \tag{3.21}$$

as a constitutive equation. Combining (3.20) and (3.21) yields

$$\rho \ddot{u}_i = C_{iklm}, \quad \frac{\partial \varepsilon_{lm}}{\partial x_k} = C_{iklm} \varepsilon_{lm,k}.$$

From the definition of strain in the strain displacement equations,

$$\varepsilon_{lm} = \frac{1}{2} \left(\frac{\partial u_l}{\partial x_m} + \frac{\partial u_m}{\partial x_l} \right) = \frac{1}{2} (u_{l,m} + u_{m,l}); \tag{3.22}$$

again, combining (3.22) and (3.21), we find the result:

$$\rho \ddot{u}_i = \frac{1}{2} C_{iklm} \left(\frac{\partial^2 u_l}{\partial x_k \partial x_m} + \frac{\partial^2 u_m}{\partial x_k \partial x_l} \right) \tag{3.23}$$

or

$$\rho \ddot{u}_i = \tfrac{1}{2} C_{iklm} (u_{l,km} + u_{m,kl}). \tag{3.24}$$

Note that C_{iklm} is symmetrical with respect to l and m, so

$$C_{iklm} = C_{ikml} = C_{kilm}.$$

Therefore, we can interchange l and m.

Let us assume plane harmonic traveling waves to see if a solution of this form is possible. Put

$$u_i = A_i \exp\{i(k_j x_j - \omega t)\}, \tag{3.25}$$

where k_j is the unit wavevector. (Note that $A_i = A\alpha_i$, where the α_i are direction cosines of particle displacement.) Substituting (3.25) into (3.24) gives an eigenvalue problem. From earlier work in harmonic motion, in considering the amplitude of the second derivative we know that $\rho \ddot{u}_i = \rho \omega^2 u_i$. Note that $k_j x_j$ is a single dot product, which is useful in differentiating to obtain

$$\rho \ddot{u}_i = \rho \omega^2 u_i = C_{iklm} k_k k_l u_m.$$

Tensor analysis can be used to establish that $C_{iklm} u_{m,kl}$ is equivalent to $C_{iklm} k_k k_l u_m$, with i the free index and with summing over k, l, and m. For one term we can show the equivalence quite easily as follows. If $u_m = A \exp i(k_1 x_1 + k_2 x_2 + k_3 x_3 - \omega t) = A \exp Q$ then it is easy to see that

$$\frac{\partial u_m}{\partial x_1} = A \exp Q k_1 \quad \text{and} \quad \frac{\partial^2 u_m}{\partial x_1 \partial x_2} = A \exp Q k_1 k_2;$$

hence $u_{m,kl} = k_k k_l u_m$ (note that $u_i = u_m \delta_{im}$). Therefore,

$$(\rho \omega^2 \delta_{im} - C_{iklm} k_k k_l) u_m = 0. \tag{3.26}$$

This is the famous Christoffel equation for anisotropic media.

The Christoffel acoustic tensor may be defined as

$$\lambda_{im} = \Gamma_{im} = C_{iklm} n_k n_l, \tag{3.27}$$

where n_k are direction cosines of the normal to the wavefront (since $k_k = k n_k$, $k_l = k n_l$, and $c^2 = \omega^2 / k^2$; we will eventually solve for the wave velocity c). Therefore, from $(C_{iklm} k_k k_l - \rho \omega^2 \delta_{im}) u_m = 0$ we have

$$(\Gamma_{im} k^2 - \rho \omega^2 \delta_{im}) u_m = 0 \quad \text{or} \quad (\Gamma_{im} - \rho c^2 \delta_{im}) u_m = 0.$$

This gives us three homogeneous equations, three real roots, and three different velocities, all from the cubic equation in c^2; this leads to an orthogonal classical eigenvalue problem.

For a nontrivial solution, we must set the determinant of the coefficient matrix equal to zero:

$$|\Gamma_{im} - \rho c^2 \delta_{im}| = 0. \tag{3.28}$$

Because Γ_{im} depends on crystal symmetry and the orientation of the waves, we have

$$\begin{vmatrix} (\lambda_{11} - \rho c^2) & \lambda_{12} & \lambda_{13} \\ \lambda_{21} & (\lambda_{22} - \rho c^2) & \lambda_{23} \\ \lambda_{31} & \lambda_{32} & (\lambda_{33} - \rho c^2) \end{vmatrix} = 0,$$

where $\lambda_{11}, \lambda_{12}, \lambda_{13}, \ldots$ are obtained from the expression for the acoustic tensor. Recall from (3.27) that $\lambda_{im} = C_{iklm} n_k n_l$.

Expand this acoustic tensor carefully for all elements of the matrix. The indices i and m are free; k and l are summers:

$$\begin{array}{cccc} & k = 1 & k = 2 & k = 3 \\ \lambda_{11} = & C_{1111} n_1 n_1 + & C_{1211} n_2 n_1 + & C_{1311} n_3 n_1 \quad l = 1 \\ & + C_{1121} n_1 n_2 + & C_{1221} n_2 n_2 + & C_{1321} n_3 n_2 \quad l = 2 \\ & + C_{1131} n_1 n_3 + & C_{1231} n_2 n_3 + & C_{1331} n_3 n_3. \quad l = 3 \end{array}$$

For ease of expansion, specify i and m as 11, 12, 13, 21, 22, 23, 31, 32, or 33 for the nine components. Consider $l = 1$, $l = 2$, and $l = 3$ for the first, second, and third rows, with $k = 1$, $k = 2$, and $k = 3$ for the first, second, and third columns.

We can simplify the results even further by converting C_{ikjl} to C_{nm} as follows: if $i = k$ then $n = i$, and if $j = l$ then $m = j$; if $i \neq k$ then $n = 9 - (i + k)$, and if $j \neq l$ then $m = 9 - (j + l)$. Therefore,

$$C_{1111} = C_{11}, \quad C_{1211} = C_{61}, \quad C_{1311} = C_{51},$$
$$C_{1121} = C_{16}, \quad C_{1221} = C_{66}, \quad C_{1321} = C_{56},$$
$$C_{1131} = C_{15}, \quad C_{1231} = C_{65}, \quad C_{1331} = C_{55}.$$

As an example,

$$\lambda_{11} = C_{11}n_1^2 + C_{16}n_2n_1 + C_{15}n_3n_1$$
$$+ C_{16}n_1n_2 + C_{66}n_2^2 + C_{56}n_3n_2$$
$$+ C_{15}n_1n_3 + C_{65}n_2n_3 + C_{55}n_3^2.$$

Continuing with these computations would allow us to plot a phase profile. Given constants and directions, various c can be calculated in the wavevector $\bar{k}$ directions. This is a tedious process that calls for an efficient computer program.

Once the phase velocity values are extracted for specific directions, we would like to see if we have a pure mode – that is, if the particle velocity direction is aligned perfectly with the chosen $\bar{k}$ direction for the phase velocity computation. To see if there are pure modes, we must expand and extract all roots of the bi-cubic equation. For a given direction $\bar{k}$ of propagation, there are three waves, $k_l = kn_l$, possibly with mutually perpendicular displacement vectors. It may be possible to find a special direction with one pure longitudinal and two pure transverse waves. Toward this end, we will find the direction cosines α_i of the particle displacements.

Substitute each value of c^2 back into the system of original homogeneous equations to calculate the eigenvectors:

$$\begin{vmatrix} (\lambda_{11} - \rho c^2) & \lambda_{12} & \lambda_{13} \\ \lambda_{21} & (\lambda_{22} - \rho c^2) & \lambda_{23} \\ \lambda_{31} & \lambda_{32} & (\lambda_{33} - \rho c^2) \end{vmatrix} \begin{Bmatrix} \alpha_1 \\ \alpha_2 \\ \alpha_3 \end{Bmatrix} = 0.$$

This yields

$$(\lambda_{11} - \rho c^2)\alpha_1 + \lambda_{12}\alpha_2 + \lambda_{13}\alpha_3 = 0,$$
$$\lambda_{21}\alpha_1 + (\lambda_{22} - \rho c^2)\alpha_2 + \lambda_{23}\alpha_3 = 0, \tag{3.29}$$
$$\lambda_{31}\alpha_1 + \lambda_{32}\alpha_2 + (\lambda_{33} - \rho c^2)\alpha_3 = 0.$$

Solve three times for each value of c^2 (the result can be checked, since A is orthogonal). We can arrange the solution as follows:

$$\begin{matrix} c_1^2 & c_2^2 & c_3^2 \\ \end{matrix}$$
$$A = \begin{bmatrix} \alpha_{11} & \alpha_{12} & \alpha_{13} \\ \alpha_{21} & \alpha_{22} & \alpha_{23} \\ \alpha_{31} & \alpha_{32} & \alpha_{33} \end{bmatrix}.$$

Note that determinant $A = +1$ for a right-handed Cartesian coordinate system.

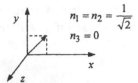

Figure 3-1. Plane waves in the [1, 1, 0] direction.

3.3.1 Sample Problem

How can we determine phase velocity in a specific direction for a given level of anisotropy? More particularly, we shall discuss two problems with the aid of C_{ij} matrices as presented in elasticity theory (for waves that are orthotropic, hexagonal, isotropic, etc.; see Appendix B).

(1) Calculate directions for a pure longitudinal wave and two pure transverse waves for a cubic crystal. *Note:* pure longitudinal, $\bar{u} \times \bar{n} = 0$; pure transverse, $\bar{u} \bullet \bar{n} = 0$.
(2) Develop equations for c_1, c_2, and c_3 as a function of the wavevector $\bar{k}$ directions in the (x_1, x_2)-plane. *Note:* $\bar{n}$ represents the direction cosine of the angle between the wavevector and the x_1, x_2, and x_3 waves. Therefore, we compute Γ_{im} for a cubic crystal. When we solve the determinant for specific $\bar{n}$, we must let $\bar{n}$ vary in the plane for small increments. In general, waves are not pure L or pure S and so are often termed quasi-longitudinal or quasi-shear. However, if $\bar{k}$ is an eigenvector of λ_{ik} then the waves are pure.

In order to solve these problems, a specific wavevector must now be considered. Therefore, imagine plane waves in the [1, 1, 0] direction for a cubic crystal (see Figure 3-1). Next, evaluate the terms of the acoustic tensor,

$$\lambda_{im} = C_{iklm} n_k n_l,$$

where i and m are free indices with double summations over k and l from 1 to 3. Using tensor-to-matrix notation yields

$$11 \to 1, \quad 22 \to 2, \quad 23, 32 \to 4, \quad 31, 13 \to 5, \quad 12, 21 \to 6.$$

The elastic constant matrix for a cubic crystal may be found in Appendix B:

$$\begin{bmatrix} C_{11} & C_{12} & C_{12} & 0 & 0 & 0 \\ C_{12} & C_{11} & C_{12} & 0 & 0 & 0 \\ C_{12} & C_{12} & C_{11} & 0 & 0 & 0 \\ 0 & 0 & 0 & C_{44} & 0 & 0 \\ 0 & 0 & 0 & 0 & C_{44} & 0 \\ 0 & 0 & 0 & 0 & 0 & C_{44} \end{bmatrix}.$$

Therefore,

$$\lambda_{11} = C_{11} n_1^2 + C_{66} n_2^2 + C_{55} n_3^2 + 2C_{56} n_2 n_3 + 2C_{15} n_1 n_3 + 2C_{16} n_1 n_2,$$

$$\lambda_{12} = C_{16}n_1^2 + C_{26}n_2^2 + C_{45}n_3^2 + (C_{46} + C_{25})n_2n_3$$
$$+ (C_{14} + C_{56})n_1n_3 + (C_{12} + C_{66})n_1n_2,$$

$\lambda_{21} = \lambda_{12}$, and so forth.

By Christoffel's equations we have $(\lambda_{im} - \rho c^2)\mu_m = 0$ and so

$$\begin{vmatrix} \lambda_{11} - \rho c^2 & \lambda_{12} & \lambda_{13} \\ \lambda_{12} & \lambda_{22} - \rho c^2 & \lambda_{23} \\ \lambda_{13} & \lambda_{23} & \lambda_{33} - \rho c^2 \end{vmatrix} = 0.$$

Using $\lambda_{11}, \lambda_{12}, \ldots$ and $n_1 = n_2 = 1/\sqrt{2}$ and $n_3 = 0$, we reach:

$$\begin{vmatrix} \frac{1}{2}(C_{11} + C_{44}) - \rho c^2 & \frac{1}{2}(C_{12} + C_{44}) & 0 \\ \frac{1}{2}(C_{12} + C_{44}) & \frac{1}{2}(C_{11} + C_{44}) - \rho c^2 & 0 \\ 0 & 0 & C_{44} - \rho c^2 \end{vmatrix} = 0;$$

therefore,

$$\left[\tfrac{1}{2}(C_{11} + C_{44}) - \rho c^2\right]\left[\tfrac{1}{2}(C_{11} + C_{44}) - \rho c^2\right]\left[C_{44} - \rho c^2\right]$$
$$- \left[\tfrac{1}{2}(C_{11} + C_{44})\right]\left[\tfrac{1}{2}(C_{12} + C_{44})\right]\left[C_{44} - \rho c^2\right] = 0.$$

Through factoring, we solve for three c; the eigenvector solution is then required to know which wave is in which direction or alignment with $\bar{k}$. Since $c_i = \sqrt{C_{44}/\rho}$,

$$\left(\tfrac{1}{2}C_{11} + \tfrac{1}{2}C_{44} - \rho c^2\right) \pm \left(\tfrac{1}{2}(C_{11} + C_{44})\right) = 0.$$

The solutions are:

$$c_1 = \sqrt{\frac{C_{11} + C_{12} + 2C_{44}}{2\rho}}, \quad c_2 = \sqrt{\frac{C_{44}}{\rho}}, \quad c_3 = \sqrt{\frac{C_{11} - C_{12}}{2\rho}}.$$

To confirm wave type and character (pure or quasi), we must solve for the three eigenvectors:

$$\begin{vmatrix} \frac{1}{2}(C_{11} + C_{44}) - \rho c^2 & \frac{1}{2}(C_{12} + C_{44}) & 0 \\ \frac{1}{2}(C_{12} + C_{44}) & \frac{1}{2}(C_{11} + C_{44}) - \rho c^2 & 0 \\ 0 & 0 & C_{44} - \rho c^2 \end{vmatrix} \begin{Bmatrix} u_x \\ u_y \\ u_z \end{Bmatrix} = 0.$$

One wave in closest alignment with $\bar{n}$ is the pure or quasi-longitudinal wave; the other two are shear or quasi-shear. Although the task is tedious, we must solve the equations three times (for u_x, u_y, and u_z) for each c_i. (*Note:* If $\bar{u}_1 \times \bar{n} = 0$ then we have a pure l wave, or use $\bar{u} \times \bar{n} = |u||n| \sin \theta$. If $\bar{u}_2 \bullet \bar{n} = 0$ and $\bar{u}_3 \bullet \bar{n} = 0$ then we have a pure transverse wave.) Consequently, if α_i are direction cosines then we can solve the following.

(1) For c_1: $\alpha_1 = \alpha_2 = 1/\sqrt{2}$ and $\alpha_3 = 0$; therefore, c_1 is a pure l wave. (In the $[1, 1, 0]$ direction, the direction cosine vector is in the same direction as the wavevector.)
(2) For c_2: $\alpha_3 = 1$ and $\alpha_1 = \alpha_2 = 0$. Therefore, when considering the dot product $[0, 0, 1] \bullet [1, 1, 0] = 0$, we see that c_2 is pure shear.
(3) Finally, for c_3: $\alpha_1 = 1/\sqrt{2}$, $\alpha_2 = -1/\sqrt{2}$, and $\alpha_3 = 0$. The dot product is again $[1, -1, 0] \bullet [1, 1, 0] = 0$; this is also a pure shear wave.

Table 3-1. *Stiffness coefficients C_{ij} for selected materials*

Material	Stiffness (10^{10} N/m²)											
	C_{11}	C_{22}	C_{33}	C_{44}	C_{55}	C_{66}	C_{12}	C_{13}	C_{14}	C_{16}	C_{23}	C_{25}
Barium titanate[ab]	15.0		14.6	4.4			6.6	6.6				
Diamond	102			49.2			25					
Lead titanate–zirconate (PZT-2)[ab]	13.5		11.3	2.22			6.79	6.81				
Quartz[a]	8.674		10.72	5.794			0.699	1.191	1.791			
Rochelle salt[a]	2.8	4.14	3.94	0.666	0.285	0.96	1.74	1.50			1.97	
Sapphire	49.4		49.6	14.5			15.8	11.4	−2.3			
Cadmium sulfide[a]	9.07		9.38	1.504			5.81	5.10				
Germanium	12.89			6.71			4.83					
Silicon	16.57			7.956			6.39					
Aluminum, crystal	10.80			2.85			6.13					
Aluminum, polycrystal	11.1			2.5								
Gold, crystal	18.6			4.20			15.7					
Titanium, crystal	16.2		18.1	4.67			9.2	6.9				
Titanium, polycrystal	16.59			4.4								
Tungsten, crystal	50.2			15.2			19.9					
Tungsten, polycrystal	58.1			13.4								
Graphite/epoxy (AS4/3501-6)	16.209	1.63			0.774		0.671				0.795	

[a]Piezoelectric.

[b]Poled ceramic. The stiffness matrix has the same form as for the hexagonal crystal system, with Z along the poling axis.

Sources: Most data from Auld (1990). Data for graphite/epoxy from Mal, Lih, and Bar-Cohen (1994).

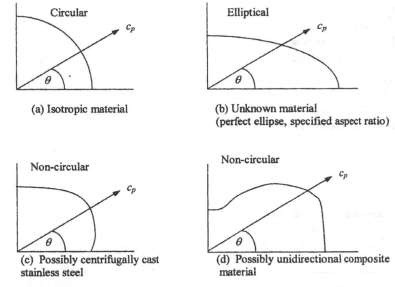

Figure 3-2. Sample longitudinal or shear wave velocity surfaces in anisotropic media (polar coordinate presentation of c_p versus θ).

These results may be confirmed by checking the eigenvector direction that leads to three mutually perpendicular directions.

In general, we can solve the eigenvector problem by solving three times for each c (and hence the direction cosines α_i for each c):

$$\alpha_1\lambda_{11} + \alpha_2\lambda_{12} + \alpha_3\lambda_{13} = \alpha_1\rho c^2,$$

$$\alpha_1\lambda_{21} + \alpha_2\lambda_{22} + \alpha_3\lambda_{23} = \alpha_2\rho c^2,$$

$$\alpha_1\lambda_{31} + \alpha_2\lambda_{32} + \alpha_3\lambda_{33} = \alpha_3\rho c^2.$$

This general solution was derived by Chistoffel many years ago.

Sample problems can be studied for a variety of different materials. Stiffness coefficients for a few selected materials can be found in Table 3-1.

3.4 On Velocity, Wave, and Slowness Surfaces

We will now contemplate several possible phase velocity profiles for anisotropic media, all nonspherical in nature. Several possible velocity surfaces are illustrated in Figure 3-2.

If we are given specific phase velocity values as a function of wavevector direction or a specific angle θ then we can visualize a slowness profile, which is simply a plot of $1/c_p$ versus $\bar{k}$ or θ. Consider the sketch in Figure 3-3, which was generated by solving the Christoffel equation discussed previously. Useful information can then be extracted from the resulting curve.

If we are given the wave surface and the $\bar{k}$ or θ direction, then we can calculate c_p and ϕ. For example, the energy velocity and group velocity vectors are the same for lossless media. This can be seen by examining a normal to the slowness profile – as illustrated in Figure 3-3, which also gives us the skew angle ϕ. If we were now able to plot (on a

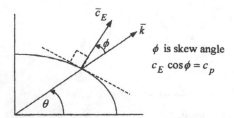

Figure 3-3. Sample slowness profile of $1/c_p$ versus θ.

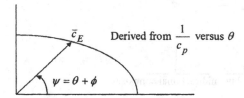

Figure 3-4. Sample wave surface showing c_E versus ψ.

point-by-point basis) a locus of points as a function of θ, we would be able to produce group velocity and skew angle as a function of θ. If we were instead to plot c_E versus the sum of θ plus ϕ (which we call ψ), then we would obtain the wave surface term illustrated in Figure 3-4. This gives us the group or energy velocity variation with angle (with no mention of launch angle if we didn't know θ and ϕ).

For some physical insight, imagine the actual wave surface propagating from an acoustic source, sending waves in one particular direction that could be sensed with some sensor for a field distribution. As a result of a superposition of all of the phase velocity contributions from different directions, if the source had some finite size then the energy velocity vector would be normal to the slowness profile for the material. See Love (1926), Musgrave (1959), or Pollard (1977) for more details.

We can now explore briefly the fact that the group velocity is normal to the slowness profile. For many different kinds of problems, we'll be extracting information similar to that found to produce a frequency spectrum of $\omega(\bar{k})$. For example, $\omega(k_x)$ is obtained for the one-dimensional string problem; $\omega(k_x, k_y)$ is the result for the anisotropic media problem in one plane. In general, $\omega(k_x, k_y, k_z)$ could be obtained for a three-dimensional problem.

Let's consider $\omega(k_x, k_y)$. This result could be plotted in three dimensions: for a particular ω value, a plane parallel to the (k_x, k_y)-plane could be used to intersect the conical-like $\omega(k_x, k_y)$-surface. For anisotropic media, an intersection plane parallel to the (k_x, k_y)-plane does not produce a circle; see Figure 3-5.

Any ω could be selected, since only a scale factor would be changed and the general shape of the curve would be the same. Note that the line drawn from the origin to a point in question would give c_p in that particular direction. From three-dimensional calculus, the normal to the tangent plane to the $\omega(k_x, k_y)$ function at the point of interest would

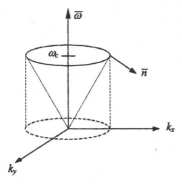

Figure 3-5. Typical intersection of the $\omega(k_x, k_y)$-surface with
a plane parallel to the (k_x, k_y)-plane, showing slowness profile
and projection onto the (k_x, k_y)-plane.

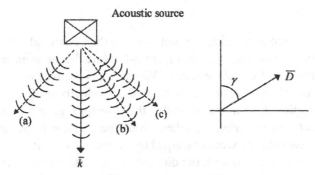

Figure 3-6. Possible directions of wave propagation from an
acoustic source in anisotropic media.

have a slope equal to $(\partial\omega/\partial k_x, \partial\omega/\partial k_y, 1)$. The intersection of the plane could achieve
the slowness curve, $(1/c_p)(k_x, k_y)$; this projection onto the (k_x, k_y)-plane is what we normally extract from the Christoffel equation solution results. The projection of $\bar{n}$ onto the
(k_x, k_y)-plane or dot product is simply $(\partial\omega/\partial k_x, \partial\omega/\partial k_y, 0)$, which we recognize as the
group velocity.

It is easy to show that this group velocity vector direction is normal to the slowness profile. Imagine an intersection of a plane parallel to the (k_x, k_y)-plane with the tangent plane
to the $\omega(k_x, k_y)$ function at the point in question. This projects as a line onto (k_x, k_y)-
space that is tangent to the curve $(1/c_p)(k_x, k_y)$. Consider $\bar{t}$ as the tangent line with slope
$(t_x, t_y, 0)$. Then $\bar{t} \bullet \bar{n} = 0$ and

$$t_x\left(\frac{\partial\omega}{\partial k_x}\right) + t_y\left(\frac{\partial\omega}{\partial k_y}\right) + 0 = 0.$$

An alternative argument is presented by Auld (1990) whereby $\bar{k} + \Delta\bar{k}$ is considered in a
limit process to show that c_g is normal to the slowness profile.

In the wave propagation diagrams of Figure 3-6, wave packets are sent out in all directions but the wavevector is in the phase velocity direction $\bar{k}$ only. If we view (via

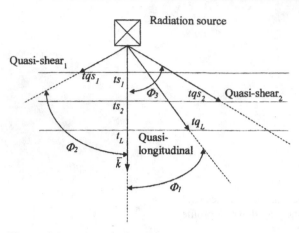

Figure 3-7. Wave propagation into an anisotropic solid in a particular wavevector direction.

Huygens's principle) the acoustic source as a series of point sources, then potential constructive interference path directions are as illustrated in Figure 3-6. Note that an infinite number of point sources is contemplated for a plane wave. Waves across the face of the transducer emanate in all directions. If the velocity surface, or phase velocity profile, is symmetric with respect to the direction $\bar{D}$ (sometimes called the *director*), giving us a sense of the orientation of an anisotropic material, then there would be no skew angle in that particular direction and the phase velocity would be equal to the energy velocity. On the other hand, if the wavevector $\bar{k}$ is *not* in line with the director of the material then the wave interference pattern becomes much more difficult to evaluate. The waves propagating from each point source are nonspherical in nature, but in this case are inclined at an angle γ. The selected wave path for energy concentration will be (a), (b), (c), or some other path, depending on the actual phase velocity profile, the inclined angle γ, and the resulting interference patterns.

Consequently, from a piston or point source (for example), three wave surfaces are produced – each with its own phase velocity variation with direction. The faster wave surface is for longitudinal waves, followed by two shear wave surfaces, possibly with different phase velocity profiles. All of this is possible owing to interference phenomena and to the changes taking place as a result of wave velocity variations with direction, which modify the interference patterns.

For a particular direction $\bar{k}$, consider the sketch in Figure 3-7. The distances illustrated to times ts_1, ts_2, and t_L are proportional to the phase velocities for the directions $\bar{k}$. The three skew angles are also shown. The distances tqs_1, tqs_2, and tq_L are proportional to the group velocities for the three wave types.

A number of interesting presentations can now be made. For example, examining in Figure 3-8 a wave surface profile of c_E versus $\theta + \phi$ shows us how to extract θ and the phase velocity value if needed. Because $\bar{k}$ is normal to the surface at the point in question for a particular $\bar{c}_E$, all information is available. If we wanted to achieve a certain energy velocity value – or, more importantly, a specific energy velocity direction – this technique shows us how to place a transducer at a specified θ or phase velocity direction, as illustrated in the figure.

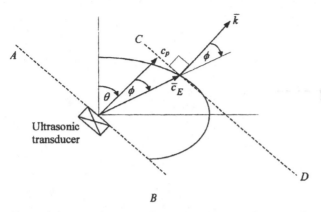

Figure 3-8. Transducer location angle calculation for a specific energy velocity magnitude or direction.

That is, if we are given $\bar{c}_E$ or the angle of $\bar{c}_E$, we can find out what transducer angle could be used to achieve this. We need only carry out the following steps, which are illustrated in Figure 3-8.

(1) Plot CD tangent to wave surface at $\bar{c}_E$ tip.
(2) Plot AB parallel to CD.
(3) We know θ and c_p (where $c_p = |\bar{c}_E| \cos \phi$) and hence the $\bar{k}$ direction from known θ.
(4) The angle ϕ between $\bar{k}$ and the $\bar{c}_E$ direction is also known and plotted.

3.5 Exercises

1. Derive the following expressions for the velocities of longitudinal and transverse waves traveling in the [1, 1, 1] direction in a cubic crystal:

$$c_1 = \sqrt{(C_{11} + 2C_{12} + 4C_{44})/3\rho},$$
$$c_2 = c_3 = \sqrt{(C_{11} - C_{12} + C_{44})/3\rho}.$$

2. For Problem 1, define the wave types c_1, c_2, c_3 as pure or quasi-longitudinal and as pure or quasi-shear.

3. Verify the formula

$$c_1 = \sqrt{C_{11}/\rho}, \quad c_2 = \sqrt{C_{44}/\rho}, \quad c_3 = \sqrt{(C_{11} - C_{12})/2\rho}$$

for longitudinal and shear waves traveling in the [1, 0, 0] direction in a hexagonal crystal.

4. Carefully list all steps necessary to produce a computer program that calculates points on a curve of phase velocity versus angle in the (x, y)-plane for a transversely isotropic material. What assumptions are made in the analysis? How can you define the modes and the pure or quasi nature of the modes? Is the acoustic tensor symmetric?

5. For a cubic tungsten crystal, plot approximately the three slowness profile curves.

6. From the cubic tungsten slowness profile, graphically estimate the wave surface of $\bar{c}_E$ versus ψ or $\theta + \phi$ for the quasi-longitudinal case.

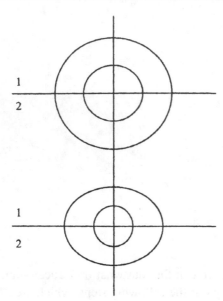

Figure 3-9. Exercise 16.

7. Graphically estimate skew angle ϕ versus θ for the quasi-longitudinal case in cubic tungsten.

8. What angle θ is required to produce a resulting energy velocity at 45° for the quasi-longitudinal case in cubic tungsten?

9. Calculate the resulting phase velocity in the [1, 1, 0] direction for transversely isotropic unidirectional graphite epoxy.

10. For a plane wave in a transversely isotropic crystal, derive an expression for the acoustic tensor and the wave velocities in the [1, 0, 0] direction.

11. For cubic silicon, develop a velocity surface of c_1, c_2, and c_3 as a function of wavevector direction in the (x_1, x_2)-plane. Use at least three points followed by interpolating estimates.

12. Calculate the phase velocity in a particular direction for transversely isotropic titanium or a barium titanate crystal.

13. Calculate the phase velocity in a particular direction for orthotropic barium sodium niobate.

14. Solve for the elastic constants of a cubic diamond material if the wave velocities were measured for plane waves in the [1, 1, 0] direction as follows: $c_1 = 17,890$ m/s; $c_2 = 11,930$ m/s; $c_3 = 10,330$ m/s.

15. For Plexiglas over tungsten, determine all critical angles. Do likewise for tungsten over Plexiglas.

16. For the slowness profiles in isotropic media 1 and 2 (shown in Figure 3-9), measure critical angles for L and S input and also for total reflection (1 to 2 and 2 to 1).

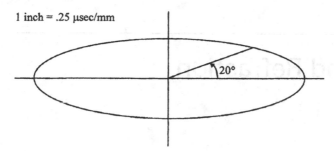

Figure 3-10. Exercise 17.

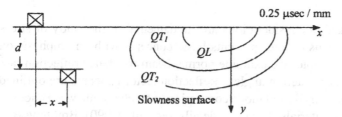

Figure 3-11. Exercise 18.

17. **(a)** For the slowness profile shown in anisotropic media (Figure 3-10), calculate the skew angle and energy velocity at the wavevector direction shown.

   (b) Estimate the energy velocity value in the wavevector direction shown.

   (c) How could this result be used to produce a wave surface?

18. Given an ultrasonic transducer emitting plane wave segments on the surface of an anisotropic medium with the slowness curves shown in Figure 3-11, determine the ideal location (of a receiving transducer in through-transmission) to receive maximum energy from quasi-longitudinal wave QL and from quasi-transverse waves QT_1 and QT_2.

19. How could you produce an entire slowness profile for an anisotropic material?

20. Plot skew angle versus $\bar{k}$ for a given $1/c_p$ curve (versus ϕ in a plane).

21. Plot energy velocity versus $\bar{k}$ for a given $1/c_p$ curve (versus ϕ in a plane).

22. Select a slowness profile for a realistic material; use any reference or research paper to obtain constants. Evaluate energy velocity and skew angle profiles.

23. Estimate or use a realistic wave surface for this problem: From a given wave surface, calcuate the phase velocity and skew angle for a particular $\bar{k}$ or wave-traveling direction.

4

Reflection and Refraction

4.1 Introduction

Wave reflection and refraction considerations are fundamental to the study of stress wave propagation in solids. This chapter presents basic concepts with an emphasis on physical phenomena. In this chapter we examine normal beam incidence reflection factors as well as computation of refraction angles. Reflection factor concepts are outlined first, followed by angle beam analysis and mode conversion as an ultrasonic wave encounters an interface between two materials. For more details, see Auld (1990), Brekhovskikh (1960), Graff (1991), or Kolsky (1963).

4.2 Normal Beam Incidence Reflection Factor

A plane wave encountering an interface between two materials is divided into two components: some energy at the interface is transmitted and some is reflected. The formula allowing us to compute reflection factor at an interface for normal incidence is presented in Figure 4-1.

This equation can be derived by matching normal stress at the interface as well as by matching displacement or particle velocity. Consider an incident harmonic plane wave σ_I traveling in an x direction to an interface between two media, as shown in Figure 4-1. Stress is reflected σ_R and transmitted σ_T. Since the elastic field is independent of the y direction, all derivatives with respect to y will vanish from the equations of motion. In this very simple case, the governing equation is the simple wave equation, which is applicable for either longitudinal or shear waves:

$$\frac{\partial^2 u_x}{\partial x^2} = \frac{1}{c_L^2} \frac{\partial^2 u_x}{\partial t^2} \tag{4.1}$$

or

$$\frac{\partial^2 u_y}{\partial x^2} = \frac{1}{c_T^2} \frac{\partial^2 u_y}{\partial t^2}, \tag{4.2}$$

where u_x and u_y are displacement vector components along the x- and y-axis, respectively.

We will now address the compressional wave. The solution for a compressional harmonic wave in equation (4.1) is

$$u_x = A_1 e^{i(kx - \omega t)} + A_2 e^{-i(kx + \omega t)}, \tag{4.3}$$

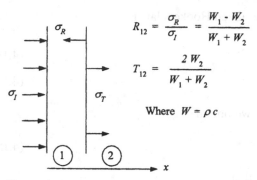

Figure 4-1. Reflection factor.

where

$$\text{wavenumber } k = \frac{\omega}{c_L}, \quad c_T^2 = \frac{\mu}{\rho}, \quad \text{and} \quad c_L^2 = \frac{\lambda + 2\mu}{\rho}. \tag{4.4}$$

The first term in (4.3) describes wave propagation in the positive x direction while the second term describes propagation in the negative direction.

Consider the incident waveform as

$$u_x^I = I e^{i(k_1 x - \omega t)}, \quad k_1 = \frac{\omega}{c_L^{(1)}}. \tag{4.5}$$

In this case, the reflected field can be written as

$$u_x^{(R)} = A_R e^{-i(k_1 x + \omega t)}. \tag{4.6}$$

The transmitted field in the second medium is

$$u_x^{(T)} = A_T e^{i(k_2 x - \omega t)}, \quad k_2 = \frac{\omega}{c_L^{(2)}}; \tag{4.7}$$

here, A_R and A_T are unknown.

We must satisfy boundary conditions on the interface. The entire elastic field in medium 1 is

$$u^1 = u_x^{(I)} + u_x^{(R)} = I e^{i(k_1 x - \omega t)} + A_R e^{-i(k_1 x + \omega t)}. \tag{4.8}$$

In medium 2,

$$u^2 = u_x^{(T)} = A_T e^{i(k_2 x - \omega t)}. \tag{4.9}$$

Boundary conditions are as follows:

$$u^{(1)}|_{x=0} = u^{(2)}|_{x=0}, \tag{4.10}$$

$$\sigma_x^{(1)}|_{x=0} = \sigma_x^{(2)}|_{x=0}, \tag{4.11}$$

$$\sigma_{xy}^{(1)}|_{x=0} = \sigma_{xy}^{(2)}|_{x=0}. \tag{4.12}$$

For the one-dimensional case, the generalized Hooke's law is

$$\sigma_x = (\lambda + 2\mu)\frac{\partial u_x}{\partial x},$$ (4.13)

$$\sigma_{xy} = 0.$$ (4.14)

By substituting (4.8) and (4.9) into (4.10), we obtain

$$I + A_R = A_T.$$ (4.15)

From (4.8), (4.9), and (4.13), it follows that

$$\sigma_x^{(1)} = i(\lambda_1 + 2\mu_1) \cdot k_1 [I e^{i(k_1 x - \omega t)} - A_R e^{-i(k_1 x - \omega t)}],$$ (4.16)

$$\sigma_x^{(2)} = i(\lambda_2 + 2\mu_2) \cdot k_2 \cdot A_T e^{i(k_2 x - \omega t)}.$$ (4.17)

Substituting (4.16) and (4.17) into (4.11) yields

$$(\lambda_1 + 2\mu_1) \cdot k_1 [I - A_R] = (\lambda_2 + 2\mu_2) k_2 A_T,$$ (4.18)

$$\text{wavenumber } k_n = \frac{\omega}{c_L^{(n)}} \quad (n = 1, 2),$$ (4.19)

$$\lambda_n + 2\mu_n = \rho_n \cdot [c_L^{(n)}]^2,$$ (4.20)

where n is the medium number. Finally, using (4.19) and (4.20), equations (4.15) and (4.18) lead to the following system:

$$I + A_R = A_T$$

$$\rho_1 c_L^{(1)} (I - A_R) = \rho_2 c_L^{(2)} \cdot A_T.$$ (4.21)

Solution of this system gives

$$A_R = \frac{\rho_1 c_L^{(1)} - \rho_2 c_L^{(2)}}{\rho_1 c_L^{(1)} + \rho_2 c_L^{(2)}} \cdot I,$$ (4.22)

$$A_T = \frac{2\rho_1 c_L^{(1)}}{\rho_1 c_L^{(1)} + \rho_2 c_L^{(2)}} \cdot I.$$ (4.23)

The reflected and transmitted wave fields can be obtained from (4.6) and (4.7) by using expressions (4.22) and (4.23).

We may now consider incident, reflected, and transmitted stresses. From (4.13) it follows that

$$\sigma_x^{(I)} = (\lambda_1 + 2\mu_1)\frac{\partial u_x^{(I)}}{\partial x} = i k_1 (\lambda_1 + 2\mu_1) e^{i(k_1 x - \omega t)}.$$ (4.24)

Similarly, the reflected stress field is obtained as

$$\sigma_x^{(R)} = -ik_1(\lambda_1 + 2\mu_1)e^{-(k_1 x - \omega t)}. \tag{4.25}$$

The transmitted wave field is presented in equation (4.17). The stress reflection and transmission coefficients are therefore obtained from (4.17), (4.24), and (4.25) as follows:

$$R = \left.\frac{\sigma_x^{(R)}}{\sigma_x^{(I)}}\right|_{x=0} = -\frac{A_R}{I} = -\frac{\rho_1 c_1 - \rho_2 c_2}{\rho_1 c_1 + \rho_2 c_2}, \tag{4.26}$$

$$T = \left.\frac{\sigma_x^{(T)}}{\sigma_x^{(I)}}\right|_{x=0} = \frac{(\lambda_2 + 2\mu_2)k_2}{(\lambda_1 + 2\mu_1)k_1} \cdot \frac{A_T}{I} = -\frac{2\rho_2 c_2}{\rho_1 c_1 + \rho_2 c_2}. \tag{4.27}$$

Defining acoustic impedance as $W = \rho c_L$, the reflection coefficient can thus be written as

$$R = \frac{W_2 - W_1}{W_1 + W_2}; \tag{4.28}$$

the transmission coefficient is

$$T = \frac{2W_2}{W_1 + W_2}. \tag{4.29}$$

The formula for energy partition at the interface into transmission and reflection modes can be derived by considering energy as proportional to the square of the pressure magnitude. The energy flow per unit of time through a unit area normal to the direction of propagation is defined as the intensity of the wave (I_I, I_R, or I_T). The intensity is evaluated over one cycle and depends on the amplitude as follows (see Pollard 1977):

$$I_R = \frac{A_R^2}{2\rho_1 c_L^{(1)}}, \quad I_T = \frac{A_T^2}{2\rho_2 c_L^{(2)}}, \quad I_I = \frac{1}{2\rho_1 c_1^{(L)}}. \tag{4.30}$$

Equations (4.26), (4.27), and (4.30) then yield

$$\frac{I_R}{I_I} = \left(\frac{B-1}{B+1}\right)^2 \quad \text{and} \quad \frac{I_T}{I_I} = \frac{4B}{(B+1)^2},$$

where $B = W_2/W_1$.

The wave velocity and acoustic impedance values of ultrasound in various materials are listed in Table 4-1; sample reflection factor results for various materials are presented in Table 4-2. (See Appendix E for a basic wave velocity experiment.) The expression for acoustic impedance, W, can be considered as a characteristic property of the material being studied:

$$W = \rho c \sim \sqrt{\rho E}.$$

The inclusion of acoustic impedance in this equation gives us some indication of material stiffness. The stiffness is often related to a Young modulus, E, of the material; in general, the stiffer the material, the higher the wave velocity.

Table 4-1. *Wave velocity values for selected material samples*

Material	Density (g/cm³)	Wave velocity (m/s) Longitudinal	Shear
Air	0.001	330	—
Aluminum	2.7–2.8	6,250–6,350	3,100
Beryllium	1.82	12,800	8,710
Bone	1.738	2,240 ± 8%	—
Brass	8.1	4,430	2,120
Bronze	8.86	3,530	2,230
Cast iron	7.7	4,500	2,400
Chocolate (dark)	1.302	2,584	960
Copper	8.9	4,660	2,260
Cork	0.24	510	—
Glass	2.23–2.51	5,570–5,770	3,430–3,440
Glycerin	1.261	1,920	—
Gold	19.3	3,240	1,200
Ice	1.00	3,980	1,990
Lead	11.4	2,160	700
Lead zirconate titanate	7.65	3,791	—
Magnesium	1.74	5,790	3,100
Nickel	8.3	5,630	2,960
Oil	0.92–0.953	1,380–1,500	—
Plexiglas	1.18	2,670	1,120
Polyethylene (typical)	0.920	2,000	—
Quartz	2.65	5,736	—
Silver	10.5	3,600	1,590
Soft tissue	1.06	1,540	—
Steel	7.8	5,850	3,230
Stainless steel	7.67–8.03	5,660–7,390	2,990–3,120
Tin	7.3	3,320	1,670
Titanium	4.54	6,100	3,120
Tungsten	19.25	5,180	2,870
Water (20°C)	1.00	1,480	—
Wood, oak (with grain)	0.4615	4,640	1,750
Wood, oak (against grain)	0.4615	1,630–2,150	1,460–1,750
Zinc	7.1	4,170	2,410

4.3   Snell's Law for Angle Beam Analysis

Several things happen when an ultrasonic wave encounters an interface between two materials at some inclined angle. First of all, refraction occurs in much the same way as it occurs for optical light waves. Consider the diagram in Figure 4-2. The refracted angle can be computed from Snell's law: $c_1 \sin \theta_2 = c_2 \sin \theta_1$.

The second thing that occurs at the interface is mode conversion. Energy is distributed into longitudinal and shear waves in the second material. Some energy is also reflected, where the angle of reflection is equal to the angle of incidence. The concept of mode conversion is illustrated in Figure 4-3. As an ultrasonic wave encounters an interface between

Table 4-2. *Reflection and transmission ratios for selected material interfaces*

Reflected surface (1)	σ_R/σ_I (2)	I_R/I_I (3)	σ_T/σ_I (4)	I_R/I_I (5)
Cork–steel	0.995	0.989	0.005	0.011
Steel–cork	−0.995	0.989	1.995	0.011
Aluminum–Plexiglas	−0.692	0.479	1.692	0.521
Plexiglas–aluminum	0.692	0.479	0.308	0.521

Notes: Column (1), material 1 to material 2; column (2), amplitude ratio of reflected to incident wave; column (3), energy ratio of reflected to incident wave; column (4), amplitude ratio of transmitted to incident wave; column (5), energy ratio of transmitted to incident wave.

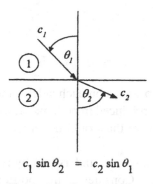

$$c_1 \sin \theta_2 = c_2 \sin \theta_1$$

Figure 4-2. Snell's law for angle beam analysis.

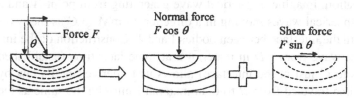

Figure 4-3. Point-source mode conversion concept.

two materials, the incident energy is portioned into different kinds of ultrasonic energy. Consider a force acting on the interface at an angle θ. This force independently generates a very complicated wave motion in the structure. The complicated wave motion can be considered as a superposition of two specific motions: one associated with a normal force and normal wave propagation, the second associated with a shear force and shear wave propagation. Keep in mind that pulse-type excitation at distances far away from the loading source causes separation of the two waveforms, since the shear wave velocity is less than the normal wave velocity. With continuous-wave excitation, however, the two wave motions are superimposed and fairly complex throughout the entire structure. A complete

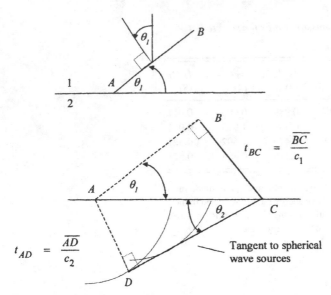

Figure 4-4. Derivation of Snell's law.

analysis of the mode conversion problem would consider two waves on each side of the interface. In this case, both normal and shear waves would be produced in both materials, and the superposition process would generate an interface wave that could travel along the interface of the two materials.

Snell's law and the qualitative aspects of mode conversion can be understood by reviewing aspects of the mathematical derivation of Snell's law. Consider a finite portion of an inclined wavefront in the plane of the paper as segment AB, traveling toward the interface between materials 1 and 2 as shown in Figure 4-4. Using Huygens's principle of spherical wave propagation, imagine a spherical wave generating from point A and a subsequent generator of spherical waves moving in a direction from A to C as the wave segment AB moves toward the interface between bodies 1 and 2. Construction of the initial wavefront AB and the wave from CD in material 2 at some later time (established when the point B travels to C just at the interface) allows us to derive Snell's law by trigonometry. Since the time traveled from B to C in material 1 must be the same as the time it takes for a wave to move from A to D in material 2, we can easily set these times equal to derive Snell's law:

$$\sin \theta_1 = \frac{\overline{BC}}{\overline{AC}} \quad \text{and} \quad \sin \theta_2 = \frac{\overline{AD}}{\overline{AC}}$$

and so $\overline{AC} = \overline{AC}$; therefore,

$$c_1 \sin \theta_2 = c_2 \sin \theta_1.$$

Snell's law, derived on wave speed analysis only, can also be used to calculate refracted shear angles in material 2. The two equations in Figure 4-5 can therefore be deduced from Snell's law.

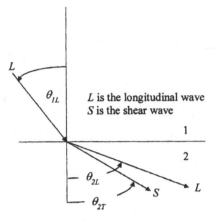

$$c_{1L} \sin \theta_{2L} = c_{2L} \sin \theta_{1L}$$

$$c_{1L} \sin \theta_{2T} = c_{2T} \sin \theta_{1L}$$

Figure 4-5. Snell's law and mode conversion.

Table 4-3. *Refraction angle values for various incident angles of a longitudinal wave*

Interface	θ_{2L} value for $\theta_I = 15°$	for $\theta_I = 30°$	for $\theta_I = 45°$	$\theta_{2L} = 90°$ $\theta_I = \theta_{cr}$
Steel–Plexiglas	6.8°	13.2°	18.8°	—[a]
Plexiglas–steel	34.5°	—	—	$\theta_{cr} = 27°$

[a]No critical angle.

Table 4-3 lists sample refraction angle values for ultrasonic waves traveling from one material to another at incidence angles of 15°, 30°, and 45°. Waves traveling from a media with slower wave speed than media 2 are bent away from the normal; this is illustrated in Figure 4-6. Waves traveling from a higher wave speed than media 2 are bent toward the normal. A basic experiment on refraction and angle beam transducers may be found in Section E.2.

4.4 Critical Angles and Mode Conversion

We will now examine some critical angle concepts, as illustrated in Figure 4-6, in order to further explore what happens as an ultrasonic wave encounters an interface between two materials.

Consideration of Snell's law indicates that two critical angles exist with respect to the refraction process.

(1) The first critical angle can be defined as

$$\theta_{cr} = \sin^{-1} \frac{c_1}{c_{2L}} \quad \text{when } \theta_2 = 90°.$$

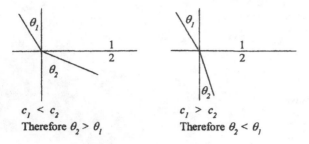

$c_1 < c_2$
Therefore $\theta_2 > \theta_1$

$c_1 > c_2$
Therefore $\theta_2 < \theta_1$

a) First critical angle can be defined as:

$$\theta_c = \sin^{-1}\frac{c_1}{c_2} \quad \text{when } \theta_2 = 90°$$

b) Second critical angle occurs when the shear refracted angle is 90°

Figure 4-6. Critical angle concepts.

In this case, all of the longitudinal energy is either reflected or converted to an interface wave. Only shear waves remain in the second material.

(2) The second critical angle occurs when the shear refracted angle is 90°. In other words, no significant energy is propagated through the second material; all of the energy is either reflected or transformed into interface wave propagation. The second critical angle can be computed by evaluating the inverse sine function of the ratio c_1/c_{2S}.

If one were to compute the critical angle between steel and Plexiglas, one would find that such an angle does not exist ($\sin\theta > 1$). In other words, refracted angles are less than incident angles if the wave velocity in the second material is slower than in the first material. On the other hand, refracted angles are greater than incident angles when wave velocities in the second material are greater than in the first material.

We may now summarize some important topics on mode conversion associated with the subject of angle beam analysis. Figure 4-7 depicts four possibilities of mode conversion. Part (a) shows the possible waveforms produced from a longitudinal incident wave at an angle θ_L. Refracted waves in a second material may be longitudinal or shear; reflected waves also may be longitudinal or shear. The two refracted and two reflected angles can be calculated from Snell's law. For longitudinal incident waves, the angle of incidence is obviously equal to the angle of reflection.

Part (b) of Figure 4-7 illustrates the situation for an incident angle wave at the first critical angle. Only shear waves are propagated into the second material, with longitudinal waves going off at an angle of 90° and so producing an interface wave. In this case, of course, both longitudinal and shear waves are reflected. Part (c) illustrates an incident longitudinal wave at the second critical angle. In this case, it is not possible to produce ultrasonic energy in the second material, because both shear and longitudinal waves are reflected. In part (d), the shear wave reflection angle is equal to the shear wave incident angle.

Figure 4-8 illustrates an interesting feature of mode conversion: an interface wave can actually be generated before the first critical angle is attained as a wave travels from material 1 to material 2. The interface wave is produced as a result of beam spreading and beam

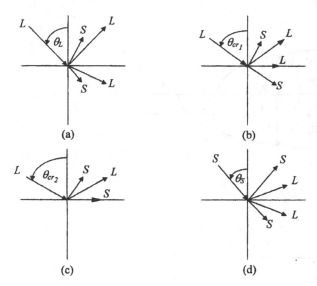

Figure 4-7. Mode conversion concepts: (a) general L input; (b) at θ_{cr_1}; (c) at θ_{cr_2}; (d) general S input.

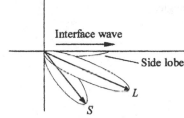

Interface wave before θ_{cr_1} (also before θ_{cr_2}) showing side lobe interaction with interface

Figure 4-8. Mode conversion and possible interface wave generation.

angle of divergence. The refracted longitudinal wave in the second material is shown with a principal lobe of ultrasonic energy surrounding it. Notice that, as the refracted longitudinal angle increases, a portion of the principal lobe of ultrasonic energy encounters the interface *before* the center point of the principal lobe reaches the 90° value.

The subject of energy partitioning is receiving attention in some areas of applied mechanics. It is often possible to evaluate what portions of the energy are converted into reflected shear and longitudinal, into refracted shear and longitudinal, and into interface waves from both the shear and longitudinal wave interaction with the interface. (Reflection and refraction factors for oblique incidence are discussed in Chapter 5.)

4.5 Slowness Profiles for Refraction and Critical Angle Analysis

We will now illustrate how slowness profiles can assist us in oblique incidence studies and subsequent critical angle analysis. If a wave encounters an interface between two anisotropic media then wave reflection and refraction will occur – in addition to any skew

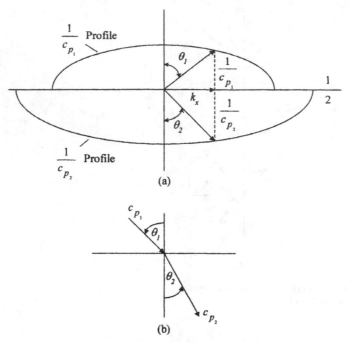

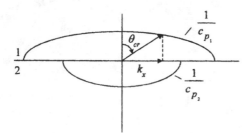

Figure 4-9. Slowness profiles for calculating refraction angle.

Figure 4-10. Slowness curves for calculating critical angles.

that may already be occurring. Let's consider symmetric but different elliptic-type slowness profiles, as illustrated in Figure 4-9. If a wave is incident to material 1 at a specific angle θ (as measured from the normal) then the wave vector component along the interface must be preserved. This is simply a restatement of Snell's law:

$$\frac{1}{c_{p1}} \sin \theta_1 = \frac{1}{c_{p2}} \sin \theta_2,$$

so $c_{p1} \sin \theta_2 = c_{p2} \sin \theta_1$. This allows us to find the refraction angle in the second material.

The same approach can also be used to evaluate critical angles. In Figure 4.9, for example, no critical angle exists; it is impossible to have θ_2 become 90°, despite the increase in θ_1, because the wave velocity in media 2 is less than that in media 1. Figure 4-10, however, shows an example where a critical angle can occur. When the incidence angle in medium 1 reaches θ_{cr}, the refraction angle in medium 2 is 90°; this is the definition of critical angle.

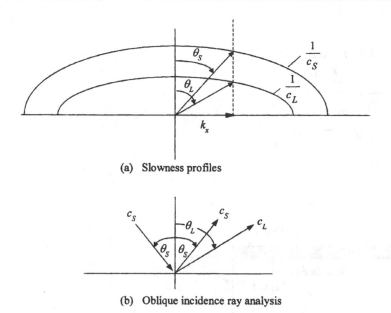

(a) Slowness profiles

(b) Oblique incidence ray analysis

Figure 4-11. Slowness curves for calculating reflection angles for shear input.

Figure 4-11 shows how slowness curves can be used to evaluate reflection angles as well as refraction and critical angles. For an incident shear wave at θ_S, the reflected shear angle will, of course, still be θ_S. Mode conversion does take place, however, and some longitudinal waves are reflected also. The reflected longitudinal wave angle will be θ_L, as illustrated in Figure 4-11, with k_x being preserved.

Numerous examples can now be studied; some are left as exercises. For realistic examples, solutions to Christoffel equations for the phase velocity functions could be combined with sample elastic constants to evaluate aspects of skew angle, energy velocity, oblique incidence, critical angles, and so forth. See Table 3-1 (p. 32) for some material constants that can be useful in problem solving.

4.6 Exercises

1. Calculate the reflection factor from steel to air. What does the negative sign mean? Think about boundary conditions on the free surface.

2. Calculate reflection factors (as in Table 4-2) for steel to air, water to Plexiglas, and Plexiglas to water.

3. Calculate refraction angle values for water to Plexiglas and Plexiglas to water at the angles indicated in Table 4-3.

4. Calculate the reflected longitudinal wave angle for an incident shear wave at 30° onto a steel–air interface.

5. Develop an equation for oblique incidence reflection angles for longitudinal or shear waves, using either longitudinal or shear waves as input.

6. Calculate the first and second critical angle from Plexiglas to steel. What special situation occurs if wave entry is limited to this range?

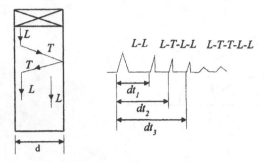

Figure 4-12. Exercise 8.

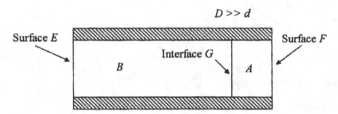

Figure 4-13. Exercise 9.

7. What would you observe when comparing reflected signals from steel to Plexiglas and from Plexiglas to steel? Explain the reason.

8. Solve for the dt values in the resulting pulse-echo rectified low-pass–filtered RF wave-front pattern shown in Figure 4-12. Assume longitudinal wave input to a rod and mode conversion along a rough outer surface.

9. In Figure 4-13, cladding (of thickness d) of material A is perfectly diffusion bonded onto a material B of length D (shading is for sound absorption). Let $c_B = 6,000$ m/s, $c_A = 2,000$ m/s, $\rho_B = 8,000$ kg/m³, and $\rho_A = 2,000$ kg/m³. If surface E is loaded with a 3-cycle 1-MHz sine wave pulse, consider the pulse-echo wave reflection pattern at E.

   (a) Calculate the pressure amplitude reflection factor at the interface G and also at the back surface F.

   (b) Considering phase variation only (not amplitude), what thickness d produces a maximum response? A minimum response? What thickness value d could be resolved with the 3-cycle 1-MHz sine wave?

   (c) Show sample results for $\lambda \ll d$ on a wave diagram for at least two reflections in D. At what time value does echo identification become difficult? Illustrate and explain.

   (d) What are the longitudinal and shear reflection factor values at the interface G for a smooth interface? (Consider G as an infinitesimally thin fluid film.)

10. What are the reflection and transmission factors for normal beam shear wave impingement onto an interface between two media?

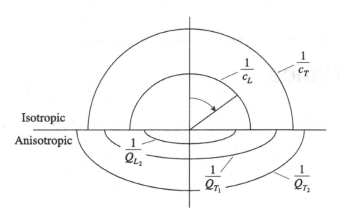

Figure 4-14. Exercise 14.

11. Suppose we were to calculate wave velocity values in the (x_1, x_2)-plane for some orthotropic media. For at least one particular direction – say, $[1, 1, 0]$ – in this plane, detail the steps of the computational process. How would you determine if the wave velocity vector had pure or quasi-longitudinal or shear components? Also, solve for one element – say, Γ_{23} – of the acoustic tensor.

12. Plot reflection and refraction angle for a specific angle of incidence of an L or S wave onto a steel–Plexiglas–steel structure. Calculate θ_1 (the first critical angle) and θ_2 (the second critical angle). Also give a graphical solution using slowness profiles.

13. Compute critical angles for waves impinging onto a silicon–steel interface that is perfectly diffusion bonded and onto a steel–silicon interface (use data from interpolation estimates if necessary). Consider L and S inputs and all possible critical angles. (Consider the definition of critical angle as a phase velocity going to 90°.)

14. For the slowness profiles illustrated in Figure 4-14, use a graphical technique to solve for all reflected and refracted angles for the incident angle shown. Also, how would you define and calculate the first, second, and third critical angles?

5

Oblique Incidence

5.1 Background

One of the most important topics associated with the subject of stress wave propagation in solid materials is the wave reflection and refraction at an interface between two different media. (For more details, see Auld 1990; Graff 1991; or Pilarski, Rose, and Balasubramaniam 1990.) The reflection (refraction) factor, or coefficient, is defined as the ratio of the amplitude of the reflected (refracted) wave to the amplitude of the incident wave. The factor depends on the angle of incidence, wave velocities, and possibly frequency, depending on the interface condition. In this chapter, we introduce two approaches for calculating these factors: a boundary condition approach and a multireflection superposition approach. We will first use the boundary condition approach for the interface between two semi-infinite medium spaces: solid–solid, solid–liquid, and liquid–solid. This approach will also be used to calculate reflection and refraction factors for a thin interface solid (and liquid) layer between two different media, following guidelines established by Jiao and Rose (1991) and from a "spring" model (Pilarski and Rose 1988a,b; Pilarski et al. 1990).

From physics and wave mechanics (Timoshenko and Goodier 1987; Graff 1991), a new wave may be generated from an incident wave at an interface, depending on the wave velocities in the two media and the angle of incidence. For example, a longitudinal wave in Plexiglas impinges the interface between Plexiglas and steel. The reflected and refracted waves could consist of both longitudinal and shear waves. At a certain angle of incidence, the refracted longitudinal wave disappears (i.e., the angle of refraction is 90°). This is the first critical angle. When the refracted shear wave disappears, the angle of incidence is the second critical angle.

What are the magnitudes of these waves in the different directions? Oblique incidence reflection and refraction factor calculations must be carried out to determine the mode conversion ratios. Several oblique incidence reflection and refraction sample problems are presented in this chapter. In the previous chapter we saw how – from Snell's law and a study of the slowness profiles between two media – we can determine all reflection and refraction angles, all critical angles, total reflection possibilities, and so forth. The problem now is to address the energy partitioning into the different modes (i.e., mode conversion).

It is possible to generate both longitudinal and shear waves in solids. However, in nonviscous liquid (e.g. water), only longitudinal waves can exist. An isotropic material has properties that are independent of direction in the material. Our emphasis is primarily on homogeneous isotropic materials; the approach would be similar for anisotropic materials (although the constitutive equations would change, and the algebra would become more cumbersome). For more information on this topic, see Auld (1990), Brekhovshikh

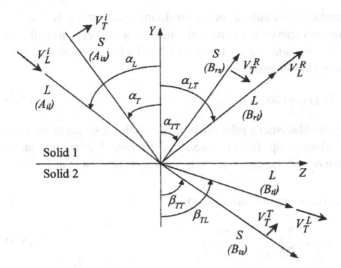

Figure 5-1. Reflection and refraction at a solid–solid media interface (β_{TT} and β_{TL} for shear input, but change to β_{LT} and β_{LL} for longitudinal input).

(1960), Henneke (1972), Jiao and Rose (1991), Pilarski and Rose (1988a,b), and Rokhlin and Marom (1986).

5.2 Reflection and Transmission Factors for Interfaces between Two Semi-Infinite Media

5.2.1 Solid–Solid Boundary Conditions

The reflection and refraction factors for a perfect solid–solid bond are derived as follows. Assume that isotropic solid media 1 and 2 are two infinite half-spaces in the coordinate system XYZ. That is, solid 1 is in the space $y > 0$; solid 2, in $y < 0$ (see Figure 5-1). For shear wave incidence, we can write expressions for particle velocities; the key to obtaining solutions is satisfying the boundary conditions. For particle velocity $\bar{V}$ and stress σ_{ij}, the boundary conditions calling for the continuity of particle velocity and stress on $y = 0$ can be written as follows for shear input:

$$
\begin{aligned}
V_y: & \ (V_T^I)_y + (V_T^R)_y + (V_L^R)_y = (V_T^T)_y + (V_L^T)_y, \\
V_z: & \ (V_T^I)_z + (V_T^R)_z + (V_L^R)_z = (V_T^T)_z + (V_L^T)_z, \\
\sigma_{yy}: & \ (\sigma_T^I)_{yy} + (\sigma_T^R)_{yy} + (\sigma_L^R)_{yy} = (\sigma_T^T)_{yy} + (\sigma_L^T)_{yy}, \\
\sigma_{yz}: & \ (\sigma_T^I)_{yz} + (\sigma_T^R)_{yz} + (\sigma_L^R)_{yz} = (\sigma_T^T)_{yz} + (\sigma_L^T)_{yz}.
\end{aligned}
\tag{5.1}
$$

Here, with V_n^m we have m as R for a reflected wave, T a transmitted wave, or I an incident wave; n is either L for a longitudinal wave or T for a transverse (shear) wave. For longitudinal input, we would consider V_L^I.

Consider now the following elastic constant–velocity relationships:

$$
C_{11} = \rho c_L = \lambda + 2\mu, \quad C_{44} = \rho c_T = \mu, \quad C_{12} = \lambda,
\tag{5.2}
$$

where C_{11}, C_{44}, and C_{12} are elastic constants, λ and μ are Lamé constants, ρ is density, and c_L and c_T are longitudinal and transverse wave velocities, respectively. Recall that the k projection on the z-axis is constant, an expression of Snell's law. We may further simplify these equations by using the identity

$$\lambda_1 + 2\mu_1 \cos^2 \alpha_L = (\lambda_1 + 2\mu_1) \cos(2\alpha_T). \tag{5.3}$$

Given the particle velocity–displacement relationship, the strain–displacement relationship, and the stress–strain relationship (from Hooke's law) for two-dimensional plane strain in isotropic media, we can obtain the stress–particle velocity relationship as follows:

$$\bar{V} = \frac{\partial \bar{u}}{\partial t} = i\omega \bar{u} \quad \text{(from harmonic motion principles)}; \tag{5.4}$$

$$\varepsilon_{yy} = \frac{\partial u_y}{\partial y}, \quad \varepsilon_{zz} = \frac{\partial u_z}{\partial z}, \quad \varepsilon_{yz} = \frac{1}{2}\left(\frac{\partial u_y}{\partial z} + \frac{\partial u_z}{\partial y}\right); \tag{5.5}$$

$$T_{yy} = \lambda(\varepsilon_{yy} + \varepsilon_{zz}) + 2\mu\varepsilon_{yy},$$
$$T_{yz} = 2\mu\varepsilon_{yz}; \tag{5.6}$$

$$T_{yy} = -\frac{i}{\omega}\lambda\left(\frac{\partial V_y}{\partial y} + \frac{\partial V_z}{\partial z} + 2\mu\frac{\partial V_y}{\partial y}\right),$$
$$T_{yz} = -\frac{i}{\omega}\mu\left(\frac{\partial V_y}{\partial z} + \frac{\partial V_z}{\partial y}\right). \tag{5.7}$$

Substituting the particle velocities and stresses into the boundary condition equations, we can find the reflection factor equations,

$$\mathbf{M}\begin{bmatrix} RTL \\ RTT \\ DTL \\ DTT \end{bmatrix} = a, \tag{5.8}$$

where a is a 4×1 matrix and M is a 4×4 matrix:

$$\mathbf{M} = \begin{bmatrix} -\cos\alpha_{LT} & \sin\alpha_{TT} & -\cos\beta_{TL} & \sin\beta_{TT} \\ -\sin\alpha_{LT} & -\cos\alpha_{TT} & \sin\beta_{TL} & \cos\beta_{TT} \\ -k_{L1}(\lambda_1 + 2\mu_1)\cos 2\alpha_{TT} & k_{T1}\mu_1\sin 2\alpha_{TT} & k_{L2}(\lambda_2 + 2\mu_2)\cos 2\beta_{TT} & -k_{T2}\mu_2\sin 2\beta_{TT} \\ -k_{L1}\mu_1\sin 2\alpha_{LT} & -k_{T1}\mu_1\cos 2\alpha_{TT} & -k_{L2}\mu_2\sin 2\beta_{TL} & -k_{T2}\mu_2\cos 2\beta_{TT} \end{bmatrix}. \tag{5.9}$$

Reflection and refraction factors are defined as

$$RTL = \frac{B_{rl}}{A_{is}}, \quad RTT = \frac{B_{rs}}{A_{is}}, \quad DTL = \frac{B_{tl}}{A_{is}}, \quad DTT = \frac{B_{ts}}{A_{is}}. \tag{5.10}$$

For each expression in (5.10), the first element in the left-hand side (LHS) is either R or D (reflected or transmitted wave), the second element is for the incident wave, T or L (transverse or longitudinal), and the third is for reflected or transmitted T or L waves. For the

RHS, A_{in} is the amplitude of the incident shear or longitudinal wave (A_{is} or A_{il}) and B_{mn} is the amplitude of the reflected or transmitted ($m = r$ or t, resp.) wave n (longitudinal or shear, resp. l or s).

For shear wave incidence,

$$a = \begin{pmatrix} \sin \alpha_T \\ \cos \alpha_T \\ -k_{T1}\mu_I \sin 2\alpha_T \\ -k_{T1}\mu_I \cos 2\alpha_T \end{pmatrix}. \tag{5.11}$$

The equation (5.8) for shear input can now be solved. For longitudinal wave incidence, a in (5.11) takes the form

$$a = \begin{bmatrix} -\cos \alpha_L \\ \sin \alpha_L \\ k_{L1}(\lambda_I + 2\mu_I) \cos 2\alpha_L \\ -k_{L1}\mu_I \sin 2\alpha_L \end{bmatrix} \tag{5.12}$$

while M remains unchanged.

The reflection and refraction factors in this case are defined as

$$RLL = \frac{B_{rl}}{A_{il}}, \quad RLT = \frac{B_{rs}}{A_{il}}, \quad DLL = \frac{B_{tl}}{A_{il}}, \quad DLT = \frac{B_{ts}}{A_{il}}. \tag{5.13}$$

These equations can now be solved for a particular input wave and given elastic constants. Solve the 4×4 matrix for RLL, RLT, DLL, and DLT.

Now consider a sample result: Figure 5-2 shows the reflection and refraction factors for oblique incident angles from $0°$ to $90°$. Note that the curves are continuous in part (a) of the figure, since there are no critical angles. In part (b), however, a discontinuity is present because of the critical angle for shear incidence.

5.2.2 Solid–Liquid Boundary Conditions

The reflection and refraction factor equations for a solid–liquid interface (see Figure 5-3) can be derived in the same way as those for solid–solid boundary conditions. The only difference is the absence of shear waves in the liquid, which leads to a zero shear stress and to continuity of normal stress and normal particle velocity at the interface.

The reflection and refraction equations under these conditions become

$$M \begin{bmatrix} RLL \\ RLT \\ DLL \end{bmatrix} = a_1 \quad \text{for longitudinal input,}$$

$$M \begin{bmatrix} RTL \\ RTT \\ DTL \end{bmatrix} = a_2 \quad \text{for shear input.} \tag{5.14}$$

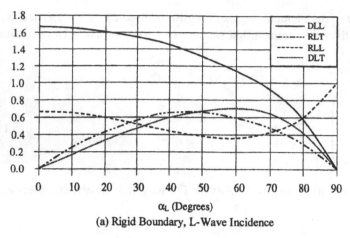

(a) Rigid Boundary, L-Wave Incidence

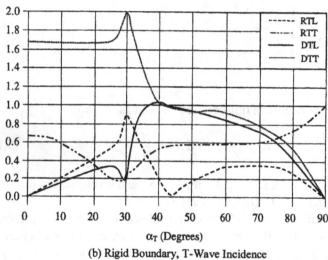

(b) Rigid Boundary, T-Wave Incidence

Figure 5-2. Reflection factor results for oblique incidence onto an aluminum–Plexiglas interface.

Both L and T exist in the solid, but only L exists in the fluid (as a natural consequence of the mode conversion process):

$$\mathbf{M} = \begin{bmatrix} -\cos\alpha_{LT} & \sin\alpha_{TT} & -\cos\beta_{TL} \\ -k_{L1}\mu(\lambda_1+2\mu_1)\cos 2\alpha_{TT} & k_{L2}\mu_1\sin 2\alpha_{TT} & 0 \\ -k_{L1}\mu_1\sin 2\alpha_{LT} & -k_{L2}\mu_1\cos 2\alpha_{TT} & 0 \end{bmatrix}; \quad (5.15)$$

$$a_1 = [\ -\cos\alpha_L \quad k_{L1}(\lambda_1+2\mu_1)\cos 2\alpha_L \quad -k_{L1}\mu_1\sin 2\alpha_L\ ]', \quad (5.16)$$

$$a_2 = [\ \sin\alpha_T \quad -k_{T1}\mu_1\sin 2\alpha_T \quad -k_{T1}\mu_1\cos 2\alpha_T\ ]'. \quad (5.17)$$

Note that primes are used to indicate the transpose of the matrix.

Let's evaluate a sample result of oblique incidence onto a solid–liquid boundary; see Figure 5-4. Again, notice that there are no critical angles for L-wave input in part (a) of that figure. A critical angle does exist in part (b).

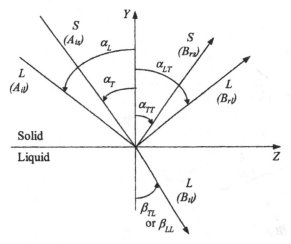

Figure 5-3. Reflection and refraction at a solid–liquid interface.

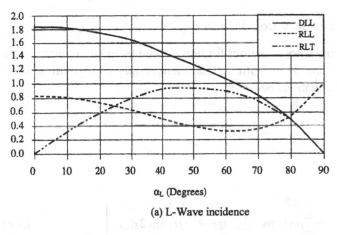

(a) L-Wave incidence

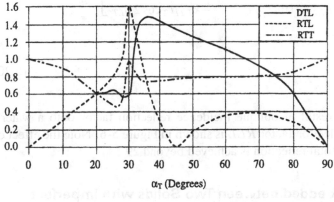

(b) T-Wave incidence

Figure 5-4. Reflection and refraction factors for oblique incidence onto a solid–water interface.

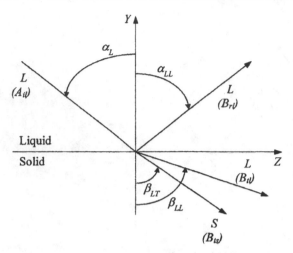

Figure 5-5. Reflection and refraction at a liquid–solid media interface.

5.2.3   Liquid–Solid Boundary Conditions

The reflection and refraction factor equations for liquid–solid boundary conditions can be obtained as before (see Figure 5-5). With liquids, there is only a longitudinal wave. The reflection and refraction equations become

$$
\mathrm{N} \begin{bmatrix} RLL \\ DLL \\ DLT \end{bmatrix} = b,
\tag{5.18}
$$

where

$$
\mathrm{N} = \begin{bmatrix} -\cos\alpha_{LL} & -\cos\beta_{LL} & \sin\beta_{LT} \\ 0 & k_{L2}(\lambda_2 + 2\mu_2)\cos 2\beta_{LT} & -k_{T2}\mu_2\sin 2\beta_{LT} \\ 0 & -k_{L2}\mu_2\sin 2\beta_{LL} & -k_{T2}\mu_2\cos 2\beta_{LT} \end{bmatrix}
\tag{5.19}
$$

and

$$
b = \begin{bmatrix} -\cos\alpha_L \\ k_{L1}\lambda_1\cos 2\alpha_L \\ 0 \end{bmatrix}.
\tag{5.20}
$$

Consider now a sample result of oblique incidence reflection factor onto a water–aluminum interface. The reflection factor RLL is shown in Figure 5-6. Note the effect of the two critical angles until total reflection is achieved beyond $29°$.

5.3   Solid Layer Embedded between Two Solids with Imperfect Boundary Conditions

In this section we model an isotropic, homogeneous, solid layer embedded between two half-space solids with imperfect interfacial conditions. We will use Thomson's matrix technique for transfer of boundary conditions (BCs) across a thin layer, based on

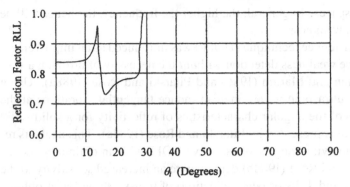

Figure 5-6. Oblique incidence reflection factor for impingement onto a water–aluminum interface.

recurrence formulas, coupled with separate normal and tangential rigidity representations for each interface. This model can be used to compute numerically the reflection and transmission coefficients for oblique incidence of longitudinal (and transverse) waves as a function of incidence angle, incident frequency, interface conditions, and material properties. Several studies of reflection factors in a three-medium case are used to provide insight on adhesively bonded structures. As a sample problem, we deal with ultrasonic quality evaluation of the commonly used aluminum–epoxy-resin–aluminum bond.

5.3.1 Introduction

In adhesively bonded structures, one of the critical defects (besides debonding and poor cohesive strength) is interfacial weakness, which is caused by an imperfection in the adhesion process between the adhesive – usually a thin layer – and the adherend (see Cagle 1972). Methods of nondestructive evaluation (NDE) for detecting adhesion weakness include ultrasonic techniques, which use the propagation of elastic waves and evaluate the associated reflection and transmission phenomena at each interface (see Segal and Rose 1980). The experimentally observed relationship between the reflection coefficient and the mechanical quality of an adhesive bond has been well explored. Several authors (Baik and Thompson 1984; Pilarski 1983; Tattersall 1973) have attempted to explain the dependence of reflectivity on adhesive bond quality, using imperfect boundary conditions in the one-dimensional case for normal incidence. However, these explanations have had only limited success with respect to subtle interfacial weakness.

An ideal connection between two solids may be described by an infinitely rigid welded boundary condition, which assumes continuity of displacements and stresses for both normal and tangential components. In contrast, interfacial weakness is modeled by a bond with a finite rigidity boundary condition. Such an approach allows us to consider a discontinuity in displacement u caused by the current stress σ at the interface.

The continuity of stresses at the interface and the linearity of the relationship between σ and u is based on the assumptions that vibrations are instantaneously transferred from one medium to another (i.e., neglecting inertial forces) and that the amplitude of the displacements is very small compared to the wavelength. This model also explains the experimentally observed frequency dependence of the reflection coefficient at an interface between two lossless semispaces. In general, the higher the frequency, the better will be the sensitivity to interfacial weakness.

The ultrasonic oblique incidence technique (UOIT) was introduced as an improvement in the sensitivity of interface weakness detection with*out* the necessity of utilizing a very high frequency. See Rokhlin and Marom (1986) and Pilarski and Rose (1988a). Using an obliquely incident bulk ultrasonic wave, one can induce tangential vibrations at the interface. The full analysis of the angular characteristics of reflectivity for a solid–solid imperfectly bonded interface is presented by Pilarski and Rose (1988a). This analysis reveals the advantages of employing the transverse wave UOIT with an optimally chosen angle of incidence. Pilarski and Rose (1988b) demonstrate an increased sensitivity to the interfacial imperfection (around 4 dB of peak reflection) of transverse waves at oblique incidence when compared with longitudinal waves in normal incidence. The UOIT shear wave approach thus opens up a new direction in the search for an appropriate tool for measuring interfacial weaknesses in adhesively bonded structures. However, configurations that are more geometrically realistic involve interference effects from a thin adhesive layer embedded between two thicker adherends. This calls for an investigation of the angular and frequency characteristics of reflectivity for a three-medium structure.

The reflection and transmission of elastic waves by a layer or multilayer medium immersed in a fluid is well known and has been studied extensively for a variety of applications; see Jackins and Guanaurd (1986) for a review of the literature. Generally, the solution to this problem is partially provided in the classical techniques used by Thomson (1950) and Haskell (1953). The reflection and transmission coefficients are determined by a method based on recurrence formulas in a form of a transformation (or propagator) matrix connecting the wave amplitudes in neighboring layers.

Many authors have also explored the reflection and transmission of bounded acoustic beams with such related phenomena as nonspecular reflection, leaky plate mode generation, and anomalous dispersion for a solid having material properties similar to those of a liquid. See Chimenti and Nayfeh (1986), Claeys and Leroy (1982), De Billy and Quentin (1984), Kundu (1988), Ngoc and Mayer (1980), and Rousseau and Gatignol (1986). Theoretical and experimental investigations have been performed for fluid-coupled plates of anisotropic, homogeneous, solid, fiber-reinforced composite layers through a derivation of formulas for the reflection coefficient (see Nayfeh and Chimenti 1988). Another approximating approach, based on a resonance formalism, is widely described in the literature (Fiorito, Madigosky, and Uberall 1979; Jackins and Gaunaurd 1986) for the case of an elastic layer or a stack of perfectly bonded elastic plates immersed in a fluid.

For all the aforementioned results, one feature is common: the incidence occurs from fluid. Hence, only longitudinal waves can be both incident and reflected on the first interaction between a fluid and solid layer, where the tangential stresses vanish. In this section we extend the existing solutions to a case where the incidence takes place from a solid semispace onto a solid layer embedded between two solids and bonded in an imperfect manner. Here, two cases of incidence are possible: one for longitudinal waves, another for transverse waves. In each case, one can consider four coefficients: two for reflection and two for transmission.

Our derivations are carried out using Thompson's matrix technique. Therefore, in the theoretical section we present a review of this classical method in terms of displacements and stresses and also give a full form of the propagator matrix. We will then develop

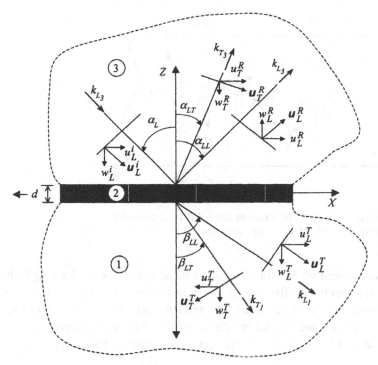

Figure 5-7. The model of the solid layer between two semi-infinite solids with relevant displacements for an incident longitudinal wave.

the set of equations for reflection and transmission coefficients for two possible incident waves and a generally imperfect bond. Most of the numerical computations will be carried out for data describing two thick aluminum plates bonded by a thin epoxy-resin glue line.

5.3.2 Theory

We will study the interaction of elastic waves with a solid plane-parallel layer 2 of thickness d separating two different solid semispaces, 3 and 1 (see Figure 5-7). The individual solids are considered to be homogeneous and isotropic, with densities ρ_n and Lamé constants λ_n and μ_n ($n = 1, 2, 3$). The flat interfaces, hereafter labeled (1) and (2), are normal to the z-axis: the lower interface (1), between layer 2 and medium 1, is located within the plane $z = 0$; the upper interface (2), between layer 2 and medium 3, is located within the plane $z = -d$. The connections between solids are assumed to be imperfect and are described by the stiffness constants: normal to the interface $KN^{(m)}$ and tangential to the interface $KT^{(m)}$ ($m = 1, 2$). This means that, at each interface, the normal and tangential components of stress are required to be continuous (contrary to the components of displacements, which can be discontinuous).

The incidence of unbounded, plane, harmonic waves – either longitudinal or transverse – occurs in the (x, y)-plane. Our case can therefore be treated as a plane strain problem for which all displacements lie in the plane of incidence, with components $u_l^k(x, z)$ and $w_l^k(x, z)$, where $l = i, r, d$ for incident, reflected, or transmitted waves (respectively) and $k = L$ or T corresponds to longitudinal or transverse waves. The wave is obliquely

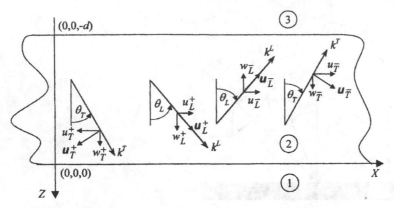

Figure 5-8. Longitudinal and transverse waves in the solid layer embedded between two solid semispaces with relevant displacements.

incident onto layer 2 from medium 3, with an angle of incidence α_L or α_T for longitudinal or transverse waves, respectively. Because of mode conversion, we have two reflected and two transmitted waves (in the upper and lower semispaces) that are reflected or refracted with relevant angles α_{ts} and β_{ts}, where the first index denotes the incident wave type (L or T) and the second index denotes the reflected or transmitted wave type (also L or T).

5.3.3 The Propagator Matrix

As a result of multiple reflections from within the boundaries of the intermediate layer, a system of longitudinal and transverse waves will exist in the layer itself, propagating in both the positive and negative directions of axis z (Figure 5-8). Here, θ_L and θ_T are the angles made with the z-axis by the longitudinal and transverse waves. The total displacement vector $\bar{u}$ in the layer is presented as a sum of the displacement vectors of all four waves – that is, of longitudinal, transverse, upward, and downward waves. Therefore, in our plane strain problem, the individual components of the displacement vector along the x and z axes can be written as

$$\sum u = u_L^+ + u_T^+ + u_L^- + u_T^-,$$
$$\sum v = 0,$$
$$\sum w = w_L^+ + w_T^+ + w_L^- + w_T^-. \tag{5.21}$$

For harmonic plane waves, the individual components in (5.21) can be obtained by suppressing the dependence on $\exp[i(k_x^{L,T} x - \omega t)]$. For horizontal displacements,

$$u_L^+ = A_L^+ \sin\theta_L \exp[ik^L \cos\theta_L z],$$
$$u_T^+ = -A_T^+ \cos\theta_T \exp[ik^T \cos\theta_T z],$$
$$u_L^- = A_L^- \sin\theta_L \exp[-ik^L \cos\theta_L z],$$
$$u_T^- = A_T^- \cos\theta_T \exp[-ik^T \cos\theta_T z]; \tag{5.22}$$

for vertical displacements,

$$w_L^+ = A_L^+ \cos\theta_L \exp[ik^L \cos\theta_L z],$$
$$w_T^+ = A_T^+ \sin\theta_T \exp[ik^T \cos\theta_T z],$$
$$w_L^- = -A_L^- \cos\theta_L \exp[-ik^L \cos\theta_L z],$$
$$w_T^- = A_T^- \sin\theta_T \exp[-ik^T \cos\theta_T z].$$

(5.23)

Here A_L and A_T are the amplitude of displacements and k^L and k^T are wavenumbers for the longitudinal and transverse waves, respectively; amplitude superscripts $+$ and $-$ correspond to upward and downward waves, respectively.

At any point within the isotropic, homogeneous layer, the stresses (both normal and tangential) can be determined using the classical stress–strain relations

$$\sigma_{ij} = \lambda\varepsilon_{kk}\delta_{ij} + 2\mu\varepsilon_{ij},$$

(5.24)

where λ and μ are the Lamé constants, ε_{ij} is the strain, and δ_{ij} is the Kronecker delta. Then, applying the following relationships between the Lamé constants, density, and velocities c_L and c_T of the longitudinal and transverse waves, we have

$$\mu = \rho c_T^2 \quad \text{and}$$

(5.25)

$$\lambda + 2\mu = \rho c_L^2,$$

(5.26)

so one can determine the general formulas for stresses σ_{zz} and σ_{xz}. These formulas, together with relationships (5.21)–(5.23) given for the displacement field, serve as a basis for further analysis.

Substituting $z = 0$ and $z = -d$ in the displacements and stresses relationships – which represent the bottom (1) and top (2) interfaces, respectively – we obtain two sets of equations with respect to the amplitudes of displacements. We can present them in matrix form as

$$\{b_i^{(n)}\} = [D_{ij}^{(n)}]\{A_j\}, \quad n = 1, 2,$$

(5.27)

where

$$\{b_i^{(n)}\} = \begin{Bmatrix} u^{(n)} \\ w^{(n)} \\ \sigma_{xz}^{(n)} \\ \sigma_{zz}^{(n)} \end{Bmatrix}$$

(5.28)

are the vectors with components of displacements and stresses at the lower ($n = 1$) and upper ($n = 2$) interfaces, but still inside the layer, and

$$\{A_j\} = \begin{Bmatrix} A_L^+ \\ A_T^+ \\ A_L^- \\ A_T^- \end{Bmatrix}$$

(5.29)

is the vector of displacement amplitudes. The matrices $[D_{ij}^{(1)}]$ and $[D_{ij}^{(2)}]$ are given by

$$[D_{ij}^{(n)}] = \begin{bmatrix} \sin\theta_L(e^{-L})^{n-1} & -\cos\theta_T(e^{-T})^{n-1} & \sin\theta_L(e^{-L})^{n-1} & \cos\theta_T(e)^{n-1} \\ \cos\theta_L(e^{-L})^{n-1} & \sin\theta_T(e^{-T})^{n-1} & -\cos\theta_L(e^{L})^{n-1} & \sin\theta_T(e^{T})^{n-1} \\ a\sin 2\theta_L(e^{-L})^{n-1} & -b\cos 2\theta_T(e^{-T})^{n-1} & -a\sin\theta_L(e^{L})^{n-1} & -b\cos 2\theta_T(e^{T})^{n-1} \\ c(e^{-L})^{n-1} & b\sin 2\theta_T(e^{T})^{n-1} & c(e^{L})^{n-1} & -b\sin 2\theta_T(e^{T})^{n-1} \end{bmatrix},$$

$$(5.30)$$

where

$$L = ik_z^L d = i(\omega/c_L)\cos\theta_L d, \qquad T = ik_z^T d = i(\omega/c_T)\cos\theta_T d;$$

$$a = i\rho(c_T^2/c_L), \quad b = i\rho c_T\omega, \quad c = i\rho\omega[c_L - 2(c_T^2/c_L)\sin^2\theta_L];$$

and $n = 1, 2$.

Solving the two matrix equations given by (5.27) and eliminating the vector of displacement amplitudes, we can obtain a relationship between the vectors $\{b_i^{(n)}\}$ at the lower and upper interface of the layer:

$$\{b_i^{(2)}\} = [C_{ij}]\{b_j^{(1)}\}. \qquad (5.31)$$

The matrix $[C_{ij}]$ is given as

$$[C_{ij}] = [D_{ij}^{(2)}][D_{ij}^{(1)}]^{-1}. \qquad (5.32)$$

Equations (5.31) and (5.32) are relationships that transfer the displacement and stress fields from one interface to another.

5.3.4 Reflection and Transmission Coefficients

We now come back to the overall picture of the thin layer separating the two solid semispaces, as shown in Figure 5-7. For the general case of an imperfect connection – represented by finite values of boundary rigidities – the boundary conditions at either of the two interfaces ($n = 1, 2$) are given as

$$\sigma_{xz}\big|_{z_n^{(+)}} = \sigma_{xz}\big|_{z_n^{(-)}} = \sigma_{xz}\big|_{z_n},$$

$$\sigma_{zz}\big|_{z_n^{(+)}} = \sigma_{zz}\big|_{z_n^{(-)}} = \sigma_{zz}\big|_{z_n},$$

$$\sigma_{xz}\big|_{z_n} = \mathrm{KT}^{(n)}(u\big|_{z_n^{(+)}} - u\big|_{z_n^{(-)}}),$$

$$\sigma_{zz}\big|_{z_n} = \mathrm{KN}^{(n)}(w\big|_{z_n^{(+)}} - w\big|_{z_n^{(-)}}),$$

$$(5.33)$$

where $\mathrm{KN}^{(n)}$ and $\mathrm{KT}^{(n)}$ are the normal and tangential components of the rigidity at the interface (n). The $(+)$ and $(-)$ superscripts represent the upper and lower sides of the interfaces, respectively. Equations (5.33) describe the continuity of both normal and tangential stresses at the interfaces and also give the linear relationships between stress and the jump of the relevant displacements at the interface. These last pairs of equations can be transferred to a more convenient form for further derivation as

$$u|_{z_2^{(-)}} = u|_{z_2^{(+)}} - \sigma_{xz}|_{z_2}(KT^{(2)})^{-1} \quad \text{and} \quad w|_{z_2^{(-)}} = w|_{z_2^{(+)}} - \sigma_{zz}|_{z_2}(KN^{(2)})^{-1} \quad (5.34)$$

for $z_2 = -d$ and as

$$u|_{z_1^{(+)}} = u|_{z_1^{(-)}} - \sigma_{xz}|_{z_1}(KT^{(1)})^{-1} \quad \text{and} \quad w|_{z_1^{(+)}} = w|_{z_1^{(-)}} - \sigma_{zz}|_{z_1}(KN^{(1)})^{-1} \quad (5.35)$$

for $z_1 = 0$. Equations (5.34) and (5.35) are expressions for the displacements in the layer at the upper and lower interfaces, respectively.

Consider now the two-dimensional case of a longitudinal wave incident with unit amplitude from the upper medium 3 (Figure 5-7). Then the components of the total displacement in this medium at the interface (2) can be written as a sum of relevant components of the incident waves and both the reflected waves, longitudinal and transverse. The sums may be given as

$$\sum u|_{z_2^{(+)}} = \sin\alpha_L e^{-L_3} + RLL\sin\alpha_{LL}e^{L_3} + RLT\cos\alpha_{LT}e^{T_3}, \quad (5.36)$$

$$\sum w|_{z_2^{(+)}} = \cos\alpha_L e^{-L_3} - RLL\cos\alpha_{LL}e^{L_3} + RLT\sin\alpha_{LT}e^{T_3}, \quad (5.37)$$

where RLL and RLT are the reflection coefficients for longitudinal and transverse waves. We also have

$$L_3 = i\omega d\cos\alpha_{LL}(c_{L_3})^{-1}, \quad (5.38)$$

$$T_3 = i\omega d\cos\alpha_{LT}(c_{T_3})^{-1}, \quad (5.39)$$

where c_L and c_T are the velocities of longitudinal and transverse waves in the upper medium 3. The normal and tangential stresses in the upper medium at the interface (2) are obtained by using the relationships (5.24)–(5.26) and equations (5.36)–(5.39):

$$\sigma_{xz}|_{z_2^{(+)}} = i\rho_3\omega\left(\frac{c_{T_3}^2}{c_{L_3}}\sin 2\alpha_{LL}e^{-L_3} - RLL\frac{c_{T_3}^2}{c_{L_3}}\right.$$
$$\left. \times \sin 2\alpha_{LL}e^{L_3} - RLTc_{T_3}\cos 2\alpha_{LT}e^{T_3}\right), \quad (5.40)$$

$$\sigma_{zz}|_{z_2^{(+)}} = i\rho_3\omega\left[\left(c_{L_3} - 2\frac{c_{T_3}^2}{c_{L_3}}\sin^2\alpha_{LL}\right)e^{-L_3}\right.$$
$$\left. + RLL\left(c_{L_3} - 2\frac{c_{T_3}^2}{c_{L_3}}\sin^2\alpha_{LL}\right)e^{L_3} - RLTc_{T_3}\sin 2\alpha_{LT}e^{T_3}\right]. \quad (5.41)$$

Because of the boundary conditions (5.33) at the interface $z_2 = -d$, the relationships (5.40) and (5.41) represent the stresses inside the layer 2 at the top interface (2). The relevant displacements are obtained using equations (5.34):

$$u|_{z_2^{(-)}} = \left(\sin\alpha_{LL} - i\rho_3\omega\frac{c_{T_3}^2}{c_{L_3}}\frac{1}{KT^{(2)}}\sin(2\alpha_{LL})\right)e^{-L_3}$$
$$+ RLL\left(\sin\alpha_{LL} + i\rho_3\omega\frac{c_{T_3}^2}{c_{L_3}}\frac{1}{KT^{(2)}}\sin(2\alpha_{LL})\right)e^{L_3}$$
$$+ RLT\left(\cos\alpha_{LT} + i\rho_3\omega c_{T_3}\frac{1}{KT^{(2)}}\cos(2\alpha_{LT})\right)e^{T_3}, \quad (5.42)$$

$$w|_{z_2^{(-)}} = \left(\cos\alpha_{LL} - i\rho_3\omega c_{L_3}\frac{1}{\mathrm{KN}^{(2)}} + 2i\rho_3\omega\frac{c_{T_3}^2}{c_{L_3}}\frac{1}{\mathrm{KN}^{(2)}}\sin^2\alpha_{LL}\right)e^{-L_3}$$

$$- RLL\left(\cos\alpha_{LL} - i\rho_3\omega c_{L_3}\frac{1}{\mathrm{KN}^{(2)}} - 2i\rho_3\omega\frac{c_{T_3}^2}{c_{L_3}}\frac{1}{\mathrm{KN}^{(2)}}\sin^2\alpha_{LL}\right)e^{L_3}$$

$$- RLT\left(\sin\alpha_{LT} + i\rho_3\omega c_{T_3}\frac{1}{\mathrm{KN}^{(2)}}\sin(2\alpha_{LT})\right)e^{T_3}. \tag{5.43}$$

A similar procedure adopted for the lower semispace 1, where two transmitted waves exist (Figure 5-7) together with the boundary conditions, provides the displacements and stresses in layer 2 at the bottom interface (1) as

$$\sigma_{xz}|_{z_1^{(+)}} = i\rho_1\omega c_{T_1}\left(\frac{c_{T_1}^2}{c_{L_1}}\sin(2\beta_{LL})DLL - \cos(2\beta_{LT})DLT\right), \tag{5.44}$$

$$\sigma_{zz}|_{z_1^{(+)}} = i\rho_1\omega\left[\left(c_{L_1} - 2\frac{c_{T_1}^2}{c_{L_1}}\sin^2\beta_{LL}\right)DLL + c_{T_1}\sin\beta_{LT}DLT\right]; \tag{5.45}$$

$$u|_{z_1^{(+)}} = DLL\left(\sin\beta_{LL} + i\rho_1\omega\frac{c_{T_1}^2}{c_{L_1}}\frac{1}{\mathrm{KT}^{(1)}}\sin(2\beta_{LL})\right)$$

$$- DLT\left(\cos\beta_{LT} + i\rho_1\omega c_{T_1}\frac{1}{\mathrm{KT}^{(1)}}\cos(2\beta_{LT})\right), \tag{5.46}$$

$$w|_{z_1^{(+)}} = DLL\left(\cos\beta_{LL} - i\rho_1\omega c_{L_1}\frac{1}{\mathrm{KN}^{(1)}} - 2i\rho_1\omega\frac{c_{T_1}^2}{c_{L_1}}\frac{1}{\mathrm{KN}^{(1)}}\sin^2\beta_{LL}\right)$$

$$+ DLT\left(\sin\beta_{LT} + i\rho_1\omega c_{T_1}\frac{1}{\mathrm{KN}^{(1)}}\sin(2\beta_{LT})\right). \tag{5.47}$$

Here, DLL and DLT represent the transmission factor scalar quantities and must not be confused with the transfer matrix $[D_{ij}]$ in (5.27).

Thus we have found the relationships for displacements and stresses in layer 2 – at the upper interface using (5.40)–(5.43) and at the lower interface using (5.44)–(5.47) – in terms of material properties of both semispaces, various angles, frequency, and the reflection and transmission coefficients. These relationships determine the vectors $\{b_i^{(n)}\}$ defined by (5.28) and (5.29), which in turn are related to each other by (5.31) with the transform matrix $[C_{ij}]$ given by (5.32) and detailed in (5.30). Equation (5.31) represents a set of four equations in matrix form with four unknown quantities, the two reflection (RLL and RLT) and two transmission (DLL and DLT) coefficients. Substituting for the individual terms in (5.31) and using the matrix technique, we find (after some straightforward though tedious operations) the final form of a nonhomogeneous set of four linear equations with respect to $\{R_j\}$:

$$[M_{ij}]\{R_j\} = \{N_i^{(L)}\}, \tag{5.48}$$

where

$$\{R_j\} = \begin{Bmatrix} RLL \\ RLT \\ DLL \\ DLT \end{Bmatrix} \tag{5.49}$$

and the individual terms of matrices $[M_{ij}]$ and $\{N_i^{(L)}\}$ are given in (5.50) as follows.

$$M_{11} = \left(\sin\alpha_{LL} + i\rho_3\omega \frac{c_{T_3}^2}{c_{L_3}} \frac{1}{KT^{(2)}} \sin(2\alpha_{LL}) \right) e^{L_3},$$

$$M_{12} = \left(\cos\alpha_{LT} + i\rho_3\omega c_{T_3} \frac{1}{KT^{(2)}} \cos(2\alpha_{LT}) \right) e^{T_3},$$

$$M_{13} = -c_{11}\left(\sin\beta_{LL} + i\rho_1\omega \frac{c_{T_1}^2}{c_{L_1}} \frac{1}{KT^{(1)}} \sin(2\beta_{LL}) \right)$$
$$- c_{12}\left(\cos\beta_{LL} + i\rho_1\omega c_{L_1} \frac{1}{KN^{(1)}} - 2i\rho_1\omega \frac{c_{T_1}^2}{c_{L_1}} \frac{1}{KN^{(1)}} \sin^2\beta_{LL} \right)$$
$$- c_{13} i\rho_1\omega \frac{c_{T_1}^2}{c_{L_1}} \sin(2\beta_{LL}) - c_{14}\left(i\rho_1\omega c_{L_1} - 2i\rho_1\omega \frac{c_{T_1}^2}{c_{L_1}} \sin^2\beta_{LL} \right),$$

$$M_{14} = c_{11}\left(\cos\beta_{LT} + i\rho_1\omega c_{T_1} \cos(2\beta_{LT}) \frac{1}{KT^{(1)}} \right)$$
$$- c_{12}\left(\sin\beta_{LT} + i\rho_1\omega c_{T_1} \sin(2\beta_{LT}) \frac{1}{KN^{(1)}} \right)$$
$$+ c_{13} i\rho_1\omega c_{T_1} \cos(2\beta_{LT}) - c_{14} i\rho_1\omega c_{T_1} \sin(2\beta_{LT});$$

$$M_{21} = \left(-\cos\alpha_{LL} - i\rho_3\omega c_{L_3} \frac{1}{KN^{(2)}} + 2i\rho_3\omega \frac{c_{T_3}^2}{c_{L_3}} \frac{1}{KN^{(2)}} \sin(2\alpha_{LL}) \right) e^{L_3},$$

$$M_{22} = \left(\sin\alpha_{LT} + i\rho_3\omega c_{T_3} \sin(2\alpha_{LT}) \frac{1}{KN^{(2)}} \right) e^{T_3},$$

$$M_{23} = -c_{21}\left(\sin\beta_{LL} + i\rho_1\omega \frac{c_{T_1}^2}{c_{L_1}} \frac{1}{KT^{(1)}} \sin(2\beta_{LL}) \right)$$
$$- c_{22}\left(\cos\beta_{LL} + i\rho_1\omega c_{L_1} \frac{1}{KN^{(1)}} - 2i\rho_1\omega \frac{c_{T_1}^2}{c_{L_1}} \frac{1}{KN^{(1)}} \sin^2\beta_{LL} \right)$$
$$- c_{23} i\rho_1\omega \frac{c_{T_1}^2}{c_{L_1}} \sin(2\beta_{LL}) - c_{24}\left(i\rho_1\omega c_{L_1} - 2i\rho_1\omega \frac{c_{T_1}^2}{c_{L_1}} \sin^2\beta_{LL} \right),$$

$$M_{24} = c_{21}\left(\cos\beta_{LT} + i\rho_1\omega c_{T_1} \cos(2\beta_{LT}) \frac{1}{KT^{(1)}} \right)$$
$$- c_{22}\left(\sin\beta_{LT} + i\rho_1\omega c_{T_1} \sin(2\beta_{LT}) \frac{1}{KN^{(1)}} \right)$$
$$+ c_{23} i\rho_1\omega c_{T_1} \cos(2\beta_{LT}) - c_{24} i\rho_1\omega c_{T_1} \sin(2\beta_{LT});$$

$$M_{31} = -i\rho_3\omega \frac{c_{T_3}^2}{c_{L_3}} \sin(2\alpha_{LL}) e^{L_3},$$

$$M_{32} = -i\rho_3\omega c_{T_3} \cos(2\alpha_{LT}) e^{T_3},$$

(5.50)

$$M_{33} = -c_{31}\left(\sin\beta_{LL} + i\rho_1\omega\frac{c_{T_1}^2}{c_{L_1}}\frac{1}{KT^{(1)}}\sin(2\beta_{LL})\right)$$

$$- c_{32}\left(\cos\beta_{LL} + i\rho_1\omega c_{L_1}\frac{1}{KN^{(1)}} - 2i\rho_1\omega\frac{c_{T_1}^2}{c_{L_1}}\frac{1}{KN^{(1)}}\sin^2\beta_{LL}\right)$$

$$- c_{33}i\rho_1\omega\frac{c_{T_1}^2}{c_{L_1}}\sin(2\beta_{LL}) - c_{34}\left(i\rho_1\omega c_{L_1} - 2i\rho_1\omega\frac{c_{T_1}^2}{c_{L_1}}\sin^2\beta_{LL}\right),$$

$$M_{34} = c_{31}\left(\cos\beta_{LT} + i\rho_1\omega c_{T_1}\cos(2\beta_{LT})\frac{1}{KT^{(1)}}\right)$$

$$- c_{32}\left(\sin\beta_{LT} + i\rho_1\omega c_{T_1}\sin(2\beta_{LT})\frac{1}{KN^{(1)}}\right)$$

$$+ c_{33}i\rho_1\omega c_{T_1}\cos(2\beta_{LT}) - c_{34}i\rho_1\omega c_{T_1}\sin(2\beta_{LT});$$

$$M_{41} = \left(i\rho_3\omega c_{L_3} - 2i\rho_3\omega\frac{c_{T_3}^2}{c_{L_3}}\sin^2\alpha_{LL}\right)e^{L_3},$$

$$M_{42} = -i\rho_3\omega c_{T_3}\sin(2\alpha_{LT})e^{T_3}$$

$$M_{43} = -c_{41}\left(\sin\beta_{LL} + i\rho_1\omega\frac{c_{T_1}^2}{c_{L_1}}\frac{1}{KT^{(1)}}\sin(2\beta_{LL})\right)$$

$$- c_{42}\left(\cos\beta_{LL} + i\rho_1\omega c_{L_1}\frac{1}{KN^{(1)}} - 2i\rho_1\omega\frac{c_{T_1}^2}{c_{L_1}}\frac{1}{KN^{(1)}}\sin^2\beta_{LL}\right) \qquad \text{(5.50)}$$
$$\text{(cont.)}$$

$$- c_{43}i\rho_1\omega\frac{c_{T_1}^2}{c_{L_1}}\sin(2\beta_{LL}) - c_{44}\left(i\rho_1\omega c_{L_1} - 2i\rho_1\omega\frac{c_{T_1}^2}{c_{L_1}}\sin^2\beta_{LL}\right),$$

$$M_{44} = c_{41}\left(\cos\beta_{LT} + i\rho_1\omega c_{T_1}\cos(2\beta_{LT})\frac{1}{KT^{(1)}}\right)$$

$$- c_{42}\left(\sin\beta_{LT} + i\rho_1\omega c_{T_1}\sin(2\beta_{LT})\frac{1}{KN^{(1)}}\right)$$

$$+ c_{43}i\rho_1\omega c_{T_1}\cos(2\beta_{LT}) - c_{44}i\rho_1\omega c_{T_1}\sin(2\beta_{LT});$$

$$N_1^{(L)} = \left(-\sin\alpha_{LL} + i\rho_3\omega\frac{c_{T_3}^2}{c_{L_3}}\frac{1}{KT^{(2)}}\sin(2\alpha_{LL})\right)e^{-L_3},$$

$$N_2^{(L)} = \left(-\cos\alpha_{LL} + i\rho_3\omega c_{L_3}\frac{1}{KN^{(2)}} - 2i\rho_3\omega\frac{c_{T_3}^2}{c_{L_3}}\frac{1}{KT^{(2)}}\sin^2\alpha_{LL}\right)e^{-L_3},$$

$$N_3^{(L)} = -i\rho_3\omega\frac{c_{T_3}^2}{c_{L_3}}\sin(2\alpha_{LL})e^{-L_3},$$

$$N_4^{(L)} = \left(-i\rho_3\omega c_{L_3} + 2i\rho_3\omega\frac{c_{T_3}^2}{c_{L_3}}\sin^2\alpha_{LL}\right)e^{-L_3};$$

$$N_1^{(T)} = \left(-\cos\alpha_{LT} - i\rho_3\omega c_{T_3}\cos(2\alpha_{LT})\frac{1}{KT^{(2)}}\right)e^{-T_3},$$

$$N_2^{(T)} = \left(-\sin\alpha_{LT} + i\rho_3\omega c_{T_3}\sin(2\alpha_{LT})\frac{1}{KN^{(2)}} \right)e^{-T_3},$$

(5.50)

$$N_3^{(T)} = i\rho_3\omega c_{T_3}\cos(2\alpha_{LT})e^{-T_3},$$

(cont.)

$$N_4^{(T)} = -i\rho_3\omega c_{T_3}\sin(2\alpha_{LT})e^{-T_3}.$$

In the foregoing expressions, recall that

$$L_3 = i\omega d\cos\alpha_{LL}(c_{L_3})^{-1} \quad \text{and} \quad T_3 = i\omega d\cos\alpha_{LT}(c_{T_3})^{-1}.$$

The superscript (L) in the RHS of (5.48) reminds us that this solution is for the problem with longitudinal wave incidence. For transverse wave incidence, only the $\{N_i^{(T)}\}$ matrix values change (see (5.50)); the matrix $[M_{ij}]$ is unaffected in form because it represents only the reflected and the transmitted waves and not the incident waves. However, for transverse incident waves, $\{R_j\}$ changes to become

$$\{R_j\} = \begin{Bmatrix} RTL \\ RTT \\ DTL \\ DTT \end{Bmatrix}.$$

(5.51)

Note that terms of the matrices $[M]$ and $\{N\}$ include material properties and velocity characteristics of both semispaces, while the intermediate layer is represented by the terms of the transform matrix $[C]$. All angles appearing here (α_{st}, β_{st}, and θ_s with $s, t = L$ or T) are mutually related through Snell's law; that is, all the x components of the wave vectors in each medium and for each kind of wave are equal. The normal (KN) and tangential (KT) rigidities of both interfaces can be treated as parameters for further numerical solutions (bond quality changes). From (5.48) and (5.50) we recognize that the reflection and transmission coefficients are frequency-dependent not only because of interference effects within the layer but also owing to the presence of imperfections at the interface.

5.3.5 Results of Numerical Computation

The set of linear equations – given in matrix form by (5.48) with the reflection and transmission coefficients as unknowns – were solved numerically in a complex domain. Only the results of the modulus of the reflection coefficients (for both longitudinal and transverse wave incidence) were presented. These calculations were carried out using three different methods. At first, the unknowns were computed as a function of the angle of incidence α_L or α_T for a specific frequency or thickness of layer and as a function of the rigidities of individual connections. In the second method, computations were carried out with the frequency as a variable and the rest of quantities held fixed. In the third group of calculations, bond rigidities were varied from zero (debonding) to very large magnitudes (ideal bond). Most of the results shown here were obtained for the aluminum–epoxy-resin–aluminum material combination, where the aluminum semispaces represent the adherends and the epoxy resin represents the adhesive. All material properties used in our computations are provided in Table 5-1.

Table 5-1. *Material properties used for computation of coefficients*

Material	Density (kg/m³)	Wave velocity (m/s)	
		Longitudinal	Shear
Water	1,000	1,500	—
Aluminum	2,700	6,320	3,080
Epoxy	1,300	2,800	1,100
Polystyrene	1,060	2,350	1,150

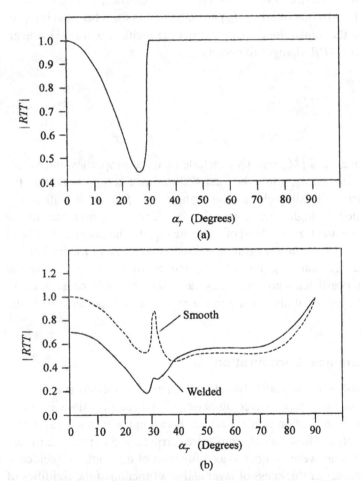

Figure 5-9. Tests for the numerical program: (a) reflection of transverse waves from a free surface of aluminum; (b) reflection coefficient RTT from interface between aluminum and epoxy for welded and smooth BCs.

At the first stage of our work, the numerical code was tested for a few simple cases and compared with well-known results from the literature; see Figure 5-9. The first case, shown in part (a) of the figure, represents the angular characteristic of modulus of the reflection RTT for a transverse wave incident from aluminum onto a free surface. Using

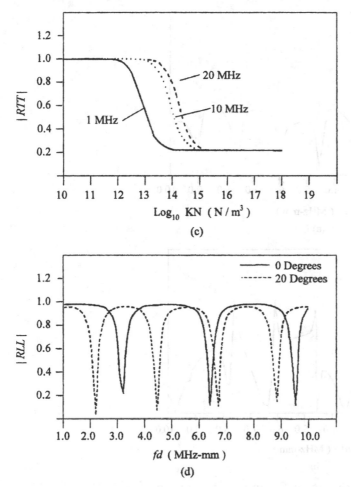

Figure 5-9 *(continued).* (c) RTT versus bond rigidities for $\alpha_T = 30°$ and $f = 1$, 10, and 30 MHz; (d) RLL versus fd product for single aluminum layer immersed in water for $\alpha_L = 0°$ and 20°.

the program for a three-medium problem, these results were obtained by treating the second and third media as air. Similar computations were carried out for the reflectivity of transverse waves from the interface between aluminum and epoxy resin, treated as two semispaces; see part (b). This was achieved by considering the second and third media as the same epoxy resin. The welded and smooth BCs were assumed alternatively for interface (2) between aluminum and a layer of epoxy resin, while interface (1) was kept as a welded one. For the first BC, so-called welded (ideal), the bond rigidities – both normal (KN) and tangential (KT) – were considered very large; for the second "smooth" BC (poor bonding), the tangential rigidity was taken as zero while KN remained large.

This last kind of boundary condition can be achieved by having two solids separated by an inviscid liquid of infinitely small thickness with an ability to completely transfer normal but not tangential stresses. As expected for this particular case, the second interface boundary condition makes no difference. Part (c) of Figure 5-9 shows the results of the modulus of the reflection coefficient for transverse waves (RTT) reflected from a similar interface but as a function of the bond rigidity KT, when KN $= 2$(KT) is kept fixed.

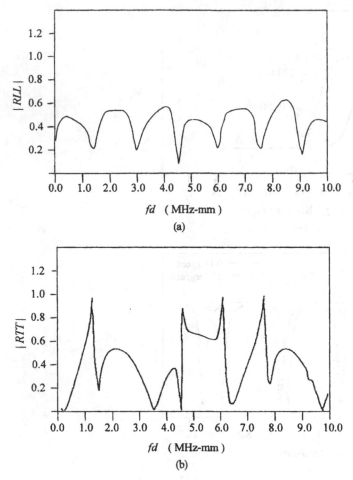

Figure 5-10. The modulus of the reflection coefficient versus fd product for (a) longitudinal wave incident at $\alpha_L = 50°$ and (b) transverse wave incident at $\alpha_T = 30°$ for an epoxy layer embedded between aluminum semispaces with perfect bonding.

These results were obtained for an angle of incidence $\alpha_T = 30°$ and for three different frequencies of 1, 10, and 20 MHz. The last sample result, shown in part (d) of the figure, illustrates the reflection coefficient of a single aluminum plate in water at two angles of incidence ($\alpha_L = 0°$ and $20°$) for the longitudinal waves versus the frequency–thickness product fd in the range $0 < fd < 10$ MHz-mm.

The foregoing results compare well with curves found in the literature and so inspire confidence in our model. Now we can present the results for more complex situations – that is, for a solid layer separating two other solid semispaces. The influence of variations in thickness of the glue line on the reflection coefficients are shown – in part (a) of Figure 5-10 for longitudinal waves incident at 50° and in part (b) for transverse waves incident at 30° – from aluminum onto the epoxy layer embedded between two aluminum semispaces. The thickness of the layer was varied within a range of 0.1–1.0 mm at a constant frequency of 10 MHz, giving an fd product range of 1–10 MHz-mm. For both types of waves, all interfaces were considered as perfect.

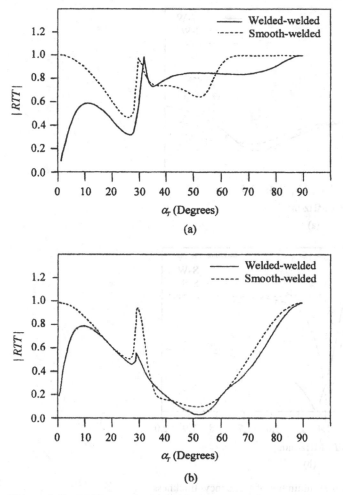

Figure 5-11. RTT versus angle of incidence α_T for epoxy layer of 0.2-mm thickness between aluminum semispaces for (a) $f = 5$ MHz and (b) $f = 10$ MHz (the continuous line corresponds to the welded–welded BC; the broken line denotes the smooth–welded BC).

Figure 5-11 shows the angular characteristics of the reflection coefficient for transverse wave incidence onto an epoxy layer of thickness 0.2 mm. In part (a) of the figure, results are obtained for a frequency of 5 MHz; in part (b), for a frequency of 10 MHz. In both cases, the calculations were provided for two situations: (i) perfect bonds at both interfaces through welded BCs; and (ii) a smooth BC for the upper interface (2) while keeping the lower interface (1) perfect. Figure 5-12 illustrates the frequency characteristics of reflection factors RLL and RTT in the vicinity of the first minimum for three BC combinations: welded–welded, smooth–welded, and smooth–smooth. The poor connection was considered as an interface whose bond rigidity values were zero.

From a practical viewpoint, the next figure (Figure 5-13) is very important. It shows the reflection coefficient RTT for transverse wave incidence at 30° as a function of bond rigidities $KT^{(2)}$, with a constant ratio $KN^{(2)} = 2KT^{(2)}$, covering the range of bond quality from poor to perfect. For part (a) of Figure 5-13, the layer of epoxy was 1.0 mm thick

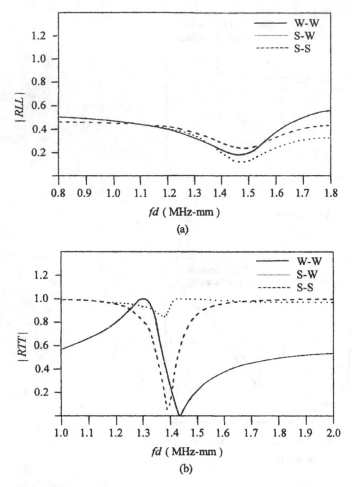

Figure 5-12. Vicinity of the first minimum of frequency–thickness characteristics of reflectivity for (a) longitudinal and (b) transverse waves incident onto the epoxy layer embedded between aluminum semispaces for $\alpha_L = 50°$ and $\alpha_T = 30°$ for three sets of boundary conditions.

and the wave frequencies were 1, 10, and 50 MHz; part (b) varies the thicknesses (0.1, 0.25, and 0.5 mm) with a single frequency of 10 MHz.

All the results presented so far have involved material combinations for which acoustic impedances of the two semispaces are larger than that of the layer. However, in Figure 5-14, the reflection factor (RLL or RTT) versus fd product is shown for the case when

$$Z_2 > \sqrt{(Z_1 \cdot Z_3)}, \tag{5.52}$$

where Z_n ($n = 1, 2, 3$) are the relevant acoustic impedances. The epoxy layer ($n = 2$) separates the aluminum ($n = 3$) as the upper semispace and the polystyrene ($n = 1$) as the lower semispace. Both interfaces were assumed to be perfect. The angle of incidence chosen for longitudinal waves was 50° and for transverse waves was 30°.

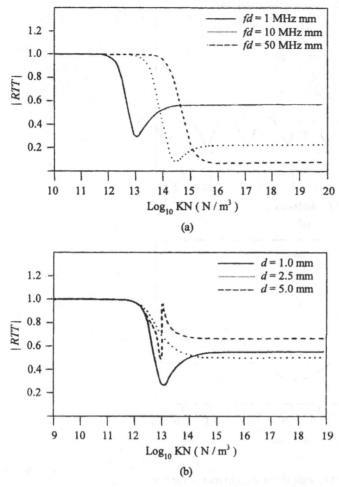

Figure 5-13. RTT versus bond rigidity for $\alpha_T = 30°$: (a) for $f = 1$, 10, and 50 MHz with $d = 1.0$ mm; (b) for $f = 10$ MHz with $d = 0.1$, 0.25, and 0.5 mm.

5.3.6   Discussion

The chosen angles of incidence for the longitudinal waves ($\alpha_L = 50°$) and for the transverse waves ($\alpha_T = 30°$) were the best possible, from a sensitivity point of view, for distinguishing between smooth and welded conditions (as established by our earlier model for the pair of semispaces of aluminum and epoxy resin; see Pilarski and Rose 1988a). But as expected, the overall picture of a layer embedded between two semi-spaces is more complex. For example, the recommended 30° angle for transverse wave incidence, which is close to the first critical angle for an aluminum–epoxy material combination, is probably not always accurate for an epoxy layer of thickness comparable with the wavelength. Comparing the angular characteristics (a) and (b) in Figure 5-11 for two frequencies (5 and 10 MHz, resp.), one finds that at exactly 30° the difference between the reflection coefficients for the two interfacial situations is the same, but it changes substantially with a slight perturbation in this incident angle. One can also see

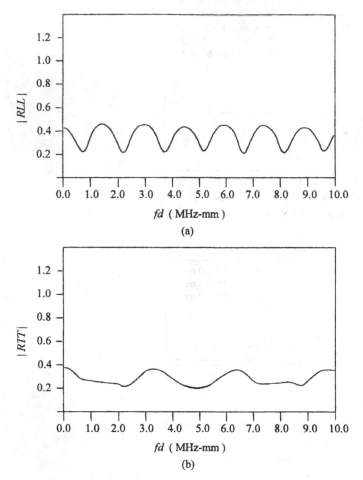

Figure 5-14. Reflectivity of (a) longitudinal and (b) transverse waves from an epoxy layer embedded between aluminum and polystyrene semispaces with perfect boundary conditions.

that, whereas 30° is the best choice for the 10-MHz frequency, 25° would be most appropriate for 5 MHz. Hence, the correct choice of incident frequency–angle combination, for a given thickness of glue line, is as important as the decision on the preferred type of wave.

Analyzing the results shown in Figure 5-12, one finds that – even for the recommended angle of incidence of the longitudinal waves – there is no observable shift in the position of the first minimum when the connection quality is varied. However, for the transverse wave incident at the preferred angle, the boundary conditions influence a displacement of the first minimum. Therefore, by choosing the frequency–thickness product around the first minimum, we can achieve an increase in sensitivity to interfacial weakness utilizing the transverse waves at 30°. Unfortunately, even small changes in the thickness of the adhesive layer can alter the situation completely. This last conclusion is nicely illustrated in Figure 5-13: for the same bond rigidity, one obtains different values of reflection coefficients depending on the frequency–thickness product. Also, the dips observed in these curves for low fd products indicates a zone of some confusion.

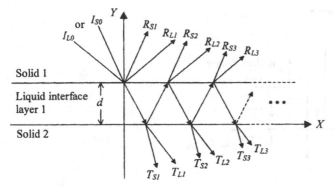

Figure 5-15. Multireflection considerations from a nonviscous liquid layer for oblique incidence.

Comparing the results shown in Figure 5-14 with those presented in Figure 5-10, we observe an exchange in the positions of minima and maxima. This confirms known results (see Brekhovskikh 1960) for normal incidence on the influence of mutual relationships between acoustic impedances of the individual media within a layered structure. Also, the minima in Figure 5-10 are more distinct than those of Figure 5-14, which is also anticipated owing to acoustic impedance combinations.

5.3.7 Conclusions

In the first part of this section we derived the transform matrix on the basis of classical Thompson–Haskell theory and so formulated a fundamental set of equations for reflection and transmission coefficients in the case of a solid layer separating two solid semispaces with imperfect boundary conditions. The details, for both longitudinal and transverse incident waves, are provided in (5.50).

Following this description, and using existing formalism for a stratified elastic medium, our solution for a single layer could easily be extended to the multilayered problem. Also, a superposition technique to handle multifrequency waveforms and more realistic ultrasonic pulses could be obtained by using the monochromatic plane wave solution and Fourier decomposition of the wave packet.

5.4 Multireflection Approach

5.4.1 Thin Liquid Layer

Reflection and transmission factors across a thin liquid layer are considered first as a sample problem of the multireflection approach. Without losing generality, we assume that the amplitude of the incident wave is of unit 1. The reflection factor across a thin liquid layer can be calculated by superposition of all like waves from the liquid layer to medium 1 (Figure 5-15). For example, a longitudinal wave reflection with respect to an incident longitudinal wave (L wave) may be a superposition of the following waves:

$$R_{L0} + R_{L1} + R_{L2} + \cdots. \tag{5.53}$$

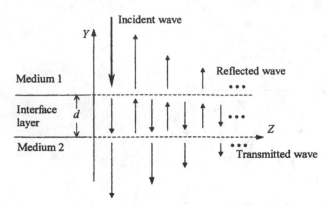

Figure 5-16. Multireflection considerations from an interface
layer for normal incidence.

First, consider the directly reflected L wave from the upper interface. Its amplitude is
easily obtained from the single interface problem between a solid and a liquid:

$$R_{L0} = (RLL)_{1-I}, \tag{5.54}$$

where RLL is the reflection factor for longitudinal incidence of an L wave reflected from
medium 1 (solid) to I (the liquid interface layer).

The first reflected L wave across the thin layer penetrates the upper interface, passes
through the liquid layer, reflects from the lower interface, then passes through the liquid
layer again, and finally leaves through the upper interface. Its amplitude is

$$R_{L2} = (DLL)_{1-I}(RLL)_{I-2}(DLL)_{I-1}. \tag{5.55}$$

We can control this process for subsequent reflections as follows:

$$R_{L3} = (DLL)_{1-I}(RLL)_{I-2}(RLL)_{I-1}(RLL)_{I-2}(DLL)_{I-1}, \tag{5.56}$$

$$R_{L4} = (DLL)_{1-I}(RLL)_{I-2}(RLL)_{I-1}(RLL)_{I-2}$$
$$\times (RLL)_{I-1}(RLL)_{I-2}(DLL)_{I-1}. \tag{5.57}$$

This process can be continued in order to calculate the overall reflection and refraction
factors for shear or longitudinal input to the layer.

The reflected wave across the liquid layer forms a geometric progression series. Recall
that $a_n = a_1 q^{n-1}$ for $n = 1, 2, 3, \ldots$ and $0 < |q| < 1$; its summation is $S = a_1/(1 - q)$.
The final reflection factor is the summation of all of the longitudinal waves.

5.4.2 A Layer for Normal Incidence

For normal incidence (Figure 5-16), if we know the reflection and refraction factors on
both upper and lower interfaces between the interface layer and semi-infinite solid media,
then the technique illustrated in Section 5.4.1 can be used to solve for reflection and transmission factors for a single embedded layer. See Brekhovskikh (1960) for an extension
of this analysis to multiple layers.

5.5 Exercises

1. Write out the general boundary condition equations for reflection and refraction at a solid–liquid interface (for shear input as well as longitudinal input).

2. Compute all reflection and refraction factors for L-wave and T-wave input to a Plexiglas–aluminum interface.

3. Write boundary conditions for oblique incidence onto a liquid–solid interface.

4. Derive the reflection and refraction factor equation for a liquid–solid interface.

5. What changes in the reflection factor derivations are necessary if we are to calculate oblique incident reflection and refraction factors between two anisotropic media?

6. What changes are expected in the derivation if two interface layers are considered?

7. How many boundary conditions are there for a four-layer structure immersed in water?

8. Use computer programs to generate the following reflection factor curves from $0°$ to $90°$ angle of incidence.

   (a) RTT versus angle of incidence curve for an epoxy–aluminum interface ($f = 10.0$ MHz).

   (b) RTT versus angle of incidence curve for an epoxy–aluminum space separated by a thin water layer of thickness 0.01 mm ($f = 10.0$ MHz).

   (c) RTT versus fd curve for an epoxy–aluminum space separated by a thin solid layer of thickness 0.01 mm ($f = 10.0$ MHz). The solid layer properties are $c_L = 1.9$ Km/s, $c_T = 0.9$ Km/s, and density $\rho = 1.05$ g/cm^3; the epoxy properties are $c_L = 2.4$ Km/s, $c_T = 1.1$ Km/s, and $\rho = 1.10$ g/cm^3; and the aluminum properties are $c_L = 6.3$ Km/s, $c_T = 3.1$ Km/s, and $\rho = 2.70$ g/cm^3.

9. An epoxy–aluminum space is separated by a thin water layer of thickness 0.1 mm. Incident longitudinal wave is normal to the interface. Generate a reflection factor curve of RLL versus fd in the frequency range 0.1–20 MHz.

10. Using the multireflection approach across a thin liquid layer, derive an expression for the overall reflection and/or refraction factor from the layer for oblique longitudinal and/or shear input. (Use the geometric progression series to develop the result.)

11. Calculate the reflection and transmission factors for a normal incident longitudinal wave onto a liquid layer between two solids. (Use a geometric progression series to develop the result.)

6

Wave Scattering

6.1 Background

The subject of wave scattering is addressed in many textbooks as well as in hundreds of research papers. The scattering of waves by an obstacle is a topic of study for wave propagation in elastodynamics, optics, acoustics, and electromagnetism. The subject is merely introduced in this chapter in order to make students aware of activity in the field. Research emphasis has been on the solution of an inverse problem, as opposed to the direct wave scattering solution. Problems exist in the former case because of the uniqueness of a solution.

As an alternative to theoretical and numerical approaches, experiments using various pattern recognition and neural net approaches have contributed to the literature on this reflector classification problem. For some interesting reading on the subject, see Chu, Askar, and Cakmak (1982), Gubernatis et al. (1977), Knopoff (1956), Lewis, Temple, and Wickham (1996), Lindsay (1968), Mal and Lee (1995), Miklowitz (1978), Nagy and Rose (1993), Rose and Richardson (1982), Thompson and Gray (1983), and Wu and Aki (1990).

6.2 A Sample Problem

In order to illustrate a method of scattering theory, we will consider scattering of shear horizontal (SH) waves by a cylindrical cavity (see Graff 1991). This is one of the simplest cases because, for SH waves, we have just one governing equation of motion for one unknown displacement $u = u_x(y, z)$ in the x direction, where $u_y = u_z = 0$. See Figure 6-1.

The equation of motion is

$$\nabla^2 u = \frac{1}{c_T^2} \frac{\partial^2 u}{\partial t^2} \quad \text{or} \tag{6.1}$$

$$\nabla^2 u = \frac{1}{r} \frac{\partial}{\partial r} \left(r \frac{\partial u}{\partial r} \right) + \frac{1}{r^2} \frac{\partial^2 u}{\partial \theta^2} + \frac{\partial^2 u}{\partial x^2}. \tag{6.2}$$

Consider the traction-free boundary conditions on the cavity surface:

$$\tau_{rr} = \tau_{r\theta} = \tau_{rx} = 0 \quad \text{at } r = a, \tag{6.3}$$

$$\tau_{rr} = \lambda \left(\frac{\partial u_r}{\partial r} + \frac{1}{r} \frac{\partial u_\theta}{\partial \theta} + \frac{u_r}{r} + \frac{\partial u_x}{\partial x} \right) + 2\mu \frac{\partial u_r}{\partial r}, \tag{6.4}$$

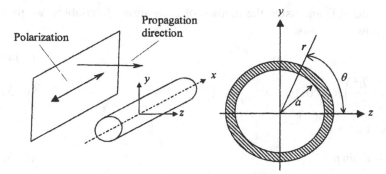

Figure 6-1. SH wave scattering from a circular cavity.

$$\tau_{r\theta} = \mu \left(\frac{1}{r} \frac{\partial u_r}{\partial \theta} + \frac{\partial u_\theta}{\partial r} - \frac{u_\theta}{r} \right), \tag{6.5}$$

$$\tau_{rx} = \mu \left(\frac{\partial u_x}{\partial r} + \frac{\partial u_r}{\partial x} \right). \tag{6.6}$$

It follows that

$$u_r = u_\theta = 0; \tag{6.7}$$

u_x does not depend on x. Therefore, from equations (6.4)–(6.6), we obtain the boundary condition:

$$\frac{\partial u}{\partial r} = 0 \quad \text{at} \quad r = a. \tag{6.8}$$

The incident SH wave propagating in the z direction is

$$u_i = U_0 e^{i(\omega t - kz)}, \quad \text{where} \tag{6.9}$$

$$k = \frac{2\pi}{\lambda} = \frac{\omega}{c_T}, \tag{6.10}$$

which satisfies the governing equation (6.1).

When the incident wave interacts with the cavity, a scattering field will occur. Because of the harmonic wave field, the scattering displacement can be written as

$$u_s = U_s(r, \theta) e^{i\omega t}. \tag{6.11}$$

The unknown amplitude $U_s(r, \theta)$ can be sought in the form

$$U_s(r, \theta) = R(r)S(\theta); \tag{6.12}$$

the total displacement field is given by

$$u(r, \theta) e^{i\omega t} = u_i + u_s \tag{6.13}$$

and should satisfy the equation of motion (6.1).

Substituting (6.13) into (6.1) and using the method of separation of variables, we find two differential equations as follows:

$$S'' + k^2 S = 0, \tag{6.14}$$

$$R'' + \frac{1}{r} R' + \left(k^2 - \frac{h^2}{r^2} \right) R = 0. \tag{6.15}$$

From (6.14), it follows that

$$S(\theta) = C \cos n\theta + D \sin n\theta. \tag{6.16}$$

The solution $S(\theta)$ should be symmetric with respect to the x-axis; therefore, $D = 0$. The condition $S(\theta) = S(\theta + 2\pi)$ requires an integer value for n.

One solution of the Bessel equation (6.15) is

$$R(r) = A H_n^{(2)}(kr) + B H_n^{(1)}(kr). \tag{6.17}$$

The Hankel function solution is chosen in order to satisfy radiation conditions. For $k \gg 1$, we have the following asymptotic representation of the Hankel functions:

$$H_n^{(1)}(kr) = \sqrt{\frac{2}{\pi kr}} \exp\left\{ i\left(kr - \frac{\pi}{4} - \frac{n\pi}{2} \right) \right\}, \tag{6.18}$$

$$H_n^{(2)}(kr) = \sqrt{\frac{2}{\pi kr}} \exp\left\{ -i\left(kr - \frac{\pi}{4} - \frac{n\pi}{2} \right) \right\}. \tag{6.19}$$

From radiation conditions, it follows that the outward propagating wave has the form $\exp\{i(\omega t - kr)\}$; hence, $B = 0$ in (6.17).

From equations (6.12), (6.16), and (6.18), we finally reach the result of the amplitude of the scattered wave:

$$U_s(r, \theta) = \sum_{n=0}^{\infty} A_n H_n^{(2)}(kr) \cos n\theta. \tag{6.20}$$

In Figure 6-1 we see that, in polar coordinates, $z = r \cos \theta$. Therefore, the expression for the incident wave equation (6.9) may be given in polar coordinates as

$$u_i = U_0 e^{-ikr \cos\theta} \cdot e^{i\omega t}. \tag{6.21}$$

Graff (1991) has shown that u_i can be expressed through Bessel functions as

$$u_i = U_0 \sum_{n=0}^{\infty} \varepsilon_n (-1)^n J_n(kr) \cos n\theta \quad (\varepsilon = 1, \ \varepsilon_n = 2), \tag{6.22}$$

where $J_n(kr)$ are the Bessel functions.

The total displacement field in (6.13) should satisfy the boundary conditions in (6.8). Substituting (6.13) into (6.8) yields

$$\frac{\partial u_i}{\partial r} + \frac{\partial u_s}{\partial r} = 0 \quad \text{at } r = a. \tag{6.23}$$

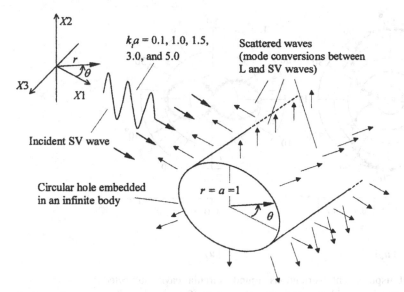

Figure 6-2. Plane SV wave scattering from a circular cavity.

Using (6.22) and (6.20), we then obtain

$$\sum_{n=0}^{\infty} \left\{ U_0 \varepsilon_n (-1)^n \frac{dJ_n(kr)}{dr} + A_n \frac{dH_n^{(2)}(kr)}{dr} \right\} \cos n\theta |_{r=a} = 0. \tag{6.24}$$

Because functions $\cos n\theta$ are orthogonal, we find the coefficient values A_n from (6.24) as follows:

$$A_n = \frac{\varepsilon_n (-1)^{n+1} U_0 J_n'(ka)}{H_n^{(2)'}(\gamma a)}. \tag{6.25}$$

The scattered field has the form of (6.20), with coefficients A_n given by (6.25). In the far field, using an asymptotic expression in (6.19), the result can be written as

$$u_s(r, \theta, t) = U_0 \sqrt{\frac{2}{\pi kr}} e^{i(\omega t - kr)} \psi_s(\theta) \quad (kr \gg 1), \tag{6.26}$$

where $\psi_s(\theta)$ is a known function. Equation (6.26) is the angular displacement distribution.

Sample results are shown in Graff (1991). Cho (1995) presents numerical results of the elastic wave scattering from a circular cavity in an infinite elastic media subjected to plane harmonic SV (shear vertical) polarization. See Figure 6.2.

If the wavelength of an incident wave is extremely small compared with the radius of a circular cavity, most of the incident energy will be reflected straight back from the cavity by the main lobe. This occurs because the cavity behaves like a flat, infinitely long, traction-free surface in comparison to the incident wavelength. Finally, this problem converges to normal incidence of plane harmonic SV waves on that surface. See the sample results in Figures 6-3 to 6-7.

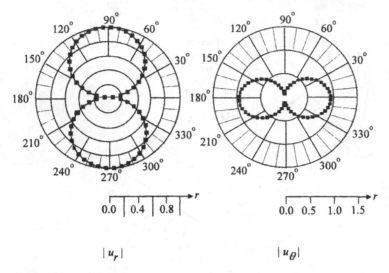

Figure 6-3. Total displacement distributions around a circular cavity subjected to plane SV wave incidence with Poisson's ratio 0.25 ($ka = 0.1$).

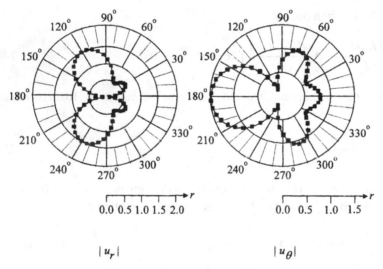

Figure 6-4. Total displacement distributions around a circular cavity subjected to plane SV wave incidence with Poisson's ratio 0.25 ($ka = 1.0$).

6.3   Other Concepts

6.3.1   General

We will now review some other basic concepts in wave scattering. Several methods used to solve scattering problems are based on formulating the boundary value problem in integral form. These techniques include the boundary element methods, the Born approximation, and the T-matrix approach. In order to attain an integral representation, we

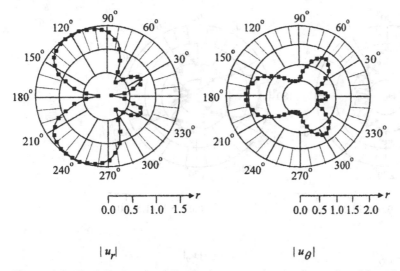

$|u_r|$ $|u_\theta|$

Figure 6-5. Total displacement distributions around a circular cavity subjected to plane SV wave incidence with Poisson's ratio 0.25 ($ka = 1.5$).

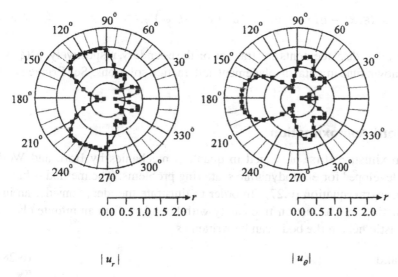

$|u_r|$ $|u_\theta|$

Figure 6-6. Total displacement distributions around a circular cavity subjected to plane SV wave incidence with Poisson's ratio 0.25 ($ka = 3.0$).

need to transform differential equations – which describe the behavior of the unknown functions inside an elastic body and on the boundary – into the boundary integral formulation. This integral equation contains only the value of the functions on the boundary. The value of the functions inside the body can be expressed through the boundary value functions. Solution of the integral equation satisfies the original differential equation with appropriate boundary conditions. The boundary integral equation without the body force contribution can be written in the form

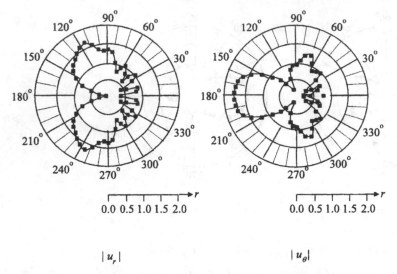

$$|u_r| \qquad\qquad\qquad |u_\theta|$$

Figure 6-7. Total displacement distributions around a circular cavity subjected to plane SV wave incidence with Poisson's ratio 0.25 ($ka = 5.0$).

$$u(\xi) = \int_\Gamma [\sigma(x)u^*(\xi, x) - u(x)\sigma^*)(\xi, x)]\, d\Gamma(x) \quad (\xi \in \Gamma), \tag{6.27}$$

where $u(x)$ and $\sigma(x)$ are displacements and traction on the boundary, respectively; $u^*(\xi, x)$ and $\sigma^*(\xi, x)$ are known fundamental displacement and traction solutions. See Eringen and Suhubi (1975).

6.3.2   The Born Approximation

The Born approximation, first presented in quantum mechanics by Born and Wolf (1965), was later developed for elastodynamic scattering problems. The method is based on the boundary integral equation (6.27). In order to illustrate the idea, consider an incident wave interacting with a traction-free cavity with boundary Γ in an infinite elastic body. The total elastic field in the body can be written as

$$u = u_s + u_i \quad \text{and} \tag{6.28}$$

$$\sigma = \sigma_s + \sigma_i, \tag{6.29}$$

where indices s and i represent the scattering and incident field, respectively. Substituting (6.28) and (6.29) into (6.27) leads to an integral boundary equation with respect to u_s and σ_s. As a result, we obtain

$$u_s(\xi) = \int_\Gamma [\sigma_s(x)u^*(\xi, x) - u_s(x)\sigma^*(\xi, x)]\, d\Gamma(x) + F(x), \tag{6.30}$$

where $F(x)$ depends on the known functions u_i and σ_i.

The boundary condition on the cavity surface is

$$(\sigma_s + \sigma_i)|_\Gamma = 0. \tag{6.31}$$

From (6.30) and (6.31) it follows that

$$u_s(\xi) = - \int_\Gamma [\sigma_i(x)u^*(\xi, x) + u_s(x)\sigma^*(\xi, x)] d\Gamma(x) + F(x). \tag{6.32}$$

This is the integral equation (6.32) with respect to the unknown function $u_s(x)$. The first step in the Born approximation is to neglect the second unknown term in the integral equation (6.32). Thus,

$$u_s(\xi) = - \int_\Gamma \sigma_i(x)u^*(\xi, x) d\Gamma(x) + F(x). \tag{6.33}$$

The second iteration can be obtained from (6.30) by replacing $u_s(x)$ and $\sigma_s(x)$ with $u_s^{(1)}$ and $\sigma_s^{(1)}$, respectively:

$$u_s^{(2)}(\xi) = \int_\Gamma [\sigma_s^{(1)}(x)u^*(\xi, x) - u_s^{(1)}(x)\sigma^*(\xi, x)] d\Gamma(x) + F(x). \tag{6.34}$$

Equation (6.34) leads to the following iterative procedure:

$$u_s^{(n+1)}(\xi) = \int_\Gamma [\sigma_s^{(n)}(x)u^*(\xi, x) + u_s^{(n)}(x)\sigma^*(\xi, x)] d\Gamma(x) + F(x). \tag{6.35}$$

Convergence for the Born method was shown numerically in Chu et al. (1982). The Born approximation becomes more accurate as the size of the cavity decreases.

6.3.3 The T-Matrix Approach

A number of wave scattering problems have also been solved by the T-matrix approach. For elastodynamics, this method is described in Varadan and Varadan (1980). In this approach, the known incident field, the unknown scattered field, and the fundamental displacement and traction are expressed as a series of a complete set of orthogonal functions. Substituting these functions into (6.27), we obtain the transition matrix (T-matrix), which expresses the unknown coefficients of the scattered field through the known incident wave:

$$\bar{b} = T\bar{a}, \quad \text{where} \tag{6.36}$$

$$\bar{a} = (a_1, a_2, \ldots) \quad \text{and} \tag{6.37}$$

$$\bar{b} = (b_1, b_2, \ldots). \tag{6.38}$$

The T-matrix depends only on the size and the shape of the obstacle. Therefore, the same T-matrix can be used for a given obstacle within different incident fields.

6.4 Exercises

1. What is the limiting case for the scattering problem when the wavelength of the incident wave is much smaller than the radius of a circular cavity?

2. Describe the iteration procedure for the Born approximation.

3. Why is the SH wave scattering problem in general easier to solve than the SV problem?

7

Surface and Subsurface Waves

7.1 Background

The existence of surface waves was predicted theoretically over a century ago. There is much literature on this subject, including for example Chadwick and Smith (1977), Farnell (1970), Pollard (1977), and Viktorov (1967). Experimental evidence was first obtained in observing wave propagation over the surface of the earth (as a result of earthquakes) and subsequent mode conversion at the earth's surface. Observations were made regarding the unusual behavior of energy decay with increased depth and the ability of waves to travel along curved surfaces.

This chapter examines surface waves on an isotropic, homogeneous, linear elastic semi-space. We take a rather classical approach to the problem, one that is based on potential functions and boundary conditions for a free surface. Assumptions of isotropy, homogeneity, and linear elastic response will also be made. For more detail, see Auld (1990), Basatskaya and Ermolov (1980), Couchman and Bell (1978), Heelan (1953), Kolsky (1963), Nikiforov and Kharitonov (1981), Pilarski and Rose (1989), Uberall (1973), and Viktorov (1967).

7.2 Surface Waves

A general derivation of the equation for surface wave propagation will be treated next. As a result of Helmholtz decomposition of the displacement vector, which is divided into two components, we have

$$\bar{u} = \nabla \phi + \nabla \times \bar{\psi}. \tag{7.1}$$

Taking the option to use grad, curl, or rot notation, we find that

$$\bar{u} = \operatorname{grad} \phi + \operatorname{rot} \bar{\psi}.$$

Navier's governing wave equation can be separated into two simple wave equations, one for dilatational waves and one for rotational or shear waves (see e.g. Kolsky 1963; Pollard 1977):

$$\nabla^2 \phi - \frac{1}{c_L^2} \ddot{\phi} = 0,$$

$$\nabla^2 \bar{\psi} - \frac{1}{c_T^2} \ddot{\bar{\psi}} = 0, \tag{7.2}$$

where c_L is longitudinal and c_T shear velocity.

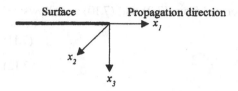

Figure 7-1. Coordinate system used (x_1, x_2, x_3 or x, y, z).

For the plane two-dimensional case,

$$\bar{u} = (u_1, 0, u_3)$$

(see the coordinate system in Figure 7-1). Therefore, the solution can be obtained from the two functions ϕ and ψ:

$$\phi = \phi(x_1, x_3, t); \tag{7.3}$$

the values of ψ_1 and ψ_3 are zero or constant because $u_2 = 0$, given

$$\bar{u} = \nabla \times \psi.$$

Hence $\psi = \psi_2 = \psi(x_1, x_3, t)$. Note that ψ_2 motion is in the (x_1, x_3)-plane only. In Lamé form, (7.1) can therefore be written as

$$u = u_1 = \phi_{,1} + \psi_{,3} = \phi_{,x} + \psi_{,z},$$
$$w = u_2 = \phi_{,3} - \psi_{,1} = \phi_{,z} - \psi_{,x}. \tag{7.4}$$

We are seeking the harmonic solution of (7.2), which represents harmonic waves traveling in the direction x_1. We have

$$\phi = D_1(z)e^{i(kx - \omega t)}, \tag{7.5}$$
$$\bar{\psi} = D_2(z)e^{i(kx - \omega t)}. \tag{7.6}$$

Substituting (7.5) and (7.6) into (7.2) (and discarding the impractical part of the solution, which doesn't attenuate), for an increase in x_3 we obtain

$$\phi = A_1 e^{-kqz} e^{ik(x-ct)}, \tag{7.7}$$
$$\psi = B_1 e^{-ksz} e^{ik(x-ct)}, \tag{7.8}$$

where

$$q = \sqrt{1 - \left(\frac{c}{c_L}\right)^2}, \quad s = \sqrt{1 - \left(\frac{c}{c_T}\right)^2}, \quad c = \frac{\omega}{k},$$

and A_1, B_1 are arbitrary constants.

From equations (7.5), (7.7), and (7.8), it then follows that

$$u = k(iA_1 e^{-qz} - sB_1 e^{-sz})e^{ik(x-ct)}, \tag{7.9}$$
$$w = -k(qA_1 e^{-qz} - iB_1 e^{-sz})e^{ik(x-ct)}. \tag{7.10}$$

On the basis of Hooke's law, stresses may be derived from (7.9) and (7.10) as follows:

$$\sigma_{33} = k^2 G(rA_1 e^{-qz} + 2isB_1 e^{-sz})e^{ik(x-ct)}, \tag{7.11}$$

$$\sigma_{13} = k^2 G(-2iqA_1 e^{-qz} + rB_1 e^{-sz})e^{ik(x-ct)}, \tag{7.12}$$

where $r = 2 - (c/c_t)^2$.

We must now satisfy the boundary conditions

$$\sigma_{33} = \sigma_{13} = 0 \quad \text{for } z = 0. \tag{7.13}$$

From (7.11)–(7.13) it follows that

$$rA_1 + 2isB_1 = 0 \quad \text{and} \quad -2iqA_1 + rB_1 = 0. \tag{7.14}$$

Equations (7.14) lead to the equation for the unknown phase velocity c:

$$r^2 - 4sq = 0. \tag{7.15}$$

From (7.14) we also obtain

$$A_1 = -\frac{ir}{2q}B_1 = -\frac{2is}{r}B_1. \tag{7.16}$$

By substituting (7.16) into (7.9) and (7.10), we find the forms of u and w that satisfy the boundary conditions:

$$u = A(re^{-qz} - 2sqe^{-sz})e^{ik(x-ct)}, \tag{7.17}$$

$$w = iAq(-re^{-qz} + 2e^{-sz})e^{ik(x-ct)}, \tag{7.18}$$

where $A = kB_1/2q$.

Equation (7.15) is the characteristic equation for this surface wave problem. In order to solve this equation, we will introduce new variables:

$$\eta = \frac{k_T}{k} = \frac{c}{c_T}, \qquad \zeta = \frac{k_L}{k_T} = \frac{c_T}{c_L} \quad \text{(since } k = \omega/c), \tag{7.19}$$

where we have used that $r = 2 - \eta^2$, $q = \sqrt{1 - \eta^2}$, and $s = \sqrt{1 - \eta^2 \zeta^2}$. Algebraic manipulation then yields

$$\eta^6 - 8\eta^4 + 8\eta^2(3 - 2\zeta^2) + 16(\zeta^2 - 1) = 0, \tag{7.20}$$

which is the Rayleigh wave velocity equation.

Note now that

$$\zeta = \frac{c_T}{c_L} = \sqrt{\frac{1 - 2\nu}{2(1 - \nu)}}; \tag{7.21}$$

since $\eta = c/c_T$, it also equals c_R/c_T with c equal to c_R, which in turn is equal to the Rayleigh *surface* wave velocity. As a result, we have three roots of (7.20) for η as a function of Poisson's ratio ν only.

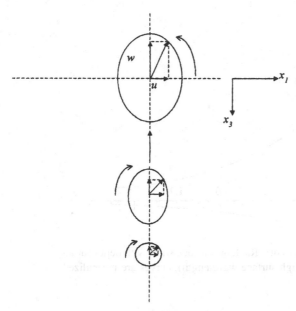

Figure 7-2. Displacement vector elliptical particle motion
with time and depth.

There is an approximate solution from Viktorov, $\eta = (0.87 + 1.12\nu)/(1 + \nu)$. Note that, in the cubic equation, there are two complex conjugate roots and one real root for $\nu > 0.263$; for $\nu \leq 0.263$, there are three real roots but only one realistic one. We note further that η does not depend on frequency. Therefore, Rayleigh surface waves are nondispersive.

We now travel to the area of displacement analysis. Because the components of the displacement vector are real, from (7.17) and (7.18) we obtain

$$\tilde{u} = \frac{u}{A} = (re^{-qz} - 2sqe^{-sz})\cos k(x - ct), \qquad (7.22)$$

$$\tilde{w} = \frac{w}{A} = q(re^{-qz} - 2e^{-sz})\sin k(x - ct), \qquad (7.23)$$

where $\tilde{u}$ and $\tilde{w}$ are normalized displacements with respect to the unknown constant A.

Consider now some aspects of the solution. From (7.22) and (7.23) it follows that, for any value of z, the vector with coordinates $(\tilde{u}, \tilde{w})$ moves along an ellipse (see Figure 7-2). We have

$$\frac{(\tilde{u})^2}{(re^{-qz} - 2sqe^{-sz})^2} + \frac{(\tilde{w})^2}{q^2(re^{-qz} - 2e^{-sz})^2} = 1, \qquad (7.24)$$

which is the equation of an ellipse. Note that the original expressions (7.9) and (7.10) show two waves, one longitudinal and one transverse, propagating along a boundary that attenuates with depth. The superposition, however, leads to the surface waves and special qualities that we now observe.

The equation in u and w on the surface is an ellipse for all time values. This means that the vector sum of u and w gives the ellipse, as shown in Figure 7-2. It can be demonstrated

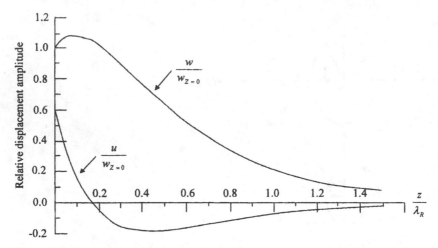

Figure 7-3. Sample displacement values of a Rayleigh surface wave with depth for an aluminum half-space (λ_R is the Rayleigh surface wavelength); curves are normalized with respect to w at $z = 0$.

that elliptical particle motion occurs as a function of depth, with the motion actually reversing after a particular depth. Analysis of the displacement equation can be used to plot these graphs. Further analysis can be used to plot displacement profiles with depth that are similar to those shown in Figure 7-3.

Figure 7-4 is a detailed presentation of particle displacements in an aluminum plate. Each displacement vector shown is the vector sum of the two normal components, $\bar{u}$ and $\bar{w}$. The figure can be considered as a snapshot of the resulting particle displacement vectors at different time periods. This gives us an idea of the displacement vector rotation along the surface and elsewhere throughout a thick aluminum plate. Note the decay with depth and the change in rotation of the particle displacement vector in space and time. Note also that, along any horizontal line above a, the particle displacement vector rotation is clockwise as you move from left to right; below a, the particle displacement vector rotation is counterclockwise. On the other hand, if we were to fix a position in space to observe the particle displacement vector rotation, it would be counterclockwise above a and clockwise below a. The vertical displacement component w is also generally larger than the horizontal component u, hence leading to an elliptical particle displacement rotation with a vertical major axis.

It is interesting to plot out the stress components with depth, again on a normalized basis. This can be done using stress equations (7.11) and (7.12); a typical result is presented in Figure 7-5.

7.3 Generation and Reception of Surface Waves

Surface waves can be generated and received through a variety of different techniques, a few of which are discussed here. A normal beam or shear wave transducer can simply be placed on a surface, as in Figure 7-6. A wedge technique might be used as in Figure 7-7, with c_{1w} the wave velocity in the wedge material and c_R the surface wave velocity in medium 2. Snell's law states that $c_1 \sin \theta_2 = c_2 \sin \theta_1$; applied to this case it yields $c_{1w} \sin 90° = c_R \sin \theta_w$, so

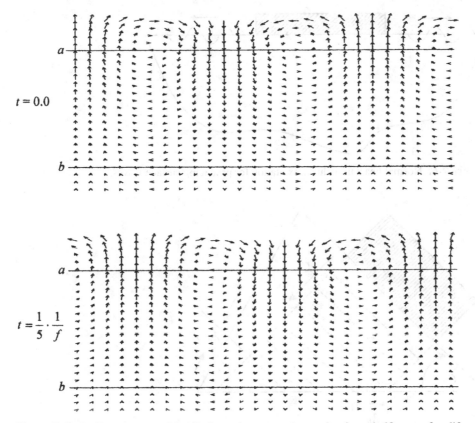

Figure 7-4. Surface wave particle displacement vector in an aluminum half-space for different time periods (line a represents the depth at which reversal in particle rotation occurs; line b represents one wavelength in depth.

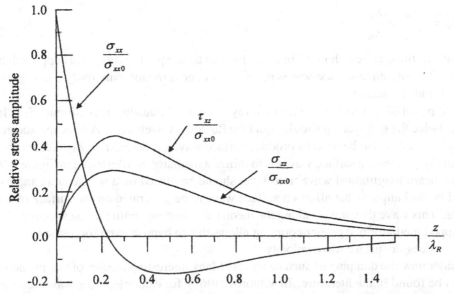

Figure 7-5. Sample stress values of a Rayleigh surface wave with depth for an aluminum half-space (λ_R is the Rayleigh surface wavelength; curves are normalized with respect to σ_{xx} at $z = 0$.

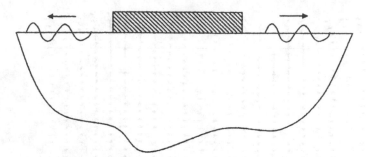

Figure 7-6. Normal beam transducer excitation for producing surface waves.

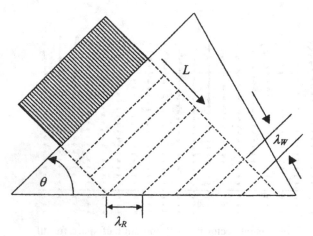

Figure 7-7. Angle beam transducer excitation for producing surface waves.

$$\sin \theta_w = \frac{c_{1w}}{c_R} = \frac{\lambda_w}{\lambda_R}.$$

Note that c_{1w} must be less than c_R in order for this technique to work. This approach is the most efficient, since surface waves propagate in one direction only (only one angle is required for all frequencies).

A third possibility is to use a periodic array or "comb" transducer; see Figure 7-8. In this case, twice the gap spacing should equal the Rayleigh wavelength. A direct piezoelectric coupling could also be used to produce surface waves, where $2a = \lambda_R$. Yet another technique for generating surface waves is to utilize a mediator, as illustrated in Figure 7-9. A normal beam longitudinal wave transducer should be placed on a wedge at an angle – the third critical angle – that allows a surface wave to be generated on the surface of the mediator. This wave then travels along the mediator to impinge onto the test specimen. A sharp-tipped mediator and proper coupling allows this to happen even for test specimens with low values of surface wave velocity.

Consider now the damping of surface waves. Many interesting studies of this phenomenon can be found in the literature. In Viktorov (1967), for example, if the wavenumbers are treated as complex then

$$\bar{k}_L = k'_L + i k''_L, \quad \bar{k}_T = k'_T + i k''_T, \quad \text{and} \quad \bar{k}_R = k'_R + i k''_R.$$

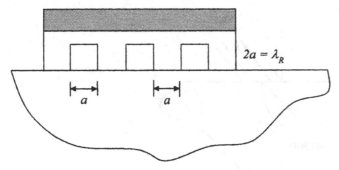

Figure 7-8. A periodic array or comb transducer for producing surface waves.

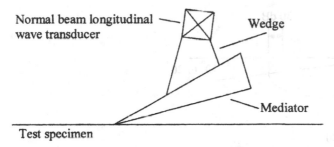

Figure 7-9. Mediator technique of generating surface waves in a test specimen.

It can be shown that the damping of Rayleigh waves is similar to that of cylindrical waves, as follows: $A_R \sim 1/\sqrt{k_R r}$ is compared to bulk waves, where $A_{L,T} \sim 1/k_{L,T} r$ (L for longitudinal and T for transverse). Rayleigh waves are less attenuated than bulk waves and are thus, for example, responsible for greater damage from earthquakes. Propagation of Rayleigh waves can be studied on curved surfaces. As an example, consider the wave equation in a system using cylindrical coordinates.

A noteworthy analysis of surface waves on anisotropic media may be found in Rose, Pilarski, and Huang (1990). These conditions give rise to variations in surface wave velocity with angle and subsequent differences in phase and in group or energy velocity, as well as to the presence of a skew angle. An inverse problem to evaluate certain composite material properties as a function of skew angle could be carried out.

7.4 Subsurface Longitudinal Waves

Subsurface elastic waves, mostly geoacoustical, are reported under many names: head waves, lateral waves, creeping longitudinal waves, or fast surface waves. Here, the term *subsurface* waves or subsurface longitudinal (SSL) waves is used to describe the field of longitudinal waves excited in a solid half-space by an angle beam transducer with an angle of incidence close to the first critical angle.

Subsurface longitudinal waves have been extensively investigated, not only theoretically but also experimentally to detect defects in subsurface layers of an isotropic material.

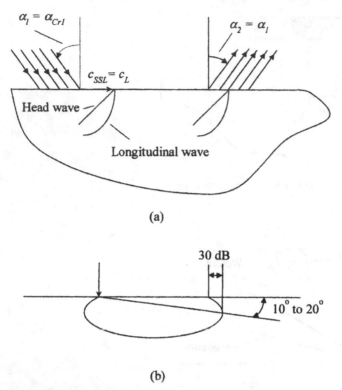

(a)

(b)

Figure 7-10. Subsurface longitudinal waves at first critical angle:
(a) coexistence of head and longitudinal waves; (b) pressure field
pattern.

It has been established that – at or near the first critical angle, for longitudinal waves incident onto an interface (liquid–solid or solid–solid) from the medium with a smaller velocity of longitudinal waves – there coexist two waves: SSL and head waves. This is shown in part (a) of Figure 7-10. The two wave types cooperatively fulfill the boundary conditions on the free surface of the solid, where all stresses are supposed to equal zero. Any disturbance on the free surface moves with a velocity equal to the velocity of longitudinal waves in the solid. The amplitude of this displacement decreases as distance increases, according to the l/r^n law, where n ranges from 1.5 to 2.0. This means that the SSL waves close to the free surface are strongly attenuated compared with the bulk waves, since the former are proportional to l/r^n with n ranging from 0.5 to 1.0. The SSL waves can be detected at some other spot on the same surface, but the receiving transducer must be inclined at an angle equal to the first critical angle.

One characteristic of subsurface waves is the distribution of the amplitude of acoustic pressure in the plane of incidence; this is shown in part (b) of Figure 7-10. The shape of the pressure field distribution reveals that the maximum sensitivity of the ray occurs at an angle of 10 to 20 degrees from the free surface. Hence, the name chosen for these waves is appropriate, since they can be utilized for the detection of subsurface defects – and especially since the SSL waves show a relatively small sensitivity to surface roughness.

The definition given here of SSL waves is related to the first critical angle, which (for isotropic media) is given by Snell's law as $\alpha_{cr} = \sin^{-1}(c_1/c_L)$, where c_1 and c_L are longitudinal wave velocities for the upper (shoe of angle beam probe) and lower media, respectively. For an anisotropic medium the critical angle must be redefined, since the phase and group velocity vectors are generally of different orientations.

7.5 Exercises

1. Show that the equation $r^2 - 4sq = 0$ becomes
$$\eta^6 - 8\eta^4 + 8(3 - 2\zeta^2)\eta^2 - 16(1 - \zeta^2) = 0;$$
let $\eta = c/c_T$ and $\zeta = c_T/c_L$.

2. Solve the Rayleigh surface wave velocity equation as a function of Poisson's ratio ν. Plot graphs. Compare with Viktorov's approximate solution, $\eta = (0.87 + 1.12\nu)/(1 + \nu)$.

3. Where does the particle velocity projection onto an ellipse reverse direction?

4. Study the damping of Rayleigh waves, comparing such media as steel, aluminum, and epoxy. Use Viktorov (1967) or other references.

5. For a wave incident on a Plexiglas specimen, assume that the longitudinal velocity is 2.71 mm/μs and the shear wave velocity is 1.38 mm/μs. Calculate the theoretical Rayleigh surface wave velocity and compare with the measured values.

6. Discuss or explain why the Rayleigh equation is a function of only Poisson's ratio for isotropic material.

7. For $t = 0$, plot the particle displacement on a 12×12 grid as a function of x and z in order to demonstrate concepts of particle motion.

8. Design a comb-type transducer to generate a surface wave in a structure.

9. For a Poisson's ratio of $\nu = 0.25$, find the surface wave velocity. Which root is correct, and why?

10. Show the development of equation (7.4) from (7.1).

11. How can one be assured of displacement being zero at a depth z of infinity?

12. An SSL wave can create secondary waves, head waves, that can coexist with SSLs. Make a sketch showing head wave formation and the proper angle of propagation into a test material.

13. How do you select the correct root from the cubic equation in a Rayleigh surface wave problem for isotropic material?

14. What is a typical particle motion profile for a surface wave as we consider particles further away from the surface in question?

15. What spacing is required in a comb transducer to produce a surface wave of specified frequency?

16. Show that attenuation for surface waves is less than that of bulk waves.

17. Compare the attenuation coefficients of bulk longitudinal waves with subsurface longitudinal waves.

18. Using an angle beam transducer to generate a surface wave in a material, how would you achieve the best result with maximum energy into the surface wave? What conditions would be necessary to generate the surface wave with the angle beam transducer?

19. Show on a sketch the SSL wave, longitudinal head wave, SST wave, transverse head wave, and surface wave due to a point source loading on a half-space.

20. Make a plot of surface wave particle displacement for a given time t.

21. How would you proceed to calculate surface wave velocities for excitation on an anisotropic material?

22. Using surface waves, discuss a procedure to identify certain anisotropic material constants as a function of measured skew angles.

8

Waves in Plates

8.1 Introduction

This chapter presents the governing equations of elastodynamics for waves in plates, along with a series of sample problems and practical discussions. The method of displacement potentials is used to obtain a solution for the case of propagation in a free plate (see e.g. Achenbach 1984 for more detail). Also, we give a brief outline of the method of partial waves (see Auld 1990). Visualization aspects of guided wave propagation in a glass plate are discussed in Appendix E, which includes several basic plate wave experiments.

The classical problem of Lamb wave propagation is associated with wave motion in a traction-free homogeneous and isotropic plate. The procedures we use to develop the governing equations and dispersion curve results of phase velocity versus frequency are similar to those used in a countless number of guided wave problems that incorporate bars, tubes, multiple layers, and anisotropic media. In this chapter we shall therefore detail the basic concepts of guided wave analysis. Interpretation procedures and mathematical analysis of phase and group velocity dispersion curves and wave structure can then be extended to a variety of different guided wave problems.

We will now briefly re-visit the fundamental differences between guided waves and bulk waves. Bulk waves travel in the bulk of the material – hence, away from the boundaries. However, often there is interaction with boundaries by way of reflection and refraction, and mode conversion occurs between longitudinal and shear waves. Although bulk and guided waves are fundamentally different, they are actually governed by the same set of partial differential wave equations. Mathematically, the principal difference is that, for bulk waves, there are no boundary conditions that need to be satisfied by the proposed solution. In contrast, the solution to a guided wave problem must satisfy the governing equations as well as some physical boundary conditions.

It is the introduction of boundary conditions that makes the guided wave problem difficult to solve analytically; in many cases, analytic solutions cannot even be found. Another interesting feature of guided wave propagation is that, unlike the finite number of modes (primarily longitudinal shear, shear perhaps being horizontal or vertical) that might be present in a bulk wave problem, there are generally an infinite number of modes associated with a given guided wave problem. That is, a finite body can support an infinite number of different guided wave modes.

Some examples of guided wave problems that have been solved – and whose solution has inherited the name of the investigator – are Rayleigh, Lamb, and Stonely waves. *Rayleigh waves* are free waves on the surface of a semi-infinite solid. Traction forces must vanish on the boundary, and the waves must decay with depth (see Figure 8-1). *Lamb waves* are waves of plane strain that occur in a free plate, and the traction force must vanish on the

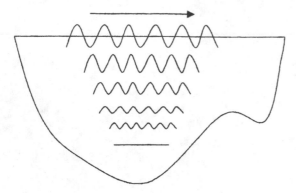

Figure 8-1. Rayleigh (surface) wave schematic.

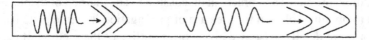

Figure 8-2. Lamb wave schematic.

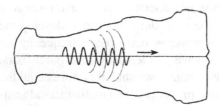

Figure 8-3. Stonely wave schematic.

upper and lower surface of the plate. Different mode structures occur from point to point as wave entry angle and frequency are varied (see Figure 8-2). *Stonely waves* are free waves that occur at an interface between two media (see Figure 8-3). Continuity of traction and displacement is required at the interface, and a radiation condition must be satisfied.

We recall the following from the theory of elasticity (here we are using Cartesian tensor notation):

$$\sigma_{ij,j} + \rho f_i = \rho \ddot{u}_i, \qquad \text{3 equations of motion } (i = 1, 2, 3); \tag{8.1}$$

$$\varepsilon_{ij} = \tfrac{1}{2}(u_{i,j} + u_{j,i}), \qquad \text{6 independent strain displacement equations;} \tag{8.2}$$

$$\sigma_{ij} = \lambda \varepsilon_{kk} \delta_{ij} + 2\mu \varepsilon_{ij}, \qquad \text{6 independent constitutive equations} \\ \text{(isotropic materials).} \tag{8.3}$$

The first two equations are valid for any continuous medium; the specific type of medium concerned is introduced via (8.3). If we eliminate the stress and strain factors from these equations, then we have

$$\mu u_{i,jj} + (\lambda + \mu) u_{j,ji} + \rho f_i = \rho \ddot{u}_i. \tag{8.4}$$

The equations of motion (8.4), which contain only the particle displacements, are the governing partial differential equations for displacement. If the domain in which a solution is sought is infinite, then these equations are sufficient. If the domain is finite, then

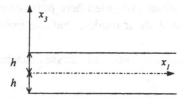

Figure 8-4. Geometry of the free plate problem.

boundary conditions are needed for a well-posed problem. The boundary conditions take the form of prescribed tractions and/or displacements on the boundaries of the domain of interest. The general forms of such boundary conditions are as follows:

$$u(x, t) = u_0(x, t) \qquad \text{on surface displacements;} \qquad (8.5)$$

$$t_i = \sigma_{ji} n_j, \qquad \text{on surface tractions;} \qquad (8.6)$$

$$u(x, t) = u_0(x, t) \text{ on } S_1$$
$$\text{and } t_i = \sigma_{ji} n_j, \text{ on } S_2 \qquad \text{as a mixed boundary condition.} \qquad (8.7)$$

8.2 The Free Plate Problem

The geometry of the free plate problem is illustrated in Figure 8-4. This problem is governed by the equations of motion (8.4), with boundary conditions of type (8.6). The surfaces at the coordinates $y = d/2 = h$ and $y = -d/2 = -h$ are considered traction-free. Ultrasonic excitation occurs at some point in the plate; as ultrasonic energy from the excitation region encounters the upper and lower bounding surfaces of the plate, mode conversions occur (L wave to T wave, and vice versa). After some travel in the plate, superpositions cause the formation of "wave packets," or what are commonly called *guided wave modes* in the plate. Based on entry angle and frequency used, we can predict how many different modes can be produced in the plate.

The exact solution of this problem has been obtained through the use of several different approaches. The most popular methods of solution are the displacement potentials and the partial wave techniques (see Achenbach 1984 and Auld 1990, respectively).

8.2.1 Solution by the Method of Potentials

If the displacement vector (field) is decomposed according to Helmholtz decomposition and the result substituted into (8.4), as demonstrated previously, we obtain two uncoupled wave equations. For plane strain, these are

$$\frac{\partial^2 \phi}{\partial x_1^2} + \frac{\partial^2 \phi}{\partial x_3^2} = \frac{1}{c_L^2} \frac{\partial^2 \phi}{\partial t^2}, \quad \text{governing longitudinal waves;} \qquad (8.8)$$

$$\frac{\partial^2 \psi}{\partial x_1^2} + \frac{\partial^2 \psi}{\partial x_3^2} = \frac{1}{c_T^2} \frac{\partial^2 \psi}{\partial t^2}, \quad \text{governing shear waves.} \qquad (8.9)$$

The case of plane strain is not the most general for the problem at hand, but the analysis is greatly simplified in this case. Achenbach (1984) shows that taking the general state of strain as a starting point results in the same set of solutions presented here plus some additional modes (infinite in number), known as horizontal shear modes, that can exist independently of the other wave modes.

As a result of our assumption of plane strain, the displacements and stresses can be written in terms of the potentials as

$$u_1 = u = \frac{\partial \phi}{\partial x_1} + \frac{\partial \psi}{\partial x_3}, \tag{8.10a}$$

$$u_2 = v = 0, \tag{8.10b}$$

$$u_3 = w = \frac{\partial \phi}{\partial x_3} - \frac{\partial \psi}{\partial x_1}; \tag{8.10c}$$

$$\sigma_{31} = \mu \left(\frac{\partial u_3}{\partial x_1} + \frac{\partial u_1}{\partial x_3} \right) = \mu \left(\frac{\partial^2 \phi}{\partial x_1 \partial x_3} - \frac{\partial^2 \psi}{\partial x_1^2} + \frac{\partial^2 \psi}{\partial x_3^2} \right), \tag{8.11a}$$

$$\sigma_{33} = \lambda \left(\frac{\partial u_1}{\partial x_1} + \frac{\partial u_3}{\partial x_3} \right) + 2\mu \frac{\partial u_3}{\partial x_3}$$

$$= \lambda \left(\frac{\partial^2 \phi}{\partial x_1^2} + \frac{\partial^2 \phi}{\partial x_3^2} \right) + 2\mu \left(\frac{\partial^2 \phi}{\partial x_3^2} - \frac{\partial^2 \psi}{\partial x_1 \partial x_3} \right), \tag{8.11b}$$

where λ and μ are Lamé constants.

We begin the analysis by assuming solutions to (8.8) and (8.9) in the form

$$\phi = \Phi(x_3) \exp[i(kx_1 - \omega t)], \tag{8.12}$$

$$\psi = \Psi(x_3) \exp[i(kx_1 - \omega t)]. \tag{8.13}$$

Note that these solutions represent traveling waves in the x_1 direction and standing waves in the x_3 direction. This is evident from the fact that, although there is a complex exponential term (hence sines and cosines) containing the time variable for the x_1 dependencies, there is only an unknown "static" function of x_3 for the x_3 dependence. This phenomenon is referred to in many texts as *transverse resonance* and is exploited in many ways to arrive at a solution. Again, these solutions represent waves that travel along the direction of the plate and that have fixed (as yet, unknown) distributions in the transverse directions.

Substitution of these assumed solutions into (8.8) and (8.9) yields equations governing the unknown functions Φ and Ψ. The solutions to these equations are

$$\Phi(x_3) = A_1 \sin(px_3) + A_2 \cos(px_3), \tag{8.14}$$

$$\Psi(x_3) = B_1 \sin(qx_3) + B_2 \cos(qx_3), \tag{8.15}$$

where

$$p^2 = \frac{\omega^2}{c_L^2} - k^2 \quad \text{and} \quad q^2 = \frac{\omega^2}{c_T^2} - k^2. \tag{8.16}$$

With these results, the displacements and stresses can be obtained directly from (8.10) and (8.11). Omitting the term $\exp[i(kx_1 - \omega t)]$ in all expressions, the results are as follows:

$$u_1 = \left[ik\Phi + \frac{d\Psi}{dx_3} \right], \tag{8.17}$$

$$u_3 = \left[\frac{d\Phi}{dx_3} - ik\Psi \right]; \tag{8.18}$$

$$\sigma_{33} = \left[\lambda \left(-k^2\Phi + \frac{d^2\Phi}{dx_3^2} \right) + 2\mu \left(\frac{d^2\Phi}{dx_3^2} - ik\frac{d\Psi}{dx_3} \right) \right], \tag{8.19}$$

$$\sigma_{31} = \mu \left(2ik\frac{d\Phi}{dx_3} + k^2\Psi + \frac{d^2\Psi}{dx_3^2} \right). \tag{8.20}$$

Now, since the field variables involve sines (resp. cosines) with argument x_3, which are odd (resp. even) functions about $x_3 = 0$, we split the solution into two sets of modes: symmetric and antisymmetric modes. Specifically, for displacement in the x_1 direction, the motion will be symmetric (with respect to the midplane of the plate) if u_1 contains cosines but will be antisymmetric if u_1 contains sines. The reverse is true for displacements in the x_3 direction. Thus, we split the modes of wave propagation in the plate into two systems:

Symmetric modes

$$\Phi = A_2 \cos(px_3),$$

$$\Psi = B_1 \sin(qx_3),$$

$$u = u_1 = ikA_2 \cos(px_3) + qB_1 \cos(qx_3),$$

$$w = u_3 = -pA_2 \sin(px_3) - ikB_1 \sin(qx_3), \tag{8.21}$$

$$\sigma_{31} = \mu[-2ikpA_2 \sin(px_3) + (k^2 - q^2)B_1 \sin(qx_3)],$$

$$\sigma_{33} = -\lambda(k^2 + p^2)A_2 \cos(px_3) - 2\mu[p^2 A_2 \cos(px_3) + ikqB_1 \cos(qx_3)];$$

Antisymmetric modes

$$\Phi = A_1 \sin(px_3),$$

$$\Psi = B_2 \cos(qx_3),$$

$$u = u_1 = ikA_1 \sin(px_3) - qB_2 \sin(qx_3),$$

$$w = u_3 = pA_1 \cos(px_3) - ikB_2 \cos(qx_3), \tag{8.22}$$

$$\sigma_{31} = \mu[2ikpA_1 \cos(px_3) + (k^2 - q^2)B_2 \cos(qx_3)],$$

$$\sigma_{33} = -\lambda(k^2 + p^2)A_1 \sin(px_3) - 2\mu[p^2 A_1 \sin(px_3) - ikqB_2 \sin(qx_3)].$$

For the symmetric modes, note that the wave structure across the thickness of the plate is symmetric for u and antisymmetric for w. On the other hand, for the antisymmetric modes, the wave structure across the thickness is symmetric for w and hence antisymmetric for u.

It should be noted that this separation of waves into symmetric and antisymmetric modes is an exception rather than a rule. In hollow cylinders, the lack of structure symmetry does not allow this separation. Plate wave modes do exist in an anisotropic plate, but the separation into symmetric and antisymmetric modes is not possible unless the wave propagates along a symmetry axis of the plate. (See Solie and Auld 1973 for an excellent discussion of plate waves in anisotropic plates.)

The constants A_1, A_2, B_1, B_2, as well as the dispersion equations, are still unknown. They can be determined by applying the traction-free boundary condition, which reduces to

$$\sigma_{31} = \sigma_{33} \equiv 0 \quad \text{at } x_3 = \pm d/2 = \pm h \text{ (for convenience)} \tag{8.23}$$

in the case of plane strain. The resulting displacement, stress, and strain fields depend upon the type of mode (i.e., symmetric or antisymmetric). However, applying the boundary conditions will give a homogeneous system of two equations for the appropriate two constants A_2, B_1 (for the symmetric case) and A_1, B_2 (antisymmetric case). For homogeneous equations we require that the determinant of the coefficient matrix vanish in order to ensure solutions other than the trivial one. From (8.23) we thus have

$$\frac{(k^2 - q^2)\sin(qh)}{2ikp(\sin(ph))} = \frac{-2\mu ikq(\cos(qh))}{(\lambda k^2 + \lambda p^2 + 2\mu p^2)\cos(ph)}. \tag{8.24}$$

After some manipulation, this may be rewritten as

$$\frac{\tan(qh)}{\tan(ph)} = \frac{4k^2 qp\mu}{(\lambda k^2 + \lambda p^2 + 2\mu p^2)(k^2 - q^2)}. \tag{8.25}$$

The denominator on the RHS of (8.25) can be further simplified by using wave velocities and the definitions of p and q from (8.16). From the definition of c_L, we obtain

$$\lambda = c_L^2 \rho - 2\mu. \tag{8.26}$$

Then

$$\lambda k^2 + \lambda p^2 + 2\mu p^2 = \lambda(k^2 + p^2) + 2\mu p^2$$
$$= (c_L^2 \rho - 2\mu)(k^2 + p^2) + 2\mu p^2 \tag{8.27}$$
$$\lambda k^2 + \lambda p^2 + 2\mu p^2 = \rho c_L^2(k^2 + p^2) - 2\mu k^2. \tag{8.28}$$

Using (8.16) and $c_T^2 = \mu/\rho$ yields

$$\lambda k^2 + \lambda p^2 + 2\mu p^2 = \rho\omega^2 - 2\rho c_T^2 k^2, \tag{8.29}$$

which therefore implies

$$\rho c_T^2 \left[\left(\frac{\omega}{c_T}\right)^2 - 2k^2 \right] = \rho c_T^2(q^2 - k^2) = \mu(q^2 - k^2). \tag{8.30}$$

Now, substituting (8.30) into an initial form of the dispersion equation (8.25), we obtain

$$\frac{\tan(qh)}{\tan(ph)} = -\frac{4k^2 pq}{(q^2 - k^2)^2} \quad \text{for symmetric modes.} \tag{8.31}$$

Proceeding along analogous lines, we can show that

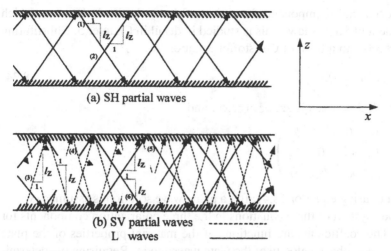

(a) SH partial waves

z

x

(b) SV partial waves - - - - - - - - - - -
L waves —————

Figure 8-5. Types of partial waves used in the isotropic problem.

$$\frac{\tan(qh)}{\tan(ph)} = -\frac{(q^2 - k^2)^2}{4k^2 pq} \quad \text{for antisymmetric modes.} \tag{8.32}$$

Recall that p and q are as defined in (8.16).

For a given ω and derived k, the displacements can be calculated using the expressions for u and w in (8.21) and (8.22). More explicit expressions are given in Auld (1990).

These equations are known as the Rayleigh–Lamb frequency relations, and they were first derived at the end of the nineteenth century. These equations can be used to determine the velocity (or velocities) at which a wave of a particular frequency (fh or fd product) will propagate within the plate. Equations of this nature are known as *dispersion relations*. Although the equations look simple, they can be solved only by numerical methods.

8.2.2 The Partial Wave Technique

Although the method just presented is quite simple and elegant, its usefulness is restricted to isotropic plates: only then will the governing equations (8.4) be in such a simple form. For the problem of plate waves in anisotropic plates, the only suitable technique is the partial wave (or transverse resonance) technique. As pointed out in Solie and Auld (1973), the partial wave technique has two major advantages over the method of displacement potentials: (1) it leads more directly to wave solutions, and (2) it provides more insight into the physical nature of the waves.

Keep in mind that the formulation of the free plate problem has in no way changed; we are merely trying another solution method. In the partial wave technique, we try to construct solutions to the problem defined by (8.4) and (8.6) from simple exponential-type waves that reflect back and forth between the boundaries of the plate (see Figure 8-5). We begin by assuming that each of the waves depicted in Figure 8-5 can be expressed as

$$u_j = a_i \exp[ik(x + l_z z)], \tag{8.33}$$

where $j = x, y, z$ and $l_z = k_z/k_x$ (we are now using x, y, and z, instead of the x_i, to denote position). Also, we are solving the more general problem, that is, with no assumption of plane strain. Note that the x component of each assumed partial wave is the same, which is exactly the statement of Snell's law. This is studied in detail in Chapter 5. Substituting these solutions in (8.33) into a form of Christoffel equations,

$$(k_{il}c_{IJ}k_{Jj} - \rho\omega^2\delta_{ij})u_j = 0, \tag{8.34}$$

where $i, j = x, y, z, I, J = xx, yy, zz, yz, xz, xy$, and

$$k_{il} = \begin{pmatrix} k_x & 0 & 0 & 0 & k_z & k_y \\ 0 & k_y & 0 & k_z & 0 & k_x \\ 0 & 0 & k_z & k_y & k_x & 0 \end{pmatrix}$$

(which is equivalent to using equation (8.4) after a plane wave solution is assumed) yields a linear homogeneous system of three equations in the three polarization components for each partial wave. The coefficients are functions of the material properties of the plate and also of the (unknown) phase velocity of the plate wave mode. Requiring the determinant to vanish for nontrivial solutions yields a sixth-order equation for l_z that defines the propagation direction of the six partial waves (see Figure 8-5).

Now we know the direction of propagation for each of the partial waves. Hence we can take a linear combination of them in the form

$$u_j = \sum_{n=1}^{6} C_n\alpha_j^{(n)} \exp[ik(x + l_z^{(n)}z)] \quad (j = x, y, z) \tag{8.35}$$

and so try to determine the coefficients C_n ($n = 1, 2, \ldots, 6$), or wave amplitudes, and thereby satisfy the traction-free boundary conditions (equation (8.6) evaluated at $z = \pm h$). Note that the traction-free condition must be satisfied along the entire upper and lower surfaces, so that the partial waves must reflect in such a manner that they reconstruct themselves after returning to the top of the plate. That is, they must be standing waves in the transverse direction (hence the term "transverse resonance"). This explains why, when using the method of displacement potentials, we assumed that the potentials (8.12) and (8.13) had not only a static dependence on x_2 but were also allowed to propagate in the x_1 direction.

The last step in our problem is to substitute this assumed linear combination of partial waves into the boundary condition equations. This gives a system of six (remember that we no longer assume plane strain, so to (8.23) we must add $\sigma_{xz} = 0$) homogeneous linear equations in which the coefficients C_n are now functions of the density, the elastic constants of the plate, and the product hk. Requiring the determinant of this "boundary condition matrix" to vanish (and thus yield nontrivial solutions for the wave amplitudes) gives us the dispersion relations that we seek.

Just as in the solution found by the method of potentials, an infinite number of modes are defined by our dispersion relations. In this case, however, we pick up the additional modes that were lost in that previous method by our assumption of plane strain. (We did not actually need to make that assumption in the method of potentials, but did so for simplicity.) The dispersion relations for these extra modes, known as shear horizontal (SH) modes, are

$$(M\pi)^2 = (\omega h/c_T)^2 - (kh)^2. \tag{8.36}$$

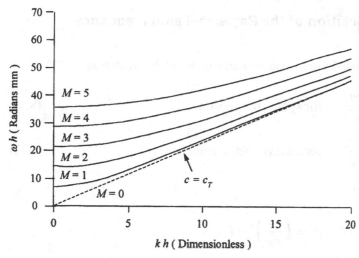

Figure 8-6. SH modes in a free copper plate ($c_T = 2.26$ mm/μs).

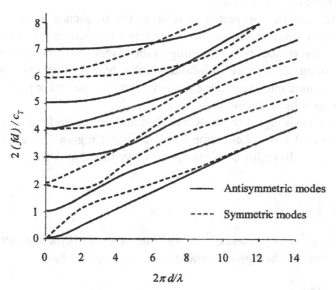

Figure 8-7. Dispersion curves for a free isotropic aluminum plate ($c_L = 6.35$ mm/μs, $c_T = 3.13$ mm/μs, $c_L/c_T = 2.0288$).

These modes are plotted in Figure 8-6. As can be seen, they are simple hyperbolas in a $(kh, \omega h)$-plane. Complete details of a horizontal shear wave solution in a plate are presented in Chapter 15.

The dispersion relations governing the symmetric and antisymmetric "in-plane" modes (known as Lamb waves) are of course the same as the ones resulting from the method of potentials, equations (8.31) and (8.32). Figure 8-7 shows a sample plot of the dispersion curves for these modes. As mentioned previously, the dispersion equations for the Lamb wave modes are simple in appearance but can be solved only by numerical methods.

8.3 Numerical Solution of the Rayleigh–Lamb Frequency Equations

Recall that the Rayleigh–Lamb frequency equations can be written as

$$\frac{\tan(qh)}{\tan(ph)} = -\frac{4k^2 pq}{(q^2 - k^2)^2} \quad \text{for symmetric modes.} \tag{8.31}$$

$$\frac{\tan(qh)}{\tan(ph)} = -\frac{(q^2 - k^2)^2}{4k^2 pq} \quad \text{for antisymmetric modes.} \tag{8.32}$$

Here p and q are given by

$$p^2 = \left(\frac{\omega}{c_L}\right)^2 - k^2 \quad \text{and} \quad q^2 = \left(\frac{\omega}{c_T}\right)^2 - k^2.$$

The wavenumber k is numerically equal to ω/c_p, where c_p is the phase velocity of the Lamb wave mode and ω is the circular frequency. The phase velocity is related to the wavelength by the simple relation $c_p = (\omega/2\pi)\lambda$.

Equations (8.31) and (8.32) can be considered as relating the frequency ω to the wavenumber k of the Lamb wave modes, resulting in the frequency spectrum, or as relating the phase velocity c_p to the frequency ω, resulting in the dispersion curves. It is known that, for any given frequency, there are an infinite number of wavenumbers that will satisfy (8.31) and (8.32). A finite number of these wavenumbers will be real or purely imaginary, while infinitely many will be complex.

It is often useful to consider various regions of the Rayleigh–Lamb equations for k compared with ω/c_L or ω/c_T (see Graff 1991). Let region 1 be $k > \omega/c_T$; region 2, $\omega/c_T > k > \omega/c_L$; and region 3, $k < \omega/c_L$. In region 1, where $c_p < c_T$, we therefore have

$$\frac{\tanh(q'h)}{\tanh(p'h)} = \left\{\frac{4p'q'k^2}{(k^2 - q'^2)^2}\right\}^{\pm 1};$$

from (8.32), $p = ip'$, $q = iq'$, $p'^2 = -p^2$, and $q'^2 = -q^2$ (the index $+1$ is for symmetric and -1 for antisymmetric modes). In region 2, where $c_T < c_p < c_L$, we have

$$\frac{\tan(qh)}{\tan(p'h)} = \pm\left\{\frac{4p'qk^2}{(k^2 - q^2)^2}\right\}^{\pm 1};$$

in region 3, where $c_p > c_L$, equations (8.31) and (8.32) are unaltered.

A key element to understanding the physical character of dispersion curves is associated with the time harmonic factor $\exp[i(kx - \omega t)]$. Equations (8.31) and (8.32) are functions of k and ω. Noting that k could possibly assume complex values, the physical meaning of k can be discussed as follows.

Let $k = k_r + ik_{im}$. Then the time-harmonic factor becomes:

$$\exp[i(k_r x - \omega t)]\exp[-k_{im}x]. \tag{8.37}$$

There are three possible signed values for k_{im}, and each has a physical interpretation:

$k_{im} < 0$,   the waves grow exponentially with distance;

$k_{im} = 0$,   the waves propagate with no damping;

$k_{im} > 0$,   the waves decay exponentially with distance.

The decaying waves are called "evanescent," which means they will disappear. Their amplitude decreases exponentially with distance from their source or a scattering center. (When studying scattering, they become important in the sense of forming a "near field.") The exponentially growing waves have not been physically observed. It can be concluded that, for the simple unloaded plate problem, only real values for k are necessary or supply information about propagating waves. An example of wave decay with distance can be seen in the water-loaded plate problem. Out-of-plane displacement at the surface of the plate produces normal loading on the fluid and hence leakage into the fluid. In this case, k_{im} would be greater than zero.

When plotting the dispersion curves, we are now interested only in the real solutions of these equations, which represent the (undamped) propagating modes of the structure. It is therefore useful to rewrite (8.31) and (8.32) so that they take on only real values for real or pure imaginary wavenumbers k. This is achieved by the following set of equations:

$$\frac{\tan(qh)}{q} + \frac{4k^2 p \tan(ph)}{(q^2 - k^2)^2} = 0 \quad \text{for symmetric modes,} \tag{8.38}$$

$$q \tan(qh) + \frac{(q^2 - k^2)^2 \tan(ph)}{4k^2 p} = 0 \quad \text{for antisymmetric modes.} \tag{8.39}$$

The numerical solution of these equations is now relatively simple. The steps in the routine may be listed as follows.

(1) Choose a frequency–thickness product $(\omega h)_0$.
(2) Make an initial estimate of the phase velocity $(c_p)_0$.
(3) Evaluate the signs of each of the left-hand sides of (8.38) or (8.39) (assuming they do not equal zero).
(4) Choose another phase velocity $(c_p)_1 > (c_p)_0$ and re-evaluate the signs of (8.38) or (8.39).
(5) Repeat steps (3) and (4) until the sign changes. Because the functions involved are continuous, a change in sign must be accompanied by a crossing through zero. Therefore, a root m exists in the interval where a sign change occurs. Assume that this happens between phase velocities $(c_p)_n$ and $(c_p)_{n+1}$.
(6) Use some sort of iterative root-finding algorithm (e.g., Newton–Raphson, bisection, ...) to locate precisely the phase velocity in the interval $(c_p)_n < c_p < (c_p)_{n+1}$ where the LHS of the required equation is close enough to zero.
(7) After finding the root, continue searching at this ωh for other roots according to steps (2) through (6).
(8) Choose another ωh product and repeat steps (2) through (7).

This procedure is performed for as many ωh as required.

Here is a helpful tip for performing the calculations: For large ωh, instead of searching for a phase velocity lying in the range $c_R < c_p < \infty$, search for $1/c_p$ in the range

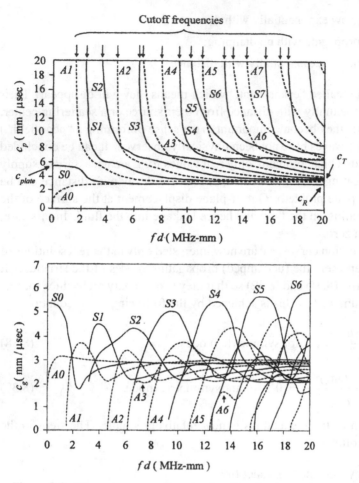

Figure 8-8. Dispersion curves for a traction-free aluminum plate.

$1/c_R < 1/c_p < 0$ (where c_R is the Rayleigh wave velocity of the material). This technique will also be useful for smaller ωh, where c_R should be replaced with the phase velocity of the lowest A0 mode. Note that alternate schemes for the root extraction process are available, including procedures for both real and imaginary roots (if both exist).

As a result of this root extraction, sample dispersion curves for an aluminum plate are presented in Figure 8-8, which depicts modes A0, S0, A1, S1, A2, S2, ... The plate velocity c_{plate}, Rayleigh surface wave velocity c_R, shear (transverse) velocity c_T, and mode cutoff regions are also shown. The classical long wavelength plate velocit, c_{plate}, is equal to $E^{1/2}[\rho(1 - \nu^2)]^{-1/2}$. Note the limiting values of c_p as fd values become large. The A0 and S0 modes converge to c_R, and the other modes converge to c_T. More will be said about cutoff frequencies in Section 8.7.4.

8.4   Group Velocity

The group velocity c_g can be found from the phase velocity c_p by use of the formula

$$c_g = \frac{d\omega}{dk}.$$

Substituting $k = \omega/c_p$ into this equation yields

$$c_g = d\omega \left[d\left(\frac{\omega}{c_p}\right) \right]^{-1}$$

$$= d\omega \left[\frac{d\omega}{c_p} - \omega \frac{dc_p}{c_p^2} \right]^{-1}$$

$$= c_p^2 \left[c_p - \omega \frac{dc_p}{d\omega} \right]^{-1}.$$

Using $\omega = 2\pi f$, the third equality can be written as

$$c_g = c_p^2 \left[c_p - (fd)\frac{dc_p}{d(fd)} \right]^{-1}, \tag{8.40}$$

where fd denotes frequency times thickness. Note that, when the derivative of c_p with respect to fd becomes zero, $c_g = c_p$. Note also that, as the derivative of c_p with respect to fd approaches infinity (i.e., at cutoff), c_g approaches zero.

8.5 Wave Structure Analysis

It is interesting to study wave structure variation as one increases the fd product along a particular mode. The symmetric mode cannot be thought of as simply an in-plane vibration mode. As one moves along the mode, the ratio of the in-plane to out-of-plane displacement changes. Of particular note are the changes on the outside surface of a structure. Similarly, the antisymmetric modes cannot be thought of as a mode with only out-of-plane displacement values.

Figures 8-9 to 8-14 depict solutions at a variety of fd values to modes S0, S1, S2, A0, A1, and A2 for an aluminum plate. The values for c_T and c_L are 3.1 mm/μs and 6.3 mm/μs, respectively. Sample points were chosen along the symmetric mode of an aluminum plate to show the variation in wave structure that occurs as one moves along the curve (see Figure 8-8 for the basic dispersion curve).

For example, the in-plane displacement is almost constant across the thickness of the plate at low fd values, but it becomes heavily concentrated at the center of the plate as the fd values increase (see Figure 8-9). In contrast, the out-of-plane displacement component w, which is initially close to zero on the outside surface for small fd values, becomes dominant on the outside surface as the fd values increase to 2 or 2.5. For in-plane distributions along the S1 mode, shown in Figure 8-10, the in-plane displacement on the outside surface is a maximum value while the out-of-plane value is zero. As fd increases to 6.0, the gradual changes show out-of-plane dominance with the in-plane component becoming zero. For the S2 mode, however, the in-plane displacement on the outer surface is close to zero for low fd values, as shown in Figure 8-11, but becomes dominant as fd changes from 5.0 to 7.0. On the other hand, the out-of-plane component goes from dominant to zero as fd changes from 5.0 to 7.0.

Similar studies are graphed for the antisymmetric modes in Figures 8-12, 8-13, and 8-14. Again, some interesting observations can be made. For example, dominant out-of-plane

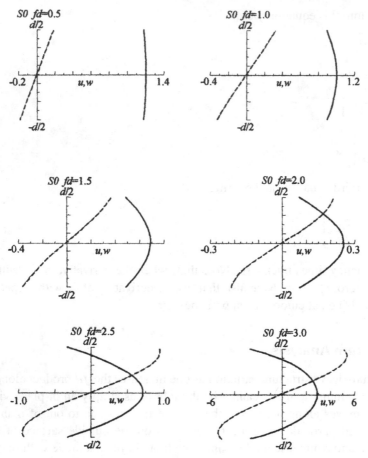

Figure 8-9. Wave structure for various points on the S0 mode of an aluminum plate, showing the in-plane (u, solid line) and out-of-plane (w, dashed line) displacement profiles across the thickness of the plate.

displacement on the surface for the A1 mode occurs at an fd value of 4.5; at this point, the in-plane displacement value is zero.

Use of wave structure can lead to increased wave penetration power along a structure – for example, by avoiding energy leakage from water loading or insulation. Improved sensitivity to certain defects can be obtained as a result of controlling in-plane or out-of-plane impingement at a certain location across the thickness of the structure. Work by Ditri, Rose, and Chen (1991) demonstrates the use of mode selection and wave structure to detect small defects on the surface of a structure. This is accomplished by getting higher energy concentrations on the outside surface. Work is currently in progress to examine other parameter distributions across the thickness of a structure for improved penetration power and/or defect detection sensitivity. Parameters including stresses, strain energy, and power – in addition to the displacement components u and w – could prove useful.

8.6 Compressional and Flexural Waves

Figure 8-15 shows a (highly exaggerated) schematic "snapshot" of the displacement vector field distribution on the surface of a plate and its effect on the shape of the plate.

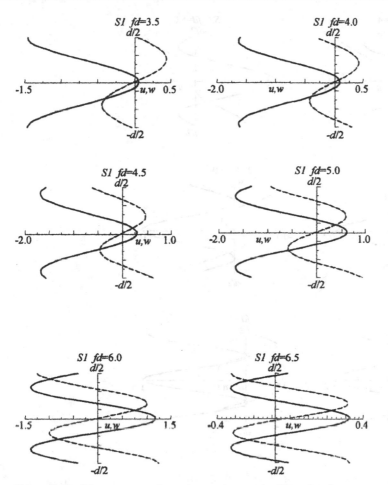

Figure 8-10. Wave structure for various points on the S1 mode of an aluminum plate.

Symmetric mode waves are often termed "compressional" and antisymmetric waves are known as "flexural." The particle displacements shown are the vector sums of the in-plane (u) and out-of-plane (w) particle displacement components.

8.7 Miscellaneous Topics

A great challenge exists with respect to the interpretation of results obtained in the Lamb wave propagation problem. Even though one might very carefully formulate the theoretical approach to the problem and develop a detailed numerical solution (for the phase velocity, group velocity, and wave structure values of displacements and stress), the ability to use these curves in a practical sense – for example, in ultrasonic nondestructive testing – requires a great deal of study and attention. A topic of great concern is associated with mode selection, isolation, and control in order to study specific wave propagation characteristics in a plate. The subject of generation and reception also becomes critical with respect to the use of the specific modes and frequencies chosen for a particular problem. This section presents miscellaneous topics associated with interpretation and utilization of some of the results associated with Lamb wave propagation. The first topic

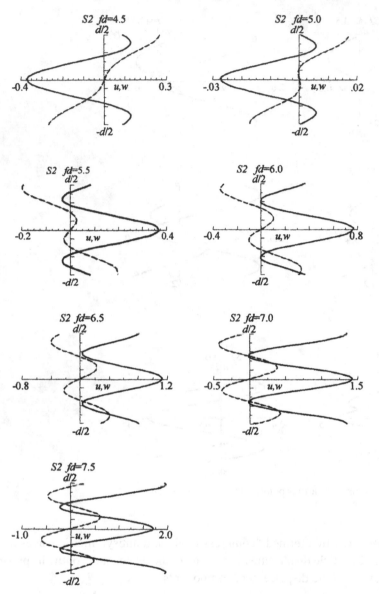

Figure 8-11. Wave structure for various points on the S2 mode of an aluminum plate.

we discuss concerns the practical reasons for locating Lamb wave propagation characteristics with a dominant longitudinal displacement. We also discuss the physical interpretation of the zeros and poles for a fluid-coupled elastic layer. Finally, we present a problem of interest in the immersion testing of a plate involving nonspecular reflection and transmission characteristics for layered media.

8.7.1 Lamb Waves with Dominant Longitudinal Displacements

Of great interest in the utilization of dispersion curves are certain points where phase velocity and frequency have particular or unusual wave structural characteristics. One

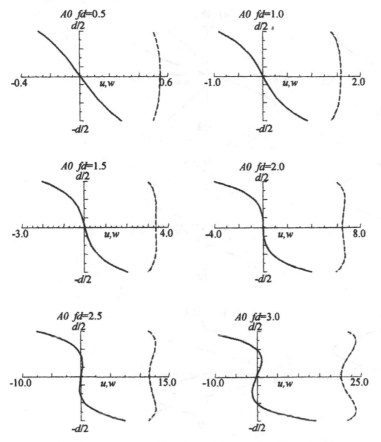

Figure 8-12. Wave structure for various points on the A0 mode of an aluminum plate.

might think of points where displacement or stress is concentrated on the outside surface or perhaps at the center section of the plate. On the other hand, one might examine the in-plane and out-of-plane displacement distributions in such a manner as to achieve dominant in-plane (or out-of-plane) characteristics on the outside surface of the plate.

Imagine, for example, a plate immersed in water. One can appreciate that energy loss would be minimal to the fluid if a dominant in-plane or longitudinal displacement were available on the outside surface of the plate. In fact, from a theoretical standpoint, the loss would be zero because we are actually trying to load the fluid with a shear force. Because shear waves cannot propagate into the fluid, the energy of propagation in the plate would be strongest at these points since the energy is retained by the plate. If we were to locate points with a dominant out-of-plane displacement, it would be easy to see that such normal pressure loading onto the fluid would propagate into the fluid as the Lamb wave propagates along the plate. This is what we call the propagation of a *leaky Lamb wave* in the structure.

In immersion nondestructive testing, we might wish to make use of this leaky Lamb wave to look at distortions in the reflection pattern as the wave travels along the plate and encounters defects or inhomogeneities in the structure. On the other hand, if we were using a pulse-echo or through-transmission technique – with the transducer actually located on the plate – then we would want strong penetration power and hence minimal leakage into the fluid; this would call for the elimination or reduction of the leaky Lamb wave component.

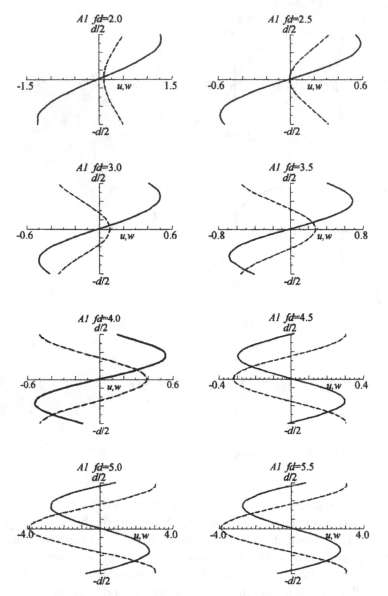

Figure 8-13. Wave structure for various points on the A1 mode of an aluminum plate.

Using elasticity concepts, one can embark upon a detailed study of the wave structural characteristics of the in-plane and out-of-plane displacement while moving along a particular mode. The continuous variations and shifting of characteristics from the in-plane and out-of-plane distribution as we move along the modes are interesting and can be useful for ultrasonic testing. Sample results were presented in Figures 8-9 to 8-14.

This section will focus on finding specific points where there is a dominant longitudinal displacement and very minimal out-of-plane component on the surface of the plate. The following discussion is taken from a paper by Pilarski, Ditri, and Rose (1993). Remarks are presented on symmetric Lamb waves with a dominant longitudinal displacement.

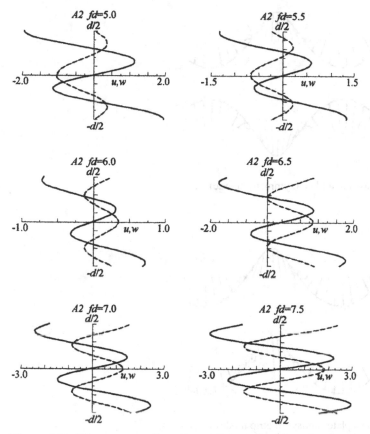

Figure 8-14. Wave structure for various points on the A2 mode of an aluminum plate.

We prove that the normal component of the particle displacement vector vanishes on the free surfaces of an isotropic, homogeneous plate for nonzero-order symmetric Lamb waves; this conclusion was stated without proof by Viktorov (1967, p. 121). We also give an expression for the frequency–thickness products where this vanishing occurs, which enables practical (i.e. experimental) selection of such modes. Other interesting features of such modes, such as the independence of their group velocity on the mode's order or its nondispersivity, are also addressed. The features discussed here are of great practical importance in the nondestructive testing (NDT) of plates loaded by liquids and in the ultrasonic tensometry of a thin plate (see Pilarski et al. 1993).

For Lamb waves in an isotropic, homogeneous solid layer immersed in a liquid, Viktorov (1967, p. 121) asserts the following: For symmetric modes, when the phase velocity reaches the value of the velocity of bulk longitudinal waves, the vertical (normal to the plate surface) component of the displacement vector vanishes on the free surfaces. Using Viktorov's notation, this can be written as

$$\lim_{r \to 0} w_s(z) = 0 \quad \text{for } z = \pm d/2 = \pm h \text{ (for convenience)}, \tag{8.41}$$

where w_s is the amplitude of the normal component of the displacement vector for a symmetric mode and d is the thickness of the plate. The quantity in this case is r, defined by

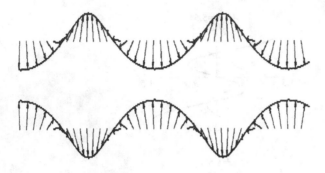

Compressional waves in a plate (symmetric mode)

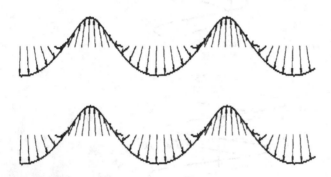

Flexural waves in a plate (antisymmetric mode)

Figure 8-15. Compressional and flexural wave particle displacement schematic.

$$r^2 = k^2 - k_L^2, \tag{8.42}$$

where k and k_L are the wavenumbers related to the phase velocity of the plate mode $c = c_p$ and to the velocity of bulk longitudinal waves c_L, respectively. The condition $r \to 0$ implies that:

$$\lim_{r \to 0} c_p = c_L. \tag{8.43}$$

We would like to (a) prove the aforementioned conclusion about the vanishing of the vertical displacement component and (b) give an expression for the frequency–thickness products where this occurs.

For NDT of plates loaded by liquids, appropriate mode selection can substantially reduce leakage of energy into the liquid owing to the vanishing normal surface displacement. For ultrasonic tensometry of a thin plate, symmetric modes that are higher than fundamental order are useful with inhomogeneously stressed plates owing to the dominance of longitudinal particle vibration, for which the values of an acoustoelastic constant (for the longitudinal applied or residual stresses) are the largest, and because the plates' field distributions vary with frequency (see Pilarski et al. 1992).

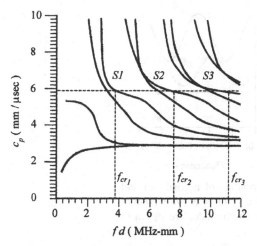

Figure 8-16. Lamb wave dispersion curves for steel, showing dominant in-plane displacement points.

We will start the analysis from the characteristic equation for symmetric modes found in Viktorov (1967, eq. II.4):

$$\Omega_s \equiv (k^2 + s^2)\cosh(rh)\sinh(sh) - 4k^2 rs\sinh(rh)\cosh(sh) = 0. \tag{8.44}$$

Here, $s^2 = k^2 - k_T^2$, where k_T is the wavenumber related to the bulk transverse wave velocity c_T. From (8.36), for r approaching zero, we arrive at the equality

$$\sinh(sh) = 0, \tag{8.45}$$

which is satisfied by

$$(sh)_n = in\pi \quad \text{for} \quad n = 0, 1, 2, \dots. \tag{8.46}$$

Finally, using that

$$s \to k_L^2 - k_T^2 \quad \text{as} \quad r \to 0,$$

we obtain the following expression for the frequency–thickness product at which the nth symmetric mode has a phase velocity equal to c_L:

$$(fd)_n = \frac{nc_T}{\sqrt{1 - (c_T/c_L)^2}} \quad \text{for} \quad n = 1, 2, \dots. \tag{8.47}$$

Note that $n = 0$ is excluded: for zero frequency, no mode exists that has a phase velocity equal to the bulk longitudinal wave velocity. Equation (8.47) has great practical significance in that it enables us to select a mode by choosing the appropriate frequency for a given plate thickness and (e.g., in the wedge technique of Lamb wave generation) by adjusting the angle of incidence of the ultrasonic wave onto the plate surface. The first such modes and their frequencies are marked on the phase velocity dispersion curve diagram (Figure 8-16) for a 1-mm-thick steel plate.

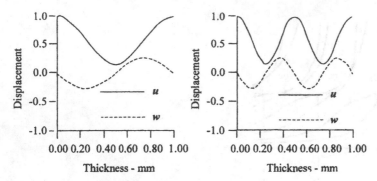

Figure 8-17. Cross-sectional normalized distribution of normal (w) and longitudinal (u) displacements for: (a) mode S1 at $(fd)_1 = 3.798$ MHz-mm and (b) mode S2 at $(fd)_2 = 7.596$ MHz-mm; both modes are shown in Figure 8-16.

To prove the validity of expression (8.41), the first of the two equations given by Viktorov (1967, eq. II.2) will be used to determine the relationship between the two constants A_s and D_s occurring in the expressions for the scalar potentials ϕ_s and ψ_s:

$$D_s = \frac{-(k^2 + s^2)\cosh(rh)}{2iks\cosh(sh)}A_s. \tag{8.48}$$

Then, given the well-known relation

$$w_s = \frac{\partial \phi_s}{\partial z} - \frac{\partial \psi_s}{\partial x}, \tag{8.49}$$

we substitute (8.48) into (8.49) and use the following equations:

$$\phi_s = A_s \cosh(rz)e^{i(kx-\omega t)},$$
$$\psi_s = D_s \sinh(sz)e^{i(kx-\omega t)} \tag{8.50}$$

(cf. eq. II.1 in Viktorov 1967).

The displacement component w_s may be determined as

$$w_s = \left(r\sinh(rz) - \frac{(k^2 + s^2)\cosh(rh)\sinh(sz)}{2s\cosh(sh)} \right)e^{i(kx-\omega t)}A_s. \tag{8.51}$$

In the limit, for r approaching zero, (8.49) can be written as

$$\lim_{r\to 0} w_s = \frac{-h(k^2 - n^2\pi^2/h^2)\sin(n\pi z/h)}{2n\pi\cos(n\pi)}e^{i(kx-\omega t)}A_s \tag{8.52}$$

(bearing in mind (8.46) and well-known relations between trigonometric and hyperbolic functions of complex variables). It is readily seen that (8.52) can be equal to zero for any n only if z is equal to $\pm h$ or zero – that is, only on the free surfaces or in the midplane. Figure 8-16 shows a sample dispersion curve for discussion purposes.

Figure 8-17 displays calculated results for both the normal w_s and the tangential (longitudinal) u_s displacement distributions across the 1-mm-thick steel plate for the first

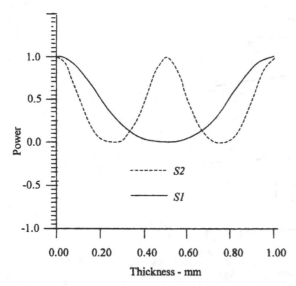

Figure 8-18. Normalized time-averaged power flux distribution for the first two nonzero-order symmetric modes of Figure 8-16: S1 at $(fd)_1 = 3.798$ MHz-mm and S2 at $(fd)_2 = 7.596$ MHz-mm.

and second symmetric modes at their respective critical frequency–thickness products. The displacement amplitudes have been normalized by the largest value of the longitudinal displacement. For both modes we can see that the normal displacement of the free surfaces vanishes and that the longitudinal displacement dominates across the entire thickness. Figure 8-18 plots the normalized time-averaged power flux distributions per unit waveguide width for the S1 and S2 modes of Figure 8-16. For these modes, the higher the order, the more the energy is concentrated near the free surfaces. This means that the higher the order of the mode, the smaller the distance between free surfaces and the first minimum displacement value.

Calculations of group velocities for frequencies near the critical frequencies of modes S1, S2, and S3 have revealed that, although for these modes the phase velocity dispersive curves look almost flat, the maxima of the relevant group velocities are shifted slightly to other frequencies. This is shown in Figure 8-19 for the first two nonzero-order symmetric modes. The maximum shift of frequency is less than $\pm 7\%$. Such a small shift from the critical frequency f_{cr} to the frequency of maximum group velocity $f_{g\,max}$ leads to a small nonzero value of the normal displacement on the surfaces, while the distribution of both displacement components remain almost unchanged (see Figure 8-20).

It is also worth noting that, for all nonzero-order symmetric modes, the values of group velocities for any of the fd products given by (8.47) are the same. This conclusion may be proved as follows. The group velocity can be found from the implicit form of the dispersion relation, equation (8.44), as

$$c_g = -\frac{\partial \Omega_s / \partial k}{\partial \Omega_s / \partial \omega}. \tag{8.53}$$

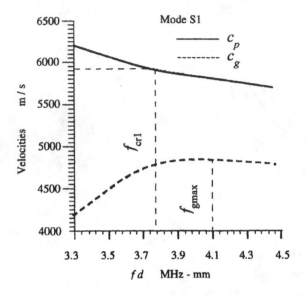

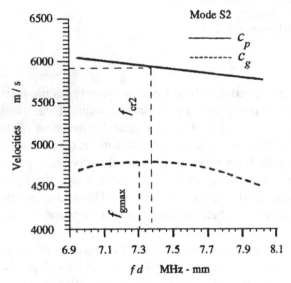

Figure 8-19. Phase and group velocities for the first two nonzero-order symmetric modes marked in Figure 8-16.

In the $r \to 0$ (i.e., $c_p \to c_L$) limit and for any of the fd products given by (8.47), we arrive at the following relation:

$$\lim_{\substack{r \to 0 \\ fd \to (fd)_n}} c_g = c_L \left(\frac{(c_T/c_L)^2 + 8P}{1 + 8P} \right), \tag{8.54}$$

where

$$P = \frac{1 - (c_T/c_L)^2}{[2 - (c_L/c_T)^2]^2}. \tag{8.55}$$

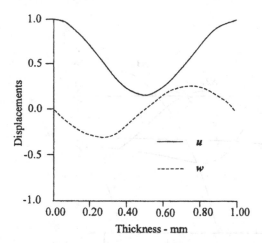

Figure 8-20. Normalized cross-sectional displacement distributions for mode S1 at $f_{g\max}$ as shown in Figure 8-19.

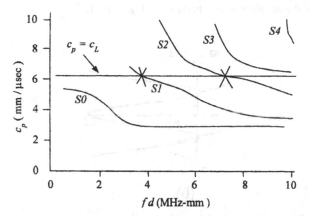

Figure 8-21. Dispersion curve for the symmetric modes of an aluminum layer, showing the intersections of the modes with a phase velocity equal to the longitudinal wave velocity.

We can see that (8.54) and (8.55) are independent of n, as stated. Note also that these two equations are written in nondimensional form, showing that the group velocity normalized by c_L is a function only of the Poisson ratio ν:

$$\frac{c_g}{c_L} = \frac{(1 - 2\nu)(\nu^2 - 4\nu + 2)}{2[\nu^2(1 - \nu) + (1 - 2\nu)^2]}. \tag{8.56}$$

When we examine the wave structure of various modes (Figure 8-18), it is obvious that there is a dramatic variation between different modes and also between different fd values of the same mode. Of particular interest on the symmetric modes are the points where $c_p = c_L$ (see Figure 8-21). Note that, in Figure 8-22, the group velocity value at these special points (where c_p equals c_L) is the same. Two of these points are S1 ($fd = 3.56$) and

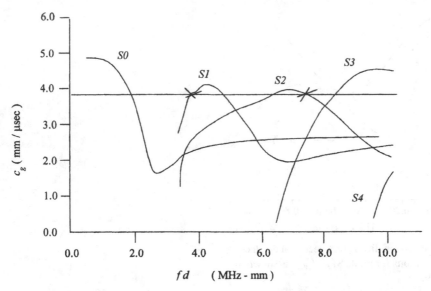

Figure 8-22. Group velocity dispersion curve for the symmetric modes of an aluminum layer, depicting the group velocity of the intersection points shown in Figure 8-21.

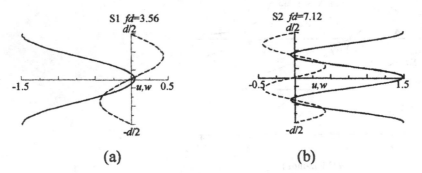

Figure 8-23. Wave structure for the S1 and S2 modes where $c_p = c_L$.

S2 ($fd = 7.12$). As expected from previous analysis, there is no out-of-plane displacement at the plate surfaces (see Figure 8-23).

8.7.2 Zeros and Poles for a Fluid-Coupled Elastic Layer

In this section we address the influence of a fluid load on an elastic layer. This problem of oblique incidence reflection factor and a corresponding physical interpretation of zeros and poles was studied by Chimenti and Rokhlin (1990). If one were to tackle the boundary value problems illustrated in Figure 8-24, using the appropriate boundary conditions at each interface, one would come up with an expression for reflection factor that would take on a special form (this expression is given by Chimenti and Rokhlin). Our emphasis will be placed on examining very small densities of the fluid with respect to the elastic layer, which represents the practical problem from an ultrasonic NDT point of view. The

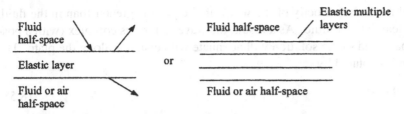

Figure 8-24. Boundary value problem for liquid-loaded layer.

form of the oblique incidence reflection factor R in (8.57) can be studied in more detail by examining the formula's poles and zeros. In its general form, R can be written as

$$R = \frac{AS - Y^2}{(S + iY)(A - iY)},$$ (8.57)

where A and S are respectively antisymmetric and symmetric Lamb mode terms (see Chimenti and Rokhlin 1990):

$$A = \frac{(q^2 - 1)^2}{q} \tan\left(kp\frac{d}{2}\right) + 4p \tan\left(kq\frac{d}{2}\right),$$

$$S = \frac{(q^2 - 1)^2}{q} \cot\left(kp\frac{d}{2}\right) + 4p \cot\left(kq\frac{d}{2}\right).$$

Here $p^2 = (c/c_L)^2 - 1$, $q^2 = (c/c_T)^2 - 1$, and $k = \omega/c$; the terms c and d denote phase velocity and thickness of the plate, respectively.

The influence of the fluid is determined by Y:

$$Y = \frac{\rho_f}{\rho}\left(\frac{c}{c_T}\right)^4 \frac{p}{qm}, \quad \text{where } m = \left(\frac{c}{c_f}\right)^2 - 1.$$

The index f is associated with the fluid parameters; hence $\rho_f =$ fluid density and $c_f =$ fluid bulk velocity ($\rho =$ plate density).

It can be shown that, for small densities, the real part of the poles may be used to produce the dispersion curves for this fluid-loaded structure, as well as expressions associated with the leaky Lamb waves. The interpretations of these results are quite interesting. For example, if the reflection factor $R = 0$ then (at least from a physical point of view) we could measure these points to produce a dispersion curve of the structure. Keep in mind, though, that the energy may go into the layer as a guided wave but could also be entirely transmitted. Or, in the most general case, a little of each is obtained – with some transmission factor and some energy going into guided waves. For small densities of the fluid with respect to the plate, the poles are close to the zero values. Note that a reflection factor of some amplitude divided by zero would equal infinity, which is not possible; on the other hand, a reflection factor of zero divided by zero is merely undefined.

In studying this material, we can recall the Cremer hypothesis (which is actually Snell's law) and so derive a relationship for oblique incidence and for energy going along the plate. Note that the poles equal to zero correspond to the leaky waves, not to the reflection factor from the plate. If the phase velocity of the waves in the plate is greater than in the fluid, then energy will leak into the fluid. As a result, the wavevector k is complex (with a positive imaginary part) and so the solution will attenuate with distance along the plate. The basic explanation is captured by

$$u = Ae^{i(k_r x - \omega t)} e^{-\alpha x}, \tag{8.58}$$

where $k = k_r + i\alpha$ and $e^{-\alpha x}$ is the decay term.

For reflection factor values of zero, the wave behaves as if the plate were completely transparent to the incident wave. But for the Lamb wave angle corresponding to the real part of the pole, we can have leaky Lamb waves on both sides of the plate. The pole equaling zero does not correspond to full reflection factor or full transmission factor; in this case, the reflection factor would have intermediate values or even a zero component.

8.7.3 Nonspecular Reflection and Transmission for Layered Media

A reflection from a flat surface is specular in nature, retaining the basic waveform shape characteristics and spatial distribution. The most familiar nonspecular reflection patterns occur when waves reflect off such unusual shapes as spheres or cylinders; the waveform is modified and the spatial distribution is also adjusted. However, nonspecular reflection from a single-layer or multilayer plate can also generate some interesting problems. Ngoc and Mayer (1980) study the nonspecular reflection pattern that can occur as an incident beam impinges onto an elastic layer immersed between two fluids (see Figure 8-25). What happens is that the reflection factor is superimposed with the leaky Lamb wave in such a manner that a null is obtained in the reflection distribution. Scanning an ultrasonic transducer along the plate, we would observe two distinct peaks and a specific null associated with that interference phenomenon.

8.7.4 Mode Cutoff Frequency

Mode cutoff values occur at specific fd values for modes higher than S0 and A0. At these points, the phase velocity approaches infinity as the group velocity approaches zero. Hence, these frequency values occur whenever standing longitudinal or shear waves are present across the thickness of the plate. See Graff (1991) for an excellent discussion on this topic.

Mode cutoff values can be calculated by examining a limiting condition of $k \rightarrow 0$. In this case, the Rayleigh–Lamb frequency equation of the form (8.24) becomes (for the symmetric case)

$$\sin qh \cos ph = 0. \tag{8.59}$$

The solutions for the fd values at cutoff frequency can therefore be calculated. For this symmetric case, let $qh = n\pi$ for $n = 0, 1, 2, \ldots$. We then have

$$qh = \frac{\omega}{c_T} \frac{d}{2} = \frac{2\pi}{c_T} \frac{fd}{2} = n\pi \tag{8.60}$$

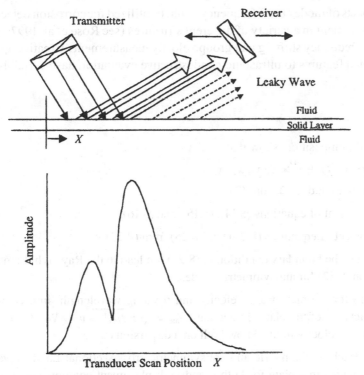

Figure 8-25. Nonspecular reflection possibility as a result of leaky wave interactions with oblique incidence reflection factor.

or $fd = nc_T$; that is,

$$fd = \{c_T, 2c_T, 3c_T, \ldots\}. \tag{8.61}$$

Alternatively, let $ph = n(\pi/2)$ for $n = 0, 1, 2, \ldots$. Then

$$ph = \frac{\omega}{c_L}\frac{d}{2} = \frac{2\pi}{c_L}\frac{fd}{2} = \frac{n\pi}{2} \tag{8.62}$$

or $fd = nc_L/2$; that is,

$$fd = \left\{ \frac{c_L}{2}, \frac{3c_L}{2}, \frac{5c_L}{2}, \ldots \right\}. \tag{8.63}$$

For the antisymmetric case, much algebraic manipulation will yield

$$fd = \left\{ \frac{c_T}{2}, \frac{3c_T}{2}, \frac{5c_T}{2}, \ldots \right\} \quad \text{and} \tag{8.64}$$

$$fd = \{c_L, 2c_L, 3c_L, \ldots\}. \tag{8.65}$$

These values of the cutoff frequency can all be found on the dispersion curves (see e.g. Figure 8-8).

Many interesting aspects of mode cutoff frequency could be utilized for corrosion detection and thickness measurement in a variety of different structures (see Rose et al. 1997a); the same can be said for frequency shifting and group velocity measurement. Creative applications of mode-related features to ultrasonic nondestructive evaluation have certainly proved useful.

8.8. Exercises

1. From Helmholtz decomposition, show that
   $$\bar{\mu} = \nabla\Phi(x_1, x_2, x_3, t) + \nabla \times \bar{\psi}(x_1, x_2, x_3, t)$$
   leads to equations (8.8) and (8.9) from (8.4).

2. Outline the development of equations (8.14), (8.15), and (8.16).

3. Show the development of equations (8.24) and (8.25) from (8.23).

4. Show how satisfying the boundary condition in (8.23) can lead to the Rayleigh–Lamb frequency equation (8.32) for antisymmetric modes.

5. Derive expressions for the plate wave velocity and the long wavelength limit compared with the thickness of the plate. Show that $c_{\text{plate}} = \sqrt{E/\rho(1 - v^2)}$. What is the high-frequency wave velocity limit? Show both on a dispersion curve.

6. Make a detailed sketch of symmetric and antisymmetric cross-sectional Lamb wave propagation possibilities in a plate for both u and w displacement components. Explain!

7. List steps for proving that the in-plane vibration is dominant (and that the out-of-plane component of particle velocity is zero) at the intersection points of the longitudinal wave velocity with higher-order symmetrical modes.

8. Why is it possible to generate dominant in-plane particle displacement on the surface of a plate with an oblique incidence wedge over a fluid film?

9. How could you distinguish ice from water as a contaminant on a test surface?

10. Calculate fd values of points of zero out-of-plane displacement on the dispersion curve for aluminum.

11. Assuming that the reflection coefficient can be written as in (8.57), how would you solve for zero reflection? How could you use this expression to produce a dispersion curve?

12. In a reflection factor problem, what does $R = 0$ imply?

13. When could the zeros of the reflection factor formula be used to produce a dispersion curve?

14. What does $T = 0$ imply? (T denotes transmission coefficient.)

15. If the roots of a characteristic equation are complex (say, $k_r + iB$), then what is physically implied with respect to wave propagation in a fluid-loaded structure and in the fluid itself?

16. How could nonspecular reflection occur from a planar object or multilayer planar structure? What instrumentation parameters affect the nonspecular reflection pattern?

17. How could you concentrate ultrasonic energy close to the surface in a Lamb wave plate experiment?

18. Compute the angle of incidence required in a Plexiglas wedge to produce a Lamb wave in a plate with phase velocity equal to 5 mm/μs.

19. How would you select a nondispersive mode from a phase velocity or group velocity dispersion diagram?

20. Make a sketch of the vibration pattern of a plate undergoing antisymmetrical (flexural) vibration motion. Show typical in-plane and out-of-plane particle displacement vectors.

21. Make a sketch of the vibration pattern of a plate undergoing symmetrical (compressional) vibration motion. Show typical in-plane and out-of-plane particle displacement vectors.

22. What might happen as a particular guided wave mode impinges onto the free edge of the plate? Why? Use a sketch to illustrate a possible scenario.

23. What are the specific boundary conditions in the surface wave problem? The plate wave problem? The two-layer plate problem?

24. Dispersion curves for a plate converge to a particular phase velocity value for large fd values. Explain.

25. Show that, for a complex wavevector, there is attenuation of the resulting displacement field. Why does this occur?

26. Plot typical wave structure (in-plane and out-of-plane distribution) for symmetric and antisymmetric modes.

27. What mode and frequency would you select if you were to use guided waves on the wing of an aircraft to detect and discriminate between ice and water? Explain.

28. Derive expressions for the Rayleigh–Lamb frequency equations for three regions: $c_p < c_T$; $c_T < c_p < c_L$; $c_p > c_L$.

29. Illustrate graphically a root extraction procedure to produce a dispersion curve.

30. List experimental methods of generating Lamb waves in a plate.

31. What angle could be used to generate a specific mode and frequency value in an aluminum plate? Give an example.

32. What are the group velocity values at the zero out-of-plane displacement points on the symmetric modes of a dispersion curve for an aluminum plate?

33. Calculate all cutoff frequency values for an aluminum plate and show the results on the dispersion curve in Figure 8-8.

34. Consider a guided wave experiment in an aluminum plate at a fixed angle and phase velocity value. Could thickness be measured by either a tone-burst frequency sweep or a Fourier transform of a shock-excited broad–frequency bandwidth transducer? Explain your answer.

35. How could cutoff frequency be used to estimate remaining wall thickness in a corrosion detection experiment?

36. In general, how does one compute the mode cutoff values?

9

Interface Waves

9.1 Introduction

A variety of interface wave types have been discussed in the literature. Some discussion has centered on seismologic interpretation as well as the utility of ultrasonic nondestructive evaluation (NDE), and studies in wave propagation will continue to address such issues. This chapter presents an introduction to two types of wave interaction: (i) at a solid–solid interface via Stoneley waves, and (ii) along a solid–liquid interface via Scholte waves. We shall derive the basic equations, but some solutions are reserved as an exercise for students.

9.2 Stoneley Waves

Consider two elastic half-spaces that are in perfect contact along the plane $z = 0$, as sketched in Figure 9-1. The terms ρ_m and λ_m, μ_m ($m = 1, 2$) denote mass density and elastic properties (resp.) for each half-space. Our goal is to determine whether harmonic wave propagation is possible along the interface, a problem that was first solved by Stoneley in 1924.

We start with the usual equations of motion of the linear theory of elasticity (see e.g. Graff 1991):

$$(\lambda + \mu)u_{j,jn} + \mu u_{n,jj} = \rho \frac{\partial^2 u_n}{\partial t^2} \quad (n = 1, 2), \tag{9.1}$$

where $u = u_1$ and $w = u_2$ are displacements in the x and z directions, respectively. To describe waves propagating along the x-axis, the displacement components of this wave should be represented in the form

$$u = u(z)e^{i(kx-\omega t)}, \tag{9.2}$$

$$w = w(z)e^{i(kx-\omega t)}, \tag{9.3}$$

where k is the wavenumber. Substituting (9.2) and (9.3) into equation (9.1), we obtain expressions for amplitude $u(z)$ and $w(z)$ (see e.g. Achenbach and Epstein 1967):

$$u(z) = \{[Ae^{-k\alpha z} + Be^{k\alpha z}] - \beta[Ce^{-k\beta z} - De^{k\beta z}]\}e^{i(\omega t - kx)}, \tag{9.4}$$

$$w(z) = i\{\alpha[Ae^{-k\alpha z} - Be^{k\alpha z}] - [Ce^{-k\beta z} + De^{k\beta z}]\}e^{i(\omega t - kx)}, \tag{9.5}$$

where

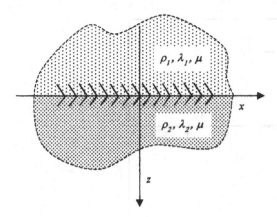

Figure 9-1. Coordinate system for interface wave studies.

$$\alpha^2 = 1 - \frac{c^2}{c_L^2} \quad \text{and} \tag{9.6}$$

$$\beta^2 = 1 - \frac{c^2}{c_T^2}. \tag{9.7}$$

As before, c_L and c_T are the velocity of the longitudinal and transverse waves, respectively. We have

$$c_L^2 = \frac{\lambda + 2\mu}{\rho} \quad \text{and} \quad c_T^2 = \frac{\mu}{\rho}, \tag{9.8}$$

where c is phase velocity of the propagating waves:

$$c = \frac{\omega}{k}. \tag{9.9}$$

Applying Hooke's law to the displacements (9.4) and (9.5), we derive the following expressions for stress:

$$\sigma_z = i\mu\{-k(1 + \beta^2)[Ae^{-k\alpha z} + Be^{k\alpha z}] + 2k\beta[Ce^{-k\beta z} - De^{k\beta z}]\}e^{i(\omega t - kx)}, \tag{9.10}$$

$$\sigma_{xz} = \mu\{-2k\alpha[Ae^{-k\alpha z} - Be^{k\alpha z}] + k(1 + \beta^2)[Ce^{-k\beta z} + De^{k\beta z}]\}e^{i(\omega t - kx)}. \tag{9.11}$$

The boundary conditions for perfect contact at $z = 0$ are

$$u^{(1)} = u^{(2)}, \tag{9.12}$$

$$w^{(1)} = w^{(2)}, \tag{9.13}$$

$$\sigma_z^{(1)} = \sigma_z^{(2)}, \tag{9.14}$$

$$\sigma_{xz}^{(1)} = \sigma_{xz}^{(2)}, \tag{9.15}$$

where the indices (1) and (2) denote the upper and lower half-spaces, respectively.

We will now find solutions for the half-spaces from the general solutions (9.4), (9.5), (9.10), and (9.11). Taking into account that the displacements and stresses must decay

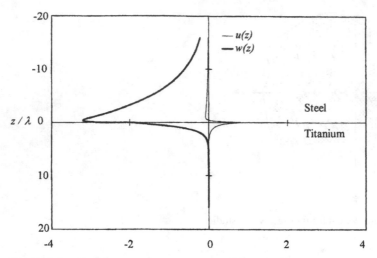

Figure 9-2. Wave structure across a steel–titanium interface.

when $|z|$ increases, we obtain the following expressions for displacement and stress in the upper half-space:

$$u^{(1)} = [B_1 e^{k\alpha_1 z} + \beta_1 D_1 e^{k\beta_1 z}]e^{i(kx - \omega t)}, \tag{9.16}$$

$$w^{(1)} = -i[\alpha_1 B_1 e^{k\alpha_1 z} + D_1 e^{k\beta_1 z}]e^{i(kx - \omega t)}, \tag{9.17}$$

$$\sigma_z^{(1)} = -ik\mu_1[(1 + \beta_1^2)B_1 e^{k\alpha_1 z} + 2\beta_1 D_1 e^{k\beta_1 z})]e^{i(kx - \omega t)}, \tag{9.18}$$

$$\sigma_{xz}^{(1)} = k\mu_1[2\alpha_1 B_1 e^{k\alpha_1 z} + (1 + \beta_1^2)D_1 e^{k\beta_1 z})]e^{i(kx - \omega t)}. \tag{9.19}$$

Analogous expressions for the lower half-space can be obtained from equations (9.16)–(9.19) by substituting A_2 for B_1, C_2 for D_1, $-\alpha_2$ for α_1, and $-\beta_2$ for β_1.

From (9.6) and (9.7) it follows that, for $n = 1, 2$,

$$\alpha_n^2 = 1 - \frac{c^2}{c_{Ln}^2}, \tag{9.20}$$

$$\beta_n^2 = 1 - \frac{c^2}{c_{Tn}^2}. \tag{9.21}$$

In equations (9.16)–(9.19), and in the solution for the upper half-space, α_k and β_k are real because of the decrease in stress and displacement. Therefore, $c < c_T$, where c_T is the smallest of c_{T1} and c_{T2} and c is less than the smallest shear velocity.

Substituting expressions (9.16)–(9.19) into the respective boundary conditions (9.12)–(9.15) yields a set of four homogeneous equations for the constants B_1, D_1, A_2, C_2. The system of equations will have a nontrivial solution when the determinant is zero:

$$\begin{vmatrix} -(1 + \beta_1^2) & -2\beta_1 & (1 + \beta_2^2)g & -2\beta_2 g \\ 2\alpha_1 & (1 + \beta_1^2) & 2\alpha_2 g & -(1 + \beta_2^2)g \\ 1 & \beta_1 & -1 & \beta_2 \\ -\alpha_1 & -1 & -\alpha_2 & 1 \end{vmatrix} = 0, \tag{9.22}$$

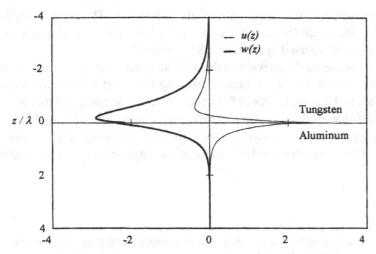

Figure 9-3. Wave structure across a tungsten–aluminum interface.

where $g = \mu_2/\mu_1$. This condition gives a dispersion equation for the unknown phase velocity c of the wave propagating along the interface.

Stoneley waves exist for the two perfectly joined materials when equation (9.22) has real roots. Note that (9.22) does not depend on frequency and so the Stoneley wave is not dispersive. It should also be noted that the domain of physical parameters for which Stoneley waves exist is small (Achenbach and Epstein 1967; Miklowitz 1984). It can be shown that, for values of c_{T1}/c_{T2} that are close to unity, the existence of real-valued roots for (9.22) is a practical criterion for establishing the presence of Stoneley waves.

Figure 9-2 displays sample results for wave structure across a perfectly bonded steel–titanium interface. A like result across a tungsten–aluminum interface is shown in Figure 9-3.

9.3 Scholte Waves

In this section we consider wave propagation along a solid–liquid interface. We can convert the general equation (9.22) for the special case when one of the half-spaces is an ideal nonviscous liquid. If the upper half-space is liquid then $\mu_1 = 0$, because liquid does not support shear motion. Taking the limit from equation (9.22) as $\mu_1 \to 0$, we obtain

$$R(c) = i\frac{\rho_L}{\rho_2}\left[\frac{c_L}{c_{L2}}\frac{\sqrt{c_{L2}^2 - c^2}}{\sqrt{c^2 - c_L^2} \cdot c_{T2}^4}\right]. \qquad (9.23)$$

The terms ρ_L and c_L denote the mass density and the longitudinal velocity in the liquid, respectively. The equation for Rayleigh waves in an elastic half-space is $R(c) = 0$, where

$$R(c) \equiv \left[2 - \left(\frac{c}{c_{T2}}\right)^2\right]^2 - 4\sqrt{1 - \left(\frac{c}{c_{T2}}\right)^2}\sqrt{1 - \left(\frac{c}{c_{L2}}\right)^2}. \qquad (9.24)$$

Viktorov (1967) shows that, for any specified parameters, equation (9.24) has only one real root, c^*, which is less than c_L, c_{L2}, and c_{T2}. It is shown that this wave, a Scholte wave, carries almost all of the energy in the liquid rather than in the solid. The wave's amplitude decreases slightly in the liquid (but rapidly in the elastic half-space) as $|z|$ increases. Hence, Scholte waves are seldom used in NDE (see Scholte 1942).

Let c_{R2} be the velocity of the Rayleigh wave in the solid half-space. If $c_L < c_{R2}$ (which is true for most materials), then waves can propagate along the interface as a Rayleigh wave, which corresponds to the complex root of equation (9.23). The energy of this wave leaks into the liquid half-space because its phase velocity is very close to c_{R2}. The attenuation is e^{-1} over approximately ten wavelengths (see Viktorov 1967). The structure and stress distribution of this wave are similar to those of the Rayleigh wave in an elastic half-space.

9.4 Exercises

1. What is a typical solution assumed for purposes of exploring the possibility of wave propagation along a solid–solid interface?

2. How many boundary conditions are there in this Stoneley wave problem? How many for the problem of a single layer on a half-space?

3. Stoneley waves are nondispersive. Provide a brief physical explanation for this characteristic.

4. How can we use the Stoneley wave solution to produce a Scholte wave solution for a fluid on a half-space?

5. How would you examine the possibility of guided wave propagation in a layer embedded between two half-spaces? Would roots of the characteristic equation be real or complex?

6. Establish a criterion for the existence of Stoneley waves.

7. Provide an explanation of the use of equation (9.23).

8. Provide an explanation of the use of equation (9.24).

9. For the case when complex roots are extracted from the Stoneley wave characteristic equation, provide a physical interpretation with respect to wave propagation along the interface.

10. Are Scholte waves dispersive?

10

Layer on a Half-Space

10.1 Introduction

Another interesting and useful problem in wave propagation and practical model analysis is related to propagation along a layer on a half-space. In this chapter we consider the case of an isotropic layer on an elastic half-space. The problem of multiple layers on a half-space and features of anisotropic cases are both important also, and they are the subject of contemporary studies. Students will be able to relate to these developments using tools developed in this text.

The dispersive character of wave propagation is useful in evaluating material properties and density gradients with depth. In this chapter we explore longitudinal and vertical shear waves as well as shear horizontal (SH) waves along an interface, often termed "Love waves." For further reading on layers on a half-space and interface waves, see Achenbach and Keshava (1967), Chadwick and Currie (1974), Love (1926), Miklowitz (1984), Nayfeh (1995), Rokhlin, Hefets, and Rosen (1981), Rokhlin and Wang (1991a,b), Rose, Nayfeh, and Pilarski (1989b), Scholte (1942), Stoneley (1924), and Viktorov (1967).

10.2 Layers on a Half-Space

We consider layers on the elastic half-space as shown in Figure 10-1, and begin by describing the waves that propagate in the x direction. Early results for such waves were published by Sezawa (1927) and Fu (1946). The components of waves propagating in the x direction have the form of equations (9.2) and (9.3). The solution for the lower half-space was constructed in Section 9.2; the solution for the layer has the form of equations (9.4), (9.5), (9.10), and (9.11). We must satisfy the following boundary conditions:

$$\sigma_z^{(1)} = 0 \quad \text{at } z = -d, \tag{10.1}$$

$$\sigma_{xz}^{(1)} = 0 \quad \text{at } z = -d, \tag{10.2}$$

$$\sigma_z^{(1)} = \sigma_z^{(2)} \quad \text{at } z = 0, \tag{10.3}$$

$$\sigma_{xz}^{(1)} = \sigma_{xz}^{(2)} \quad \text{at } z = 0, \tag{10.4}$$

$$u^{(1)} = u^{(2)} \quad \text{at } z = 0, \tag{10.5}$$

$$w^{(1)} = w^{(2)} \quad \text{at } z = 0. \tag{10.6}$$

Substituting expressions for displacement and stress in the layer and the half-space into equations (10.1)–(10.6), we obtain a system of six homogeneous equations. Setting the determinant of the system equal to zero,

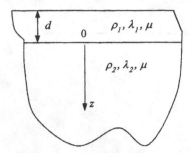

Figure 10-1. Coordinate system for studies of a layer on a half-space.

$$
\begin{vmatrix}
a_{11} & a_{12} & a_{13} & a_{14} & a_{15} & a_{16} \\
a_{21} & a_{22} & a_{23} & a_{24} & a_{25} & a_{26} \\
a_{31} & a_{32} & a_{33} & a_{34} & a_{35} & a_{36} \\
a_{41} & a_{42} & a_{43} & a_{44} & a_{45} & a_{46} \\
a_{51} & a_{52} & a_{53} & a_{54} & a_{55} & a_{56} \\
a_{61} & a_{62} & a_{63} & a_{64} & a_{65} & a_{66}
\end{vmatrix} = 0, \tag{10.7}
$$

we find the dispersive equation for the phase velocity c (see Achenbach and Epstein 1967).

The matrix in (10.7) contains an inner dashed rectangle, which represents the determinant (9.22) for a Stoneley wave for two perfectly joined elastic half-spaces. Elements of (10.7) include:

$$a_{11} = -(1 + \beta_1^2)e^{k\alpha_1 d}, \tag{10.8}$$

$$a_{12} = 2\beta_1 e^{k\beta_1 d}, \tag{10.9}$$

$$a_{21} = -2\alpha_1 e^{k\alpha_1 d}, \tag{10.10}$$

$$a_{22} = (1 + \beta_1^2)e^{k\beta_1 d}, \tag{10.11}$$

$$a_{31} = -(1 + \beta_1^2), \tag{10.12}$$

$$a_{32} = 2\beta_1, \tag{10.13}$$

$$a_{41} = -2\alpha_1, \tag{10.14}$$

$$a_{42} = 1 + \beta_1^2, \tag{10.15}$$

$$a_{51} = 1, \tag{10.16}$$

$$a_{52} = -\beta_1, \tag{10.17}$$

$$a_{53} = 1, \tag{10.18}$$

$$a_{54} = \beta_1, \tag{10.19}$$

$$a_{55} = -1, \tag{10.20}$$

$$a_{56} = \beta_2, \tag{10.21}$$

$$a_{61} = \alpha_1, \tag{10.22}$$

$$a_{62} = -1,$$ (10.23)

$$a_{63} = -\alpha_1,$$ (10.24)

$$a_{64} = -1,$$ (10.25)

$$a_{65} = -\alpha_2,$$ (10.26)

$$a_{66} = 1;$$ (10.27)

as usual,

$$k = \frac{\omega}{c}.$$ (10.28)

Equation (10.7) is a transcendental equation whose elements depend on frequency. Hence its solution c, a real number, also depends on frequency and is dispersive. When the fd product approaches zero, (10.7) degenerates to the Rayleigh equation for the half-space (see Achenbach and Epstein 1967). Therefore, small values of the fd phase velocity of the lowest mode approach the velocity of the Rayleigh wave c_{R2} for the half-space. When fd increases, the wavelength decreases and the phase velocity approaches the velocity of the Rayleigh wave c_{R1} for the layer material. It is shown that c is always less than c_{T2}. The number of propagation modes increases as fd increases if $c_{T1} < c_{T2}$. If $c_{T1} > c_{T2}$, no more than two modes can exist. The velocity of the two lowest modes for larger values of fd approach c_{R1} and the velocity of the Stoneley wave on the interface. If Stoneley waves do not exist, then the second mode from the lowest mode approaches c_{T1}.

In Figures 10-2A and 10-2B, the phase velocities of several modes are shown for a layer and half-space in perfect contact. For fd values greater than 0.7, real roots do not exist; the roots are complex and describe waves that attenuate. For our data, when c_{T1} is larger than c_{T2}, there exists only one propagation wave for a special range of fd. For an fd value larger than 0.74, the propagating wave degenerates because the phase velocity cannot exceed c_{T2}. Some practical applications of this problem can be found in Bray (1988) and Grewal (1996).

10.3 Love Waves

The Rayleigh wave that propagates in a half-space contains partial displacement in the plane of propagation only. Seismology observations show that, during an earthquake, an SH wave can appear. Displacement in a direction perpendicular to the plane of wave propagation is possible. Love showed that this wave can exist in a half-space covered by a layer with different elastic properties. The coordinate system used in his study is the same as that used in Figure 10-1.

We will consider only the displacement v, which is perpendicular to the plane of wave propagation, with $u = w = 0$. In this case, the governing equation for SH waves is:

$$\frac{\partial^2 v_k}{\partial x^2} + \frac{\partial^2 v_k}{\partial z^2} = \frac{1}{(c_{Tk})^2} \frac{\partial^2 v_k}{\partial t^2} \quad (k = 1, 2).$$ (10.29)

Let

$$c_{T1}^2 = \frac{\mu_1}{\rho_1} \quad \text{and} \quad c_{T2}^2 = \frac{\mu_2}{\rho_2};$$ (10.30)

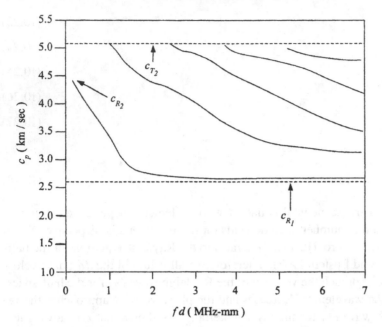

Figure 10-2A. Phase velocity dispersion curves for $c_{T1} < c_{T2}$, where
$c_{L1} = 4.7$ km/s, $c_{T1} = 3$ km/s, $c_{L2} = 8$ km/s, $c_{T2} = 5$ km/s, $G_1 = 3$ GPa,
and $G_2 = 7$ GPa (soft layer on a half-space).

v_1 and v_2 are displacements in the layer and the half-space, respectively. The resulting solutions of (10.29), which represents harmonic waves propagating in the x direction, are

$$v_1 = (Ae^{-k\beta_1 z} + Be^{k\beta_1 z})e^{i(\omega t - kx)} \quad (-d \le z \le 0) \tag{10.31}$$

for a layer and

$$v_2 = (Ne^{-k\beta_1 z})e^{i(\omega t - kx)} \quad (z \ge 0) \tag{10.32}$$

for a half-space, where

$$\beta_1^2 = 1 - \frac{c^2}{c_{T1}^2}, \tag{10.33}$$

$$\beta_2^2 = 1 - \frac{c^2}{c_{T2}^2}. \tag{10.34}$$

The displacement v_2 should decay as z increases. We thus have the restriction $\beta_2 > 0$, from which it follows that $c < c_{T2}$. Boundary conditions for this problem are

$$\sigma_{zy}^{(1)} = 0 \quad \text{at } z = -d, \tag{10.35}$$

$$\sigma_{zy}^{(1)} = \sigma_{zy}^{(2)} \quad \text{at } z = 0, \tag{10.36}$$

$$v^{(1)} = v^{(2)} \quad \text{at } z = 0. \tag{10.37}$$

By substituting the generalized solution into the boundary conditions, we discover a set of three algebraic homogeneous equations that have a nontrivial solution if the determinant,

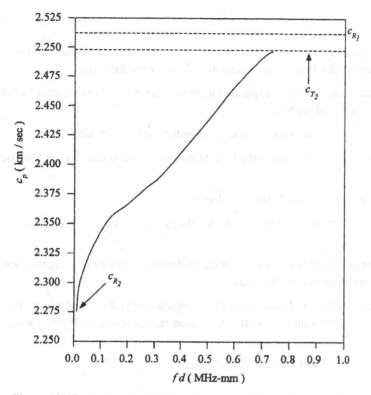

Figure 10-2B. Phase velocity dispersion curves for $c_{T1} > c_{T2}$, where $c_{L1} = 6.3$ km/s, $c_{T1} = 3.1$ km/s, $c_{L2} = 4$ km/s, $c_{T2} = 2.5$ km/s, $G_1 = 2.61$ GPa, and $G_2 = 1.5$ GPa (hard layer on a half-space).

$$\tan\left[\sqrt{\frac{1}{c_{T1}^2} - \frac{1}{c^2}} \cdot \omega d\right] = \frac{\mu_2\sqrt{1 - c^2/c_{T2}^2}}{\mu_1\sqrt{1 - c^2/c_{T1}^2}}, \tag{10.38}$$

is equal to zero.

Real roots of equation (10.38) exist if $c_{T2} > c_{T1}$ (see Miklowitz 1984). Solutions to (10.38) satisfy the following condition:

$$c_{T1} < c < c_{T2}. \tag{10.39}$$

Because (10.38) depends on frequency, the phase velocity of Love waves is dispersive. The lowest mode approaches c_{T2} as fd approaches infinity. Displacement functions for Love waves are

$$v_1 = D\cosh[\beta_1 k(z + d)]e^{i(\omega t - kx)} \quad \text{at} \ -d \leq z \leq 0, \tag{10.40}$$

$$v_2 = D\cosh(\beta_1 kd)e^{-\beta_2 kz}e^{i(\omega t - kx)} \quad \text{at} \ z \geq 0, \tag{10.41}$$

where D is an arbitrary constant. Note that the displacement has a maximum at $z = -d$.

10.4   Exercises

1. Plot a possible dispersion curve for a layer on a half-space. Consider higher and lower bulk wave velocities for the layer compared with values for the half-space.

2. What order of determinant must be expanded to obtain the characteristic equation for two isotropic layers on a half-space?

3. What is the low-frequency limit phase velocity value for a layer on a half-space?

4. For a soft layer on a half-space, what is the high-frequency or small wavelength limit on phase velocity?

5. What, physically, are Love waves? Are they dispersive?

6. Show that the Love wave modes approach c_T for the layer material as fd approaches infinity.

7. Show that the upper limit of the Love wave velocity is the shear velocity in the half-space. Provide a physical explanation of this result.

8. Compute wave structure for the Lamb-type wave propagation problem for a layer on a half-space, using estimated values from the dispersion curves shown in Figures 10-2A and 10-2B.

9. Estimate group velocity c_g curves for Figure 10-2A.

11

Waves in Rods

11.1 Background

For over half a century, the subject of wave propagation in rodlike structures has been addressed by many investigators, who report varied theoretical approaches, approximations, analyses, and experiments. Here we present a basic approach to this problem, and include an experiment on waves in rods in Section E.5. For more details, see Achenbach (1984), Graff (1991), Kolsky (1963), Meitzler (1965), Miklowitz (1984), Mindlin and McNiven (1960), Onoe, McNiven, and Mindlin (1962), and Pao and Mindlin (1960).

11.2 Longitudinal Waves in Thin Rods

Before proceeding to the general problem of wave propagation in a rod, we present a brief introduction to longitudinal waves in thin rods. We shall use a "strength of materials" approach with approximations that neglect lateral inertia. This assumption is valid for long wavelengths, but subsequent sections will show that, for smaller wavelengths and with lateral inertia, the results are dispersive in nature. For now, we proceed with the simplified case; see Figure 11-1.

Neglecting lateral inertia, we have

$$\sum F_x = ma_x \quad \text{and}$$

$$-\sigma A + \left(\sigma + \frac{\partial \sigma}{\partial x} \, dx \right) A + qA \, dx = \rho A \, dx \, \frac{\partial^2 u}{\partial t^2}.$$

By Hooke's law in one dimension, $\sigma = E\varepsilon$, where ε denotes axial strain and so $\varepsilon = \partial u / \partial x$. Therefore,

$$\frac{\partial \sigma}{\partial x} + q = \rho \frac{\partial^2 u}{\partial t^2}$$

and so

$$\frac{\partial}{\partial x} \left(E \frac{\partial u}{\partial x} \right) + q = \rho \frac{\partial^2 u}{\partial t^2}.$$

If the rod is homogeneous then neither E nor ρ is a function of x; hence

$$E \frac{\partial^2 u}{\partial x^2} + q = \rho \frac{\partial^2 u}{\partial t^2}.$$

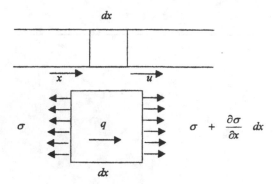

Figure 11-1. Differential element of a rod.

Thus,

$$\frac{\partial^2 u}{\partial x^2} = \frac{1}{c_0^2}\frac{\partial^2 u}{\partial t^2}, \quad \text{where} \quad c_0 = \sqrt{\frac{E}{p}}; \tag{11.1}$$

c_0 is known as the traditional *bar velocity*.

Equation (11.1) is applicable for long wavelengths – that is, for wavelengths that are greater than the diameter of the rod. This bar velocity for long wavelengths is valid for rods of any cross-sectional area. Note that the basic concepts for a string (including D'Alembert's solution) are also applicable when there is no dispersion in a thin rod, so we have u and σ propagation (where $\sigma = E(\partial u/\partial x)$).

11.3 Waves in an Infinite Rod

We may now investigate the propagation of elastic harmonic waves in an infinite rod. It is most convenient to solve this problem using cylindrical coordinates, where the z-axis is along the axis of the rod. The equation of motion can be written in full as follows, using aspects of Navier's equation in cylindrical coordinates (see Kolsky 1963 for more details):

$$(\lambda + 2\mu)\frac{\partial \phi}{\partial r} - \frac{2\mu}{r}\frac{\partial \omega_z}{\partial \theta} + 2\mu\frac{\partial \omega_\theta}{\partial z} = \rho\frac{\partial^2 u_r}{\partial t^2}, \tag{11.2}$$

$$(\lambda + 2\mu)\frac{1}{r}\frac{\partial \phi}{\partial \theta} - 2\mu\frac{\partial \omega_r}{\partial z} + 2\mu\frac{\partial \omega_z}{\partial r} = \rho\frac{\partial^2 u_\theta}{\partial t^2}, \tag{11.3}$$

$$(\lambda + 2\mu)\frac{\partial \phi}{\partial z} - \frac{2\mu}{r}\frac{\partial}{\partial r}(r\omega_\theta) + \frac{2\mu}{r}\frac{\partial \omega_r}{\partial \theta} = \rho\frac{\partial^2 u_z}{\partial t^2}, \tag{11.4}$$

where ϕ is the dilatation in cylindrical coordinates and $\omega_r, \omega_\theta, \omega_z$ represent elements of the rotation tensor. Hence

$$\phi = \frac{1}{r}\frac{\partial(ru_r)}{\partial r} + \frac{1}{r}\frac{\partial u_\theta}{\partial \theta} + \frac{\partial u_z}{\partial z},$$

$$2\omega_r = \frac{1}{r}\frac{\partial u_z}{\partial \theta} - \frac{\partial u_\theta}{\partial z},$$

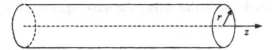

Figure 11-2. Cylindrical coordinates for a solid cylindrical rod.

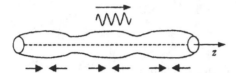

Figure 11-3. Longitudinal modes in a solid cylindrical rod.

$$2\omega_\theta = \frac{\partial u_r}{\partial z} - \frac{\partial u_z}{\partial r},$$

$$2\omega_z = \frac{1}{r}\left[\frac{\partial(r u_\theta)}{\partial r} - \frac{\partial u_r}{\partial \theta}\right].$$

At the surface of the rod, three stress components (σ_{rr}, $\sigma_{r\theta}$, σ_{rz}) must vanish. From the stress–deformation relation, we find

$$\sigma_{rr} = \lambda\phi + 2\mu\frac{\partial u_r}{\partial r},$$

$$\sigma_{r\theta} = \mu\left[\frac{1}{r}\frac{\partial u_r}{\partial \theta} + r\frac{\partial}{\partial r}\left(\frac{u_\theta}{r}\right)\right],$$

$$\sigma_{rz} = \mu\left(\frac{\partial u_r}{\partial z} + \frac{\partial u_z}{\partial r}\right).$$

We now consider propagation of harmonic waves along a rod. As follows from the equation of motion (see also Graff 1991), for the general case of vibration we have the following displacements:

$$u_r = U(r)\cos n\theta e^{i(kz-\omega t)}, \tag{11.5}$$

$$u_\theta = V(r)\sin n\theta e^{i(kz-\omega t)}, \tag{11.6}$$

$$u_z = W(r)\cos n\theta e^{i(kz-\omega t)}, \tag{11.7}$$

where n can be zero or an integer. We shall examine three types of vibration in a cylindrical rod: longitudinal, torsional, and flexural.

11.3.1 Longitudinal Modes in a Solid Cylindrical Rod

Imagine a solid circular cylindrical rod (see Figure 11-2). Longitudinal waves are axially symmetric, with displacement components in the radial and axial directions. Longitudinal waves correspond to the case $n = 0$ in equations (11.5)–(11.7); the modes are represented schematically in Figure 11-3.

It is convenient to employ the potentials Φ and Ψ that satisfy the wave equations:

$$\nabla^2\Phi = \frac{1}{c_L^2}\frac{\partial^2\Phi}{\partial t^2}, \tag{11.8}$$

$$\nabla^2\Psi = \frac{1}{c_T^2}\frac{\partial^2\Psi}{\partial t^2}; \tag{11.9}$$

$$c_L^2 = \frac{\lambda + 2\mu}{\rho}, \qquad c_T^2 = \frac{\mu}{\rho}.$$

Because of symmetry, the solution with respect to the z-axis is

$$\nabla^2 = \frac{\partial^2}{\partial r^2} + \frac{1}{r}\frac{\partial}{\partial r} + \frac{\partial^2}{\partial z^2}. \tag{11.10}$$

The scalar components of the displacement vector $\bar{u} = (u_r, 0, u_z)$ are given by

$$u_r = \frac{\partial\Phi}{\partial r} + \frac{\partial^2\Psi}{\partial r\,\partial z}, \tag{11.11}$$

$$u_z = \frac{\partial\Phi}{\partial z} - \frac{\partial^2\Psi}{\partial r^2} - \frac{1}{r}\frac{\partial\Psi}{\partial r}. \tag{11.12}$$

The stresses are given by Hooke's law as

$$\sigma_{rr} = 2\mu\frac{\partial u_r}{\partial r} + \lambda\left(\frac{u_r}{r} + \frac{\partial u_r}{\partial r} + \frac{\partial u_z}{\partial z}\right), \tag{11.13}$$

$$\sigma_{rz} = \mu\left(\frac{\partial u_r}{\partial z} + \frac{\partial u_z}{\partial r}\right). \tag{11.14}$$

The boundary conditions for the problem will be given by

$$\sigma_{rr} = \sigma_{rz} = 0 \quad \text{at } r = a. \tag{11.15}$$

The harmonic waves propagate in a cylinder along the z-axis. Thus, we consider solutions of (11.8) and (11.9) to be of the general form

$$\Phi = G_1(r)e^{i(kz-\omega t)}, \tag{11.16}$$

$$\Psi = G_2(r)e^{i(kz-\omega t)}. \tag{11.17}$$

When (11.16) and (11.17) are substituted into the wave equations (11.8) and (11.9), respectively, we obtain ordinary differential equations for $G_j(r)$ ($j = 1, 2$):

$$\frac{d^2G_j}{dr^2} + \frac{1}{r}\frac{dG_j}{dr} + \left(\frac{\omega^2}{c_j^2} - k^2\right)G_j = 0 \quad (j = 1, 2). \tag{11.18}$$

Assume that

$$\alpha^2 = \frac{\omega^2}{c_L^2} - k^2 \quad \text{and} \tag{11.19}$$

$$\beta^2 = \frac{\omega^2}{c_T^2} - k^2. \tag{11.20}$$

Equation (11.18) is Bessel's equation, whose solutions are

$$G_1(r) = AJ_0(\alpha r), \tag{11.21}$$

$$G_2(r) = BJ_0(\beta r). \tag{11.22}$$

The second solution – $Y_0(\alpha r)$ and $Y_0(\beta r)$, where Y_0 is the Bessel function of the second kind – has been discarded because of its singular behavior at the origin. Substituting (11.21) and (11.22) into (11.16) and (11.17), we discover that

$$\Phi = AJ_0(\alpha r)e^{i(kz-\omega t)}, \tag{11.23}$$

$$\Psi = BJ_0(\beta r)e^{i(kz-\omega t)}. \tag{11.24}$$

Substituting (11.23) and (11.24) into (11.11) and (11.12) then yields

$$u_r = [AJ_0'(\alpha r) + BikJ_0'(\beta r)]e^{i(kz-\omega t)}, \tag{11.25}$$

$$u_z = [AikJ_0(\alpha r) + \beta^2 BJ_0(\beta r)]e^{i(kz-\omega t)}, \tag{11.26}$$

where $J_0'(\alpha r) = (d/dr)[J_0(\alpha r)]$; hence,

$$J_0'(x) = -J_1(x). \tag{11.27}$$

From (11.27), (11.25), and (11.26), it follows that

$$u_r = [-\alpha AJ_1(\alpha r) - ik\beta BJ_1(\beta r)]e^{i(kz-\omega t)}, \tag{11.28}$$

$$u_z = [ikAJ_0(\alpha r) + \beta^2 BJ_0(\beta r)]e^{i(kz-\omega t)}. \tag{11.29}$$

Let $C = \beta B$. Then (11.21) and (11.22) lead to

$$u_r = [-aAJ_1(\alpha r) + ikCJ_1(\beta r)]e^{i(kz-\omega t)}, \tag{11.30}$$

$$u_z = [ikAJ_0(\alpha r) + \beta CJ_0(\beta r)]e^{i(kz-\omega t)}. \tag{11.31}$$

At the cylindrical surface ($r = a$), the stresses must be zero. Substituting (11.30) and (11.31) into (11.13), and setting the resulting expression for σ_{rr} equal to zero at $r = a$, we find that

$$\left[-\frac{1}{2}(\beta^2 - k^2)J_0(\alpha a) + \frac{\alpha}{a}J_1(\alpha a)\right]A + \left[-ik\beta J_0(\beta a) + \frac{ik}{a}J_1(\beta a)\right]C = 0. \tag{11.32}$$

From the condition $\sigma_{rz} = 0$ at $r = a$,

$$[-2ik\alpha J_1(\alpha a)]A - (\beta^2 - k^2)J_1(\beta a)C = 0. \tag{11.33}$$

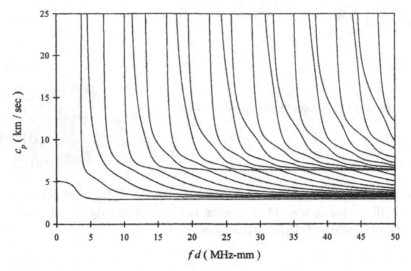

Figure 11-4. Phase velocity dispersion curves for an aluminum rod
($c_L = 6.3$ km/s, $c_T = 3.1$ km/s).

The requirement that the determinant of the coefficients must vanish presents us with the frequency equation as follows:

$$\frac{2\alpha}{a}(\beta^2 + k^2)J_1(\alpha a)J_1(\beta a) - (\beta^2 - k^2)^2 J_0(\alpha a)J_1(\beta a)$$

$$- 4k^2\alpha\beta J_1(\alpha a)J_0(\beta a) = 0. \tag{11.34}$$

This expression is known as the *Pochhammer frequency equation* for the longitudinal modes. It was first published in 1876 but, owing to its complexity, detailed calculations of the roots did not appear until the 1940s. Figure 11-4 depicts the phase velocities of longitudinal modes in terms of fd, where f is frequency and $d = 2a$ is the diameter of the rod. (The axially symmetric longitudinal modes for the rod of diameter $d = 2a$ and symmetric modes for the plate of thickness d are quite similar.)

As $fd \to \infty$, the phase velocity of the lowest mode approaches the velocity of Rayleigh waves, while velocities of the higher modes approach c_T. Over a short range of frequencies near the cutoff frequency, the behavior of the second dispersion curve is unusual; see Figure 11-5. This branch of the dispersion curve includes a range of frequencies where the group velocity and phase velocity have opposite signs. Such wave motions carry energy in one direction but appear to propagate in the other direction: the wave troughs and crests appear to move against the energy flux. This phenomenon of backward wave transmission in rods and plates was investigated by Meitzler (1965). Figure 11-6 shows graphs for these group velocity dispersion curves.

For each k_n root of (11.33), from equation (11.32) we find

$$C_n = -\frac{2ik_n\alpha_n J_n(\alpha_n a)}{(\beta_n^2 - k_n^2)J_1(\beta_n a)}A_n \quad (n = 1, 2, \ldots). \tag{11.35}$$

Substituting (11.35) into (11.30) and (11.31) yields the following displacements:

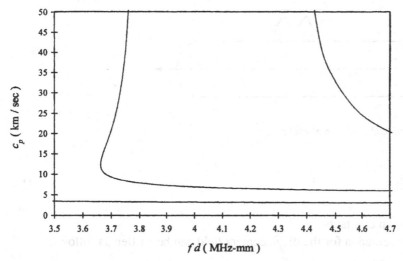

Figure 11-5. Magnified portion of the phase velocity dispersion curves for an aluminum rod ($c_L = 6.3$ km/s, $c_T = 3.1$ km/s).

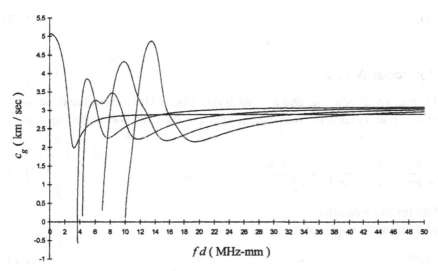

Figure 11-6. Group velocity dispersion curves for an aluminum rod ($c_L = 6.3$ km/s, $c_T = 3.1$ km/s).

$$u_r^n = \alpha_n[(\beta_n^2 - k_n^2)J_1(\alpha_n r)J_1(\beta_n a) + 2k_n^2 J_1(\alpha_n a)J_1(\beta_n r)]e^{i(k_n z - \omega t)}, \tag{11.36}$$

$$u_z^n = ik_n[(\beta_n^2 - k_n^2)J_0(\alpha_n r)J_1(\beta_n a) - 2\alpha_n \beta_n J_1(\alpha_n a)J_0(\beta_n r)]e^{i(k_n z - \omega t)}; \tag{11.37}$$

$$D_n = -\frac{A_n}{(\beta_n^2 - k_n^2)J_1(\beta_n a)}; \tag{11.38}$$

$$\alpha_n^2 = \frac{\omega^2}{c_L^2} - k_n^2, \tag{11.39}$$

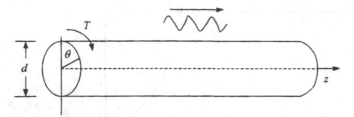

Figure 11-7. Torsional modes in a solid rod.

$$\beta_n^2 = \frac{\omega^2}{c_T^2} - k_n^2;$$ (11.40)

D_n denotes unknown constants.

A general representation for the displacement field can be written as follows:

$$u_r = \sum_{n=1}^{\infty} D_n u_r^n,$$ (11.41)

$$u_z = \sum_{n=1}^{\infty} D_n u_z^n.$$ (11.42)

11.3.2 Torsional Waves

Torsional waves result when u_r and u_z vanish (see Figure 11-7); from the equation of motion, it follows that u_θ must be independent of θ. For torsional waves, the equation of motion is

$$\frac{\partial^2 u_\theta}{\partial r^2} + \frac{1}{r}\frac{\partial u_\theta}{\partial r} - \frac{u_\theta}{r^2} + \frac{\partial u_\theta}{\partial z^2} = \frac{1}{c_T^2}\frac{\partial^2 u_\theta}{\partial t^2}$$ (11.43)

(see Graff 1991 for more details).

We consider harmonic waves of the form

$$u_\theta = V(r)e^{i(kz-\omega t)}.$$ (11.44)

Substituting (11.44) into (11.43) and solving the differential equation for the unknown function $V(r)$, we obtain

$$u_\theta = \frac{1}{\beta}BJ_1(\beta r)e^{i(kz-\omega t)},$$ (11.45)

where B is arbitrary.

Of the three boundary conditions,

$$\sigma_{rr} = \sigma_{rz} = \sigma_{r\theta} = 0 \quad \text{at } r = a,$$ (11.46)

only the condition

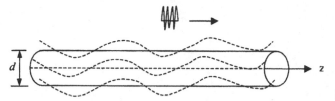

Figure 11-8. Flexural modes in a rod.

$$\sigma_{r\theta} = 0 \quad \text{at} \quad r = a \tag{11.47}$$

is nontrivial. This condition yields the dispersion frequency transcendental equation,

$$(\beta a) J_0(\beta a) - 2 J_1(\beta a) = 0, \tag{11.48}$$

whose first three roots are

$$\beta_1 = 0, \quad \beta_2 a = 5.136, \quad \beta_3 a = 8.417.$$

Taking the limit of (11.41) as $\beta \to 0$ yields:

$$u_\theta = \tfrac{1}{2} B r e^{i(kz - \omega t)}. \tag{11.49}$$

This displacement represents the lowest torsional mode. In the lowest mode, the amplitude of u_θ is proportional to the radius, and both u_r and u_z are zero. The motion corresponding to this solution is a rotation of each cross-section of the cylinder as a whole about its center. Note that, since $\beta = 0$ implies that the phase velocity equals c_T, the lowest torsional mode is not dispersive. The higher modes are dispersive and the resulting frequency spectrum has the same shape as that of SH waves in a plate.

11.3.3 Flexural Waves

Flexural waves depend on the circumferential angle θ through the trigonometric functions shown in equations (11.5)–(11.7). A schematic representation of flexural waves is shown in Figure 11-8. Of the flexural modes, the family defined by $n = 1$ is most important. Through the use of (11.5)–(11.7), we obtain

$$u_r = U(r) \cos\theta e^{i(kz - \omega t)}, \tag{11.50}$$

$$u_\theta = V(r) \sin\theta e^{i(kz - \omega t)}, \tag{11.51}$$

$$u_z = W(r) \cos\theta e^{i(kz - \omega t)}. \tag{11.52}$$

Substituting these equations into (11.2)–(11.4), we find the system of three differential equations containing $U(r)$, $V(r)$, $W(r)$. Without going into details of the solution, we present the final form:

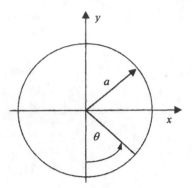

Figure 11-9. Cross-section of the rod.

$$U(r) - A\frac{\partial}{\partial r}J_1(\alpha r) + \frac{B}{r}J_1(\beta r) + ikCJ_2(\beta r), \tag{11.53}$$

$$V(r) = -\frac{A}{r}J_1(\alpha r) + ikCJ_2(\beta r) - B\frac{\partial}{\partial r}J_1(\beta r), \tag{11.54}$$

$$W(r) = ikAJ_1(\alpha r) - \frac{C}{r}\frac{\partial}{\partial r}[rJ_2(\beta r)] - \frac{C}{r}J_2(\beta r). \tag{11.55}$$

In order to illustrate the motions represented by displacement distributions (11.53)–(11.55), we choose the (y, z)-plane (the vertical plane) as the one from which θ is measured (see Figure 11-9). It now follows from (11.54) that for points in the vertical plane, the u_θ component vanishes, so that these points remain in the vertical plane. In the (x, z)-plane (the horizontal plane), where $\theta = \pm\pi/2$, the displacements u_r, u_z vanish. Points in the horizontal plane have purely vertical oscillations, which suggests the terminology "flexural" waves.

To determine the frequency equation, the displacements (11.50)–(11.52) must be substituted into the stress expressions, and $\sigma_{rr}, \sigma_{rz}, \sigma_{r\theta}$ must all be set equal to zero at $r = a$. This leads to a system of three homogeneous equations for A, B, and C. The requirement that the determinant of the coefficients must vanish yields the frequency equation (see Graff 1991). This frequency equation was examined in Pao and Mindlin (1960).

11.4 Exercises

1. Derive the governing equation of motion for an inhomogeneous rod where the modulus $E = E(1 + \varepsilon x^2)$.

2. How would you develop a characteristic or frequency equation for torsional, longitudinal, or flexible modes in a rod? Clearly outline all the necessary steps.

3. What phase velocity value is produced for a long wavelength limit in a plate wave problem? In a rod problem?

4. For longitudinal waves in a solid rod, what are the assumed generalized particle displacement functions?

5. What are the assumed generalized particle displacement functions for torsional waves in a rod?

6. What is the displacement distribution for the first torsional mode in a rod?

7. Explain how to obtain the bar velocity by using an exact solution for an infinite rod.

8. Derive an equation for the rod cutoff frequencies for (a) longitudinal and (b) torsional modes.

12

Waves in Hollow Cylinders

12.1 Introduction

Waves in hollow cylinders – such as piping and tubing – have long been a topic of considerable interest from the viewpoints of mechanics and ultrasonic inspection. Guided wave inspection using circumferential or longitudinal modes has received a great deal of attention. From a mechanics point of view, the problem can be tackled in a manner similar to that used for rods and plates. From the viewpoint of a governing wave equation and its assumed solutions and satisfied boundary conditions, dispersion curves and practical wave structure information can be generated. In this chapter we evaluate both circumferential and longitudinal waves. We also study a water-loaded hollow cylinder. Some basic experiments on guided waves in tubes are presented in Section E.6.

12.2 Circumferential Guided Waves in an Elastic Hollow Cylinder

We will now consider waves propagating in a circumferential direction within an elastic cylinder; see Figure 12-1. This problem of circumferential wave propagation was studied by Viktorov (1967) and Qu, Berthelot, and Li (1996).

We consider time-harmonic ($e^{-i\omega t}$) waves propagating in the θ direction in the (r, θ)-plane. The axial direction z is perpendicular to the cylinder cross-section, and the elastic field does not depend on the coordinate z. The equation of motion can be obtained from the governing Navier wave equations in cylindrical coordinates: simply eliminate all terms that depend on the z value. The analog of Lamb waves for a plate is found for this problem by considering a solution of the boundary value problem for an elastic cylinder. Hence, the unknown displacement field can be written as

$$u_r = u_r(r, \theta), \quad u_\theta = u_\theta(r, \theta), \quad u_z = 0. \tag{12.1}$$

The displacement components u_r, u_θ are given in terms of the potentials ϕ and ψ as

$$u_r = \frac{\partial \phi}{\partial r} + \frac{1}{r} \frac{\partial \psi}{\partial \theta},$$

$$u_\theta = \frac{1}{r} \frac{\partial \phi}{\partial r} - \frac{\partial \psi}{\partial r}. \tag{12.2}$$

Here, ϕ and ψ satisfy the wave equations

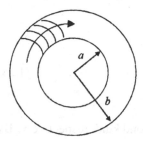

Figure 12-1. Cross-section of the cylinder
with inner radius a and outer radius b.

$$\left(\frac{\partial^2}{\partial r^2} + \frac{1}{r}\frac{\partial}{\partial r} + \frac{1}{r^2}\frac{\partial^2}{\partial \theta^2}\right)\phi + \frac{\omega^2}{c_L^2}\phi = 0,$$

$$\left(\frac{\partial^2}{\partial r^2} + \frac{1}{r}\frac{\partial}{\partial r} + \frac{1}{r^2}\frac{\partial^2}{\partial \theta^2}\right)\psi + \frac{\omega^2}{c_T^2}\psi = 0,$$

(12.3)

where c_L and c_T are (resp.) longitudinal and shear wave velocities for the elastic medium.
The stresses are given by Hooke's law as follows:

$$\sigma_{rr} = \lambda\left(\frac{\partial u_r}{\partial r} + \frac{u_r}{r} + \frac{1}{r}\frac{\partial u_\theta}{\partial \theta}\right) + 2\mu\frac{\partial u_r}{\partial r},$$

$$\sigma_{\theta\theta} = \lambda\left(\frac{\partial u_r}{\partial r} + \frac{u_r}{r} + \frac{1}{r}\frac{\partial u_\theta}{\partial \theta}\right) + 2\mu\left(\frac{u_r}{r} + \frac{1}{r}\frac{\partial u_\theta}{\partial \theta}\right),$$

$$\sigma_{r\theta} = \mu\left(\frac{\partial u_\theta}{\partial r} - \frac{u_\theta}{r} + \frac{1}{r}\frac{\partial u_r}{\partial \theta}\right).$$

(12.4)

On the inner and outer surfaces of the cylinder, we have the traction-free boundary
conditions

$$\sigma_{rr} = \sigma_{r\theta} = 0 \quad \text{at } r = a \text{ and } r = b.$$

(12.5)

We may consider circumferential waves as an analog of Lamb waves in a two-dimensional
layer. Therefore, waves propagating in the θ direction depend on the angular coordinate
according to $e^{ikb\theta}$. Here, k is called the "angular" wavenumber and is determined by satisfying boundary conditions.

Propagation characteristics may be established by considering harmonic waves of the
form

$$\phi = \Phi(r)e^{i(kb\theta - \omega t)} \quad \text{and} \quad \psi = \Psi(r)e^{i(kb\theta - \omega t)}.$$

(12.6)

The time term $e^{-i\omega t}$ is omitted in the following development.

Substitution of (12.6) into the governing equation (12.3) gives

$$\Phi'' + \frac{1}{r}\Phi' + \left[\left(\frac{\omega}{c_L}\right)^2 - \left(\frac{kb}{r}\right)^2\right]\Phi = 0,$$

$$\Psi'' + \frac{1}{r}\Psi' + \left[\left(\frac{\omega}{c_T}\right)^2 - \left(\frac{kb}{r}\right)^2\right]\Psi = 0.$$

(12.7)

The general solution of (12.7) has the form

$$\Phi(r) = A_1 J_{kb}\left(\frac{\omega r}{c_L}\right) + A_2 Y_{kb}\left(\frac{\omega r}{c_L}\right),$$

$$\Psi(r) = B_1 J_{kb}\left(\frac{\omega r}{c_T}\right) + B_2 Y_{kb}\left(\frac{\omega r}{c_T}\right),$$

(12.8)

where $J_M(z)$ and $Y_M(z)$ are Bessel functions of the first and second kind, respectively. By substituting (12.8) into the stress equations (12.4), we obtain

$$\sigma_{rr}(r,\theta) = \frac{\mu e^{ikb\theta}}{r^2}[\chi^2 r^2 \Phi'' + (\chi^2 - 2)r\Phi' \\ - (\chi^2 - 2)k^2 b^2 \Phi + 2ikb(r\Psi' - \Psi)],$$

$$\sigma_{r\theta}(r,\theta) = \frac{\mu e^{ikb\theta}}{r^2}[-r^2\Psi'' + r\Psi' - k^2 b^2 \Psi + 2ikb(r\Phi' - \Phi)],$$

(12.9)

where $\chi = c_L/c_T$.

The four unknown constants A_1, A_2, B_1, B_2 from (12.8) are determined by satisfying the four traction-free boundary conditions (12.5) on the inner and outer surfaces of the cylinder. The result is a system of four linear homogeneous equations with respect to A_1, A_2, B_1, B_2. In order to find a nontrivial solution, the determinant of the system of equations must vanish. This leads to the dispersion equation

$$\|D_{nm}\| = 0 \quad (n, m = 1, \ldots, 4),$$

(12.10)

where $h = b - a$, $M = kh/(1 - \gamma)$, $\varepsilon = \omega h/(c_T(1 - \gamma))$, and $\gamma = a/b$. Specifically:

$$D_{11} = \left[J_{M-2}\left(\frac{\varepsilon}{\chi}\right) + J_{M+2}\left(\frac{\varepsilon}{\chi}\right) - 2(\chi^2 - 1)J_M\left(\frac{\varepsilon}{\chi}\right)\right]\chi^{-2},$$

$$D_{12} = i[J_{M-2}(\varepsilon) - J_{M+2}(\varepsilon)],$$

$$D_{13} = \left[Y_{M-2}\left(\frac{\varepsilon}{\chi}\right) + Y_{M+2}\left(\frac{\varepsilon}{\chi}\right) - 2(\chi^2 - 1)Y_M\left(\frac{\varepsilon}{\chi}\right)\right]\chi^{-2},$$

$$D_{14} = i[Y_{M-2}(\varepsilon) - Y_{M+2}(\varepsilon)];$$

$$D_{21} = i\left[J_{M-2}\left(\frac{\varepsilon}{\chi}\right) - J_{M+2}\left(\frac{\varepsilon}{\chi}\right)\right]\chi^{-2},$$

$$D_{22} = -[J_{M-2}(\varepsilon) + J_{M+2}(\varepsilon)],$$

(12.11)

$$D_{23} = i\left[Y_{M-2}\left(\frac{\varepsilon}{\chi}\right) - Y_{M+2}\left(\frac{\varepsilon}{\chi}\right)\right]\chi^{-2},$$

$$D_{24} = -[Y_{M-2}(\varepsilon) + Y_{M+2}(\varepsilon)];$$

$$D_{31} = \left[J_{M-2}\left(\frac{\gamma\varepsilon}{\chi}\right) - J_{M+2}\left(\frac{\gamma\varepsilon}{\chi}\right) - 2(\chi^2 - 1)J_M\left(\frac{\gamma\varepsilon}{\chi}\right)\right]\gamma^2\chi^{-2},$$

$$D_{32} = i[J_{M-2}(\gamma\varepsilon) - J_{M+2}(\gamma\varepsilon)]\gamma^2,$$

$$D_{33} = \left[Y_{M-2}\left(\frac{\gamma\varepsilon}{\chi}\right) + Y_{M+2}\left(\frac{\gamma\varepsilon}{\chi}\right) - 2(\chi^2 - 1)Y_M\left(\frac{\gamma\varepsilon}{\chi}\right)\right]\gamma^2\chi^{-2},$$

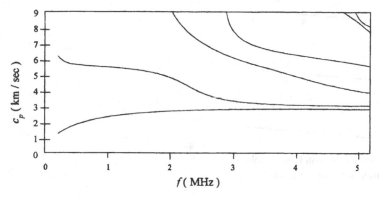

Figure 12-2. Circumferential phase velocity dispersion curves for an aluminum cylinder.

$$D_{34} = i[Y_{M-2}(\gamma\varepsilon) - Y_{M+2}(\gamma\varepsilon)]\gamma^2;$$

$$D_{41} = i\left[J_{M-2}\left(\frac{\gamma\varepsilon}{\chi}\right) - J_{M+2}\left(\frac{\gamma\varepsilon}{\chi}\right)\right]\gamma^2\chi^{-2},$$

$$D_{42} = -[J_{M-2}(\gamma\varepsilon) + J_{M+2}(\gamma\varepsilon)]\gamma^2,$$

$$D_{43} = i\left[Y_{M-2}\left(\frac{\gamma\varepsilon}{\chi}\right) - Y_{M+2}\left(\frac{\gamma\varepsilon}{\chi}\right)\right]\gamma^2\chi^{-2},$$

$$D_{44} = -[Y_{M-2}(\gamma\varepsilon) + Y_{M+2}(\gamma\varepsilon)]\gamma^2.$$

(12.11)
(*cont.*)

The variable M in the dispersion equation (12.10) is the order of the Bessel functions. The harmonic solution for wave propagation in the θ direction depends on the time t and angle θ according to $\exp[i(kb\theta - \omega t)]$. Therefore, the angular phase velocity can be calculated as

$$\alpha = \frac{\omega}{kb}. \qquad (12.12)$$

Equation (12.12) allows us to define the linear phase velocity of the circumferential waves for a given radius value r:

$$c_p = r\alpha = (\omega/k) * (r/b). \qquad (12.13)$$

Figure 12-2 shows circumferential phase velocity dispersion curves for an aluminum cylinder ($c_L = 6.3$ km/s, $c_T = 3.1$ km/s; outer radius $a = 20$ mm, inner radius $b = 19$ mm). We can observe two significant differences for the hollow cylinder compared with the behavior of the dispersion curves for a single two-dimensional single plate. (1) For lower frequencies, the dispersion curves are different. (2) For the two-dimensional plate, A0 and S0 converge to the surface wave velocity value for large fd values, whereas these lower-order modes differ significantly in the case of circumferential waves.

12.3 Longitudinal Guided Waves in an Elastic Hollow Cylinder

Now consider guided wave propagation in an infinitely long hollow cylinder, as shown in Figure 12-3. The traction-free boundary conditions are

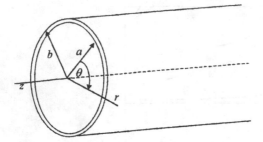

Figure 12-3. A traction-free, infinitely long hollow
cylinder with inner radius a and outer radius b.

$$\sigma_{rr} = \sigma_{rz} = \sigma_{r\theta} = 0 \quad \text{at } r = a \text{ and } r = b. \tag{12.14}$$

(Exact solutions for this boundary value problem were first published by Gazis 1959b; see
also Gazis 1959a.) The assumed particle displacement components would therefore be

$$u_r = U_r(r) \cos n\theta \cos(\omega t + kz),$$

$$u_\theta = U_\theta(r) \sin n\theta \cos(\omega t + kz), \tag{12.15}$$

$$u_z = U_z(r) \cos n\theta \sin(\omega t + kz),$$

where the circumferential order is $n = 0, 1, 2, 3, \dots$. The terms u_r, u_θ, u_z are the displacement components in the radial, circumferential, and axial directions, respectively;
U_r, U_θ, U_z are the corresponding displacement amplitudes composed of Bessel functions
(or modified Bessel functions, depending on the arguments).

When considering stress wave propagation in a hollow cylinder, three different mode
types should be studied separately: longitudinal, torsional, and flexural modes. Consider
all modes propagating in the z-axis direction (Figure 12-3). The longitudinal modes and
the torsional modes are axisymmetric modes, but the flexural modes are not axisymmetric. In the expressions for the displacement components, the displacements for $n = 0$
correspond to the axisymmetric modes and the displacements for $n = 1, 2, 3, \dots$ correspond to flexural modes, which contain sinusoidal functions with the argument $n\theta$. It is
convenient to follow the notation of Meitzler (1961), Zemanek (1972), and Silk and Bainton (1979) as follows:

longitudinal modes: $L(0, m)$ (axisymmetric modes),

torsional modes: $T(0, m)$ (axisymmetric modes),

flexural modes: $F(n, m)$ (non-axisymmetric modes).

Here the circumferential order $n = 1, 2, 3, \dots$ and the mode $m = 1, 2, 3, \dots$.

There are an infinite number of torsional modes and an infinite number of longitudinal modes for $n = 0$. For $n = 1, 2, 3, \dots$ there are an infinite number of modes for each
circumferential order n, so that – as noted by Ditri (1994c) – there are a doubly infinite
number of flexural modes. Usually, a number m of modes can be plotted in dispersion
curves for a given nth order. In this chapter we present dispersion curves for the longitudinal and flexural modes up to sixth order; a torsional mode sample problem in steel is
also presented.

12.4 Longitudinal Axisymmetric Modes

The dispersion, characteristic, or frequency equation for axisymmetric modes, $n = 0$, in hollow cylinders is

$$
\begin{vmatrix}
c_{11} & c_{12} & c_{13} & c_{14} & c_{15} & c_{16} \\
c_{21} & c_{22} & c_{23} & c_{24} & c_{25} & c_{26} \\
c_{31} & c_{32} & c_{33} & c_{34} & c_{35} & c_{36} \\
c_{41} & c_{42} & c_{43} & c_{44} & c_{45} & c_{46} \\
c_{51} & c_{52} & c_{53} & c_{54} & c_{55} & c_{56} \\
c_{61} & c_{62} & c_{63} & c_{64} & c_{65} & c_{66}
\end{vmatrix} = 0.
\tag{12.16}
$$

The first three rows of matrix elements are as follows:

$$
\begin{aligned}
c_{11} &= [2n(n-1) - (\beta^2 - \xi^2)a^2]Z_n(\alpha_1 a) + 2\lambda_1\alpha_1 a Z_{n+1}(\alpha_1 a), \\
c_{12} &= 2\xi\beta_1 a^2 Z_n(\beta_1 a) - 2\xi a(n+1)Z_{n+1}(\beta_1 a), \\
c_{13} &= -2n(n-1)Z_n(\beta_1 a) + 2\lambda_2 n\beta_1 a Z_{n+1}(\beta_1 a), \\
c_{14} &= [2n(n-1) - (\beta^2 - \xi^2)a^2]W_n(\alpha_1 a) + 2\alpha_1 a W_{n+1}(\alpha_1 a), \\
c_{15} &= 2\lambda_2\xi\beta_1 a^2 W_n(\beta_1 a) - 2(n+1)\xi a W_{n+1}(\beta_1 a), \\
c_{16} &= -2n(n-1)W_n(\beta_1 a) + 2n\beta_1 a W_{n+1}(\beta_1 a); \\
c_{21} &= 2n(n-1)Z_n(\alpha_1 a) - 2\lambda_1 n\alpha_1 a Z_{n+1}(\alpha_1 a), \\
c_{22} &= -\xi\beta_1 a^2 Z_n(\beta_1 a) + 2\xi a(n+1)Z_{n+1}(\beta_1 a), \\
c_{23} &= -[2n(n-1) - \beta^2 a^2]Z_n(\beta_1 a) - 2\lambda_2\beta_1 a Z_{n+1}(\beta_1 a), \\
c_{24} &= 2n(n-1)W_n(\alpha_1 a) - 2n\alpha_1 a W_{n+1}(\alpha_1 a), \\
c_{25} &= -\lambda_2\xi\beta_1 a^2 W_n(\beta_1 a) + 2\xi a(n+1)W_{n+1}(\beta_1 a), \\
c_{26} &= -[2n(n-1) - \beta^2 a^2]W_n(\beta_1 a) - 2\beta_1 a W_{n+1}(\beta_1 a); \\
c_{31} &= 2n\xi a Z_n(\alpha_1 a) + 2\lambda_1\xi\alpha_1 a^2 Z_{n+1}(\alpha_1 a), \\
c_{32} &= -n\beta_1 a Z_n(\beta_1 a) + (\beta^2 - \xi^2)a^2 Z_{n+1}(\beta_1 a), \\
c_{33} &= n\xi a Z_n(\beta_1 a), \\
c_{34} &= -2n\xi a W_n(\alpha_1 a) + 2\xi\alpha_1 a^2 W_{n+1}(\alpha_1 a), \\
c_{35} &= -\lambda_2 n\beta_1 a W_n(\beta_1 a) + (\beta^2 - \xi^2)a^2 W_{n+1}(\beta_1 a), \\
c_{36} &= n\xi a W_n(\beta_1 a).
\end{aligned}
\tag{12.17}
$$

The terms Z_n and W_n represent the Bessel functions (or the modified Bessel functions, depending on the arguments). The proper selections of the Bessel functions are indicated in Table 12-1. The values of λ_1 and λ_2 are 1 when the Bessel functions J and Y are used and -1 when the modified Bessel functions I and K are used. Integer n represents the

Table 12-1. *Bessel functions used in equation (12.17)*

Interval	Functions
$c_L < c_p$ or $\alpha^2, \beta^2 > 0$	$J_n(\alpha r)$ and $Y_n(\alpha r)$, $\quad J_n(\beta r)$ and $Y_n(\beta r)$
$c_L > c_p > c_T$ or $\alpha^2 < 0$, $\beta^2 > 0$	$I_n(\alpha_1 r)$ and $K_n(\alpha_1 r)$, $\quad J_n(\beta r)$ and $Y_n(\beta r)$
$c_p < c_T$ or $\alpha^2 < 0$, $\beta^2 < 0$	$I_n(\alpha_1 r)$ and $K_n(\alpha_1 r)$, $\quad I_n(\beta_1 r)$ and $K_n(\beta_1 r)$

circumferential order of the guided waves in hollow cylinders. The remaining matrix elements, c_{41} to c_{66}, are the same as elements c_{11} to c_{36} except that b replaces a in equations (12.17). Note that $\alpha^2 = \omega^2/c_L^2 - k^2$ and $\beta^2 = \omega^2/c_T^2 - k^2$; also,

$$\alpha_1 r = |\alpha r| \quad \text{and} \quad \beta_1 r = |\beta r|. \tag{12.18}$$

If $n = 0$ then the modes are axially symmetric, and the frequency equation (Gazis 1959b) can be decomposed as the product of subdeterminants:

$$D = D_1 \bullet D_2 = 0, \tag{12.19}$$

where

$$D_1 = \begin{vmatrix} c_{11} & c_{12} & c_{14} & c_{15} \\ c_{31} & c_{32} & c_{34} & c_{35} \\ c_{41} & c_{42} & c_{44} & c_{45} \\ c_{61} & c_{62} & c_{64} & c_{65} \end{vmatrix} \quad \text{and} \quad D_2 = \begin{vmatrix} c_{23} & c_{26} \\ c_{53} & c_{56} \end{vmatrix}. \tag{12.20}$$

The solutions of $D_1 = 0$ and $D_2 = 0$ correspond to the longitudinal and torsional modes, respectively. The polarization vector of the particle displacement in the longitudinal modes is in the (r, z)-plane (Figure 12-3), so that there is no circumferential component of particle displacement, $u_\theta = 0$. Torsional modes are assumed to contain only the u_θ component of the particle displacement. In practical applications, longitudinal modes are often preferred to torsional modes owing to such experimental aspects as excitability and repeatability of the modes. Usually, longitudinal modes are also preferred over flexural modes for excitation, because of the symmetry that allows us to inspect 360° along the circumference of hollow cylinders. Reflections from defects, on the other hand, are generally non-axisymmetric in nature owing to the finite size of the defect along the circumference of the hollow cylinder. See Shin and Rose (1998a).

Theoretical results of the longitudinal modes can be represented by phase velocity and group velocity dispersion curves. Sample calculations for an inconel (nickel-based alloy) tube are shown in Figure 12-4; a sample result for torsional modes in a stainless steel tube is shown in Figure 12-5. Sample wave structure results for a different inconel tube for modes $L(0, 1)$ to $L(0, 4)$ are illustrated in Figures 12-6 to 12-9, respectively. In these figures, the sample hollow cylinder is an inconel tube with 8.23-mm inner radius and 9.45-mm outer radius. The longitudinal and shear wave velocities of the tube are 6.29 km/s and 3.23 km/s, respectively; u_r and u_z indicate the radial and axial components (resp.) of the sample tube's particle displacement.

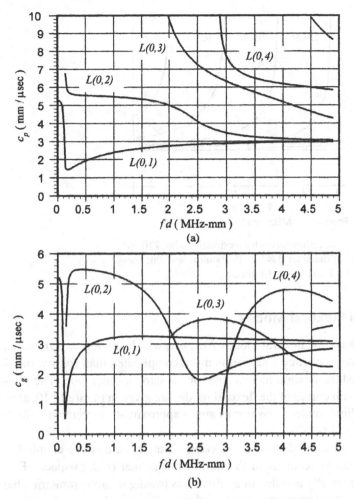

Figure 12-4. (a) Phase velocity and (b) group velocity dispersion curves for longitudinal modes, $L(0, m)$, of an inconel tube (inner diameter 16.46 mm, thickness 1.22mm, $c_L = 6.29$ km/s, $c_T = 3.23$ km/s).

Many interesting observations can be made from the figures. Notice, for example, the dominant out-of-plane displacement component on the inner and outer surface for $L(0, 1)$ over the frequency range 0.6–5.0 MHz. For $L(0, 2)$, on the other hand, the in-plane displacement component is dominant on the inner and outer surface for low frequency (0.5 MHz) but gradually reverses to a dominant out-of-plane displacement component at 1.8 MHz. For $L(0, 3)$, almost zero excitation at the center of the hollow cylinder at 2.3 MHz increases to a dominant out-of-plane distribution on the center line at 3.5 MHz. Mode $L(0, 4)$ presents a dominant in-plane displacement on the inner and outer surfaces (with both displacements zero at the tube center line) onto a dominant in-plane on the inside surface, dominant in-plane at the center line, and dominant out-of-plane on the outside surface at 5.0 MHz. These studies can be used to optimize penetration power in water-loaded tubes and for defect detection across the thickness of the tube.

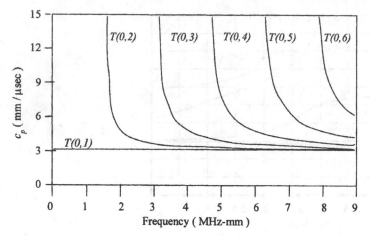

Figure 12-5. Phase velocity dispersion curves for torsional modes, $T(0, m)$, of a stainless steel tube (outer diameter $3/4" = 19.05$ mm, wall thickness $0.065" = 1.651$ mm; $c_L = 5.8$ km/s, $c_T = 3.1$ km/s).

12.5   Longitudinal Flexural Modes

If $n = 1, 2, 3, \ldots$ we have the flexural modes $F(n, m)$, which are non-axisymmetric; in this case, solution of the frequency equation is more complicated than for symmetric modes. We have calculated the flexural modes also for the same inconel tube. The phase and group velocity dispersion curves of the flexural modes are given in Figure 12-10, along with those of the longitudinal modes. The curves are in approximate agreement with the results of Cooper and Naghdi (1957).

In the flexural modes, all three displacement components exist and are coupled together; hence, the polarization vector is in the three-dimensional (r, θ, z)-space. Even though flexural modes are usually avoided in applications (owing to nonsymmetric characteristics and complexity of the wave structure), understanding the physical characteristics of flexural modes is necessary for advanced applications involving wave reflections from defects. For instance, it can be difficult experimentally to generate pure longitudinal modes over a given frequency range; hence, there are usually at least two modes in a waveguide. As shown in the dispersion curves, many flexural modes exist close to the longitudinal modes. Reflected echoes from defects are generally non-axisymmetric, so understanding such wave propagation can assist in defect characterization and sizing studies.

12.6   Leaky Guided Waves from a Water-Loaded Hollow Cylinder

Water loading of components is an important consideration in NDE applications. Water and other types of loads on a component provide leakage paths for ultrasonic energy, so energy that would normally be reflected or refracted to a receiving transducer can be lost. Water-loading effects are evaluated here to discover whether or not any particular guided wave propagation modes are insensitive to such loading. We find that there do exist certain modes that may be used for component inspection with or without minimum leakage.

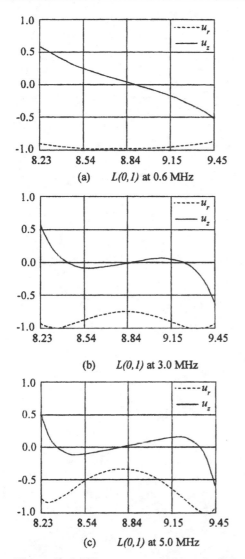

Figure 12-6. Wave structure problem for $L(0, 1)$.

The mathematical treatment essentially unfolds along the same lines as the development of dispersion equations for the unloaded plate or tube, but there are two additional considerations. (1) We must review a potential function Φ_w for longitudinal wave propagation in the water, along with an additional displacement field equation; and (2) attenuation due to leakage of energy from certain modes into the water must be accounted for. We achieve the latter by using a complex propagation number k such that $k = k_{Re} + i k_{Im}$; the imaginary part of k is equivalent to an attenuation factor. Thus we have

$$u_z = A e^{i(kx - \omega t)} = A e^{i[(k_r + i\alpha)x - \omega t]}$$
$$= A e^{i(k_r x - \omega t)} e^{-\alpha x}.$$

Complex k occurs in the water-loaded case; attenuation in the displacement function results from the leaky modes that occur as energy leaks from the hollow cylinder to the fluid.

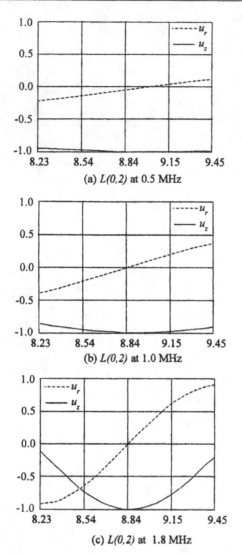

(a) $L(0,2)$ at 0.5 MHz

(b) $L(0,2)$ at 1.0 MHz

(c) $L(0,2)$ at 1.8 MHz

Figure 12-7. Wave structure problem for $L(0, 2)$.

The coordinate reference system for the problem is shown in Figure 12-11, along with the leakage concept. The interior of the cylinder and the bottom of the plate are each considered to be a medium of infinite acoustic impedance (i.e., a vacuum). Both the plate and the cylinder are assumed to have thickness d.

The expressions for the water scalar potential and water displacement for a plate are given by

$$\Phi_w = A_5 e^{-ik_w y} e^{i(kx-\omega t)},$$

$$k_w = \sqrt{(\omega/c_w)^2 - k^2},$$ (12.21)

$$\bar{U} = \bar{\nabla}\Phi_w,$$

where c_w is the longitudinal wave velocity in water. Using these equations yields the displacements

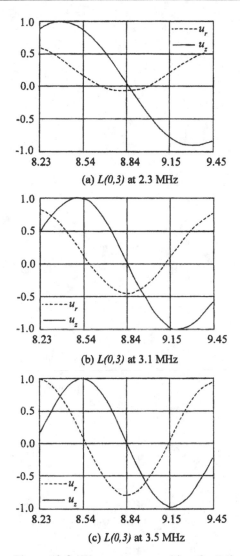

(a) $L(0,3)$ at 2.3 MHz

(b) $L(0,3)$ at 3.1 MHz

(c) $L(0,3)$ at 3.5 MHz

Figure 12-8. Wave structure problem for $L(0, 3)$.

$$u_{x(\text{water})} = \frac{\partial \Phi_w}{\partial x} = [ike^{-ik_w y}A_5]e^{i(kx-\omega t)},$$

$$u_{y(\text{water})} = \frac{\partial \Phi_w}{\partial y} = [-ik_w e^{-ik_w y}A_5]e^{i(kx-\omega t)}. \tag{12.22}$$

The normal stress component on the water side is

$$\sigma_{yy} = \lambda_w \Delta_w = [\lambda_w(-k_w^2 - k^2)e^{-ik_w y}A_5]e^{i(kx-\omega t)}, \tag{12.23}$$

and the boundary conditions are

$$\sigma_{xy} = 0, \qquad \sigma_{yy} = \sigma_{yy(\text{water})};$$

$$u_y = u_{y(\text{water})} \quad \text{at} \ y = d/2, \tag{12.24}$$

$$\sigma_{xy} = \sigma_{yy} = 0 \quad \text{at} \ y = -d/2.$$

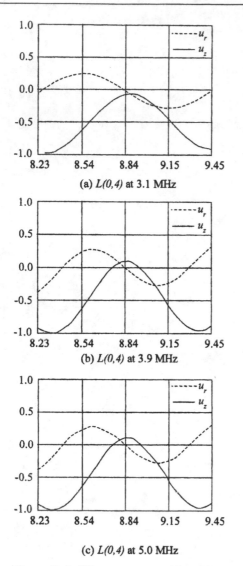

(a) $L(0,4)$ at 3.1 MHz

(b) $L(0,4)$ at 3.9 MHz

(c) $L(0,4)$ at 5.0 MHz

Figure 12-9. Wave structure problem for $L(0, 4)$.

Using the equations for displacement and stress with the boundary conditions (12.24), five equations can be written as

$$
\begin{bmatrix}
a_{11} & a_{12} & a_{13} & a_{14} & 0 \\
a_{21} & a_{22} & a_{23} & a_{24} & a_{25} \\
a_{31} & a_{32} & a_{33} & a_{34} & a_{35} \\
a_{41} & a_{42} & a_{43} & a_{44} & 0 \\
a_{51} & a_{52} & a_{53} & a_{54} & 0
\end{bmatrix}
\begin{bmatrix}
A_1 \\
A_2 \\
A_3 \\
A_4 \\
A_5
\end{bmatrix}
=
\begin{bmatrix}
0 \\
0 \\
0 \\
0 \\
0
\end{bmatrix}.
\tag{12.25}
$$

Solving the equations that result from setting the determinant equal to zero will yield the complex propagation number k. The phase velocity and attenuation curves can then be generated using the real and imaginary parts of k:

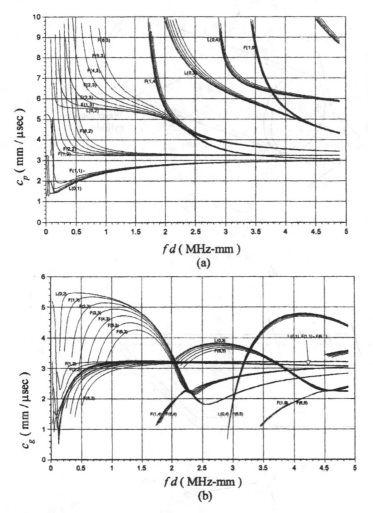

Figure 12-10. (a) Phase velocity and (b) group velocity dispersion curves for longitudinal and flexural modes (up to $n = 6$) of the inconel tube.

$$c_p = \omega/k_{Re}, \qquad \alpha = k_{Im}$$

(for decibels, use $20 \log \alpha$).

The cylinder problem is addressed in a similar manner, although all equations are cast in a cylindrical coordinate system. Because of this, special functions – called *Hankel functions* – arise in the solutions of the water-loaded cylinder problem. Special care must be exercised when dealing with these functions owing to their asymptotic behavior. Physically, asymptotic behavior is characteristic of waves that radiate away from the cylinder at great distances. One would thus expect that, at large distances, the radiated waves would disappear and that no re-radiation of these waves back toward the cylinder would occur. Intuitively, the waves leaving the cylinder are diverging, which implies that their energy density is steadily decreasing. To satisfy this assumption, solutions for the radiation portion of the problem must behave according to the so-called Sommerfeld radiation condition (see Sommerfeld 1964).

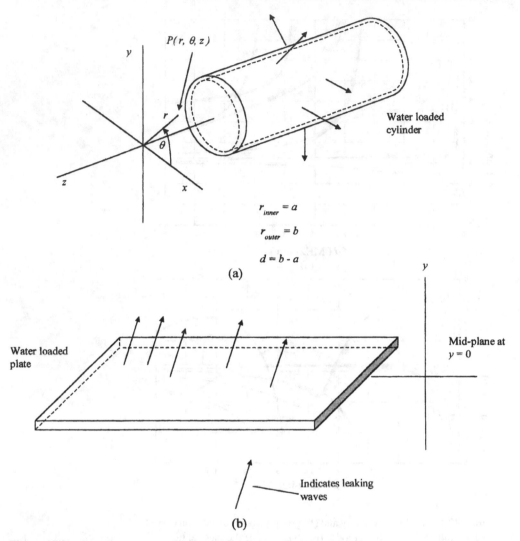

Figure 12-11. Coordinate system for water-loaded (a) cylinder and (b) plate.

A zero-order Hankel function of the second kind satisfies this condition. It has the asymptotic behavior

$$H_0^{(2)} \sim \frac{ie^{-ik_w r}}{\sqrt{k_w r}} \equiv f(k_w r). \tag{12.26}$$

The Sommerfeld radiation condition (for time-harmonic functions) is

$$\lim_{r \to \infty} \left[r \frac{\partial f}{\partial r} + ik_w r f \right] = 0. \tag{12.27}$$

Using f for $H_0^{(2)}$ as defined in (12.26), we have

$$r\frac{\partial f}{\partial r} = r\left[\frac{e^{-ik_wr}}{r} - \frac{ie^{-ik_wr}}{k_wr^2}\right] = e^{-ik_wr} - \frac{ie^{-ik_wr}}{k_wr},$$

$$ik_wrf = ik_wr \cdot \frac{ie^{-ik_wr}}{k_wr} = -e^{-ik_wr}, \tag{12.28}$$

$$r\frac{\partial f}{\partial r} + ik_wrf = -\frac{ie^{-ik_wr}}{k_wr},$$

since

$$\lim_{r\to\infty}\left[\frac{-ie^{-ik_wr}}{k_wr}\right] = 0. \tag{12.29}$$

Hence the Sommerfeld condition for diverging time-harmonic waves is satisfied. The actual calculation of the Hankel functions is performed through the calculation of Bessel and Neuman functions and adding them together:

$$H_0^{(2)} = J_0 - iN_0. \tag{12.30}$$

Again, phase velocity and attenuation are obtained as

$$c_p = \omega/k_{\text{Re}}, \qquad \alpha = k_{\text{Im}}$$

(for decibels, use $20\log\alpha$).

Considering cylindrical coordinates for the hollow cylinder, we can proceed as follows:

$$\phi = \Phi(r)e^{i(\omega t - kz)} = [B_1 J_0(k_l r) + B_2 Y_0(k_l r)]e^{i(\omega t - kz)}, \tag{12.31}$$

$$\bar{\psi} = \Psi(r)e^{i(\omega t - kz)} = [B_3 J_1(k_t r) + B_4 Y_1(k_t r)]e^{i(\omega t - kz)}\bar{e}_\theta, \tag{12.32}$$

where

$$k_l^2 = \left(\frac{\omega}{c_L}\right)^2 - k^2, \qquad k_t^2 = \left(\frac{\omega}{c_T}\right)^2 - k^2.$$

In order to satisfy the radiation condition of no incoming wave at infinity, we now introduce the appropriate outgoing second kind of Hankel function for a scalar potential of water:

$$\Phi_w = [B_5 H_0^{(2)}(k_w r)]e^{i(\omega t - kz)}, \tag{12.33}$$

where

$$k_w^2 = \left(\frac{\omega}{c_w}\right)^2 - k^2, \qquad H_0^{(2)}(k_w r) = J_0(k_w r) - iY_0(k_w r).$$

Substituting (12.31) and (12.32) into the simple wave equation for Φ and $\bar{\Psi}$ with the differential operator for cylindrical coordinates, we find:

$$u_r = \frac{\partial\Phi}{\partial r} - \frac{\partial\Psi}{\partial z}, \tag{12.34}$$

$$u_z = \frac{\partial\Phi}{\partial z} + \left(\frac{\partial\Psi}{\partial r} + \frac{\Psi}{r}\right). \tag{12.35}$$

Dilatation is

$$\Delta = \bar{\nabla} \bullet \bar{U}$$
$$= -(k_l^2 + k^2)[B_1 J_0(k_l r) + B_2 Y_0(k_l r)]e^{i(\omega t - kz)} \tag{12.36}$$

and the strain displacement relation is

$$\varepsilon_{ij} = \tfrac{1}{2}(u_{ij} + u_{ji}). \tag{12.37}$$

Hooke's law is

$$\sigma_{ij} = \lambda \Delta \delta_{ij} + 2G\varepsilon_{ij} \quad (i, j = r, \theta, z), \tag{12.38}$$

where σ_{ij} is the stress tensor, δ_{ij} the Kronecker delta, and ε_{ij} the strain tensor.

For water, substituting (12.33) into $\bar{U}_{\text{water}} = \bar{\nabla}\Phi_{\text{water}}$ yields the normal displacement

$$u_{r(\text{water})} = -k_w H_1^{(2)}(k_w r) B_5 e^{i(\omega t - kz)}. \tag{12.39}$$

The pressure of water at an interface between a waveguide and water is

$$\sigma_{rr(\text{water})} = \lambda_w \Delta$$
$$= \lambda_w [-(k_w^2 + k^2) H_0^{(2)}(k_w r)] B_5 e^{i(\omega t - kz)}. \tag{12.40}$$

Applying traction-free continuity and stress and displacement continuity to the inner and outer surfaces of a waveguide, respectively, leads to five boundary conditions and a determinant of order 5. For nontrivial solutions of the unknown amplitudes, the characteristic equation is

$$\det \begin{bmatrix} B_{11} & B_{12} & B_{13} & B_{14} & 0 \\ B_{21} & B_{22} & B_{23} & B_{24} & 0 \\ B_{31} & B_{32} & B_{33} & B_{34} & B_{35} \\ B_{41} & B_{42} & B_{43} & B_{44} & 0 \\ B_{51} & B_{52} & B_{53} & B_{54} & B_{55} \end{bmatrix} = 0. \tag{12.41}$$

This is the dispersion equation, where

$$B_{11} = -\{\lambda(k_l^2 + k^2) + 2Gk_l^2\}J_0(k_l a) + 2Gk_l\frac{J_1(k_l a)}{a},$$

$$B_{12} = -\{\lambda(k_l^2 + k^2) + 2Gk_l^2\}Y_0(k_l a) + 2Gk_l\frac{Y_1(k_l a)}{a},$$

$$B_{13} = -\{2iGk\}\left\{k_t J_0(k_t a) - \frac{1}{a}J_1(k_t a)\right\}, \tag{12.42}$$

$$B_{14} = -\{2iGk\}\left\{k_t Y_0(k_t a) - \frac{1}{a}Y_1(k_t a)\right\};$$

$$B_{21} = \{2iGkk_l\}\{J_1(k_l a)\}, \qquad B_{22} = \{2iGkk_l\}\{Y_1(k_l a)\},$$

$$B_{23} = G\{k^2 - k_t^2\}\{J_1(k_t a)\}, \qquad B_{24} = G\{k^2 - k_t^2\}\{Y_1(k_t a)\};$$

$$B_{31} = -\{\lambda(k_l^2 + k^2) + 2Gk_l^2\}J_0(k_lb) + 2Gk_l\frac{J_1(k_lb)}{b},$$

$$B_{32} = -\{\lambda(k_l^2 + k^2) + 2Gk_l^2\}Y_0(k_lb) + 2Gk_l\frac{Y_1(k_lb)}{b},$$

$$B_{33} = \{2iGk\}\left\{k_t J_0(k_tb) - \frac{1}{b}J_1(k_tb)\right\},$$

$$B_{34} = \{2iGk\}\left\{k_t Y_0(k_tb) - \frac{1}{b}Y_1(k_tb)\right\},$$

$$B_{35} = \lambda_w[(k_w^2 + k^2)H_0^{(2)}(k_wb)];$$

(12.42)
(*cont.*)

$$B_{41} = \{2iGkk_l\}\{J_1(k_lb)\}, \qquad B_{42} = \{2iGkk_l\}\{Y_1(k_lb)\},$$

$$B_{43} = G\{k^2 - k_t^2\}\{J_1(k_tb)\}, \qquad B_{44} = G\{k^2 - k_t^2\}\{Y_1(k_tb)\};$$

$$B_{51} = -k_l J_1(k_lb), \qquad B_{52} = -k_l Y_1(k_lb),$$

$$B_{53} = ikJ_1(k_tb), \qquad B_{54} = ikY_1(k_tb),$$

$$B_{55} = k_w H_1^{(2)}(k_wb).$$

A variety of numerical techniques could be used to extract the complex roots from this characteristic equation; one of the most suitable is the method of Müller (1956). Müller's method uses three initial approximations, z_0, z_1, z_2, and determines the next by an intersection z_3 of the parabolic curve through the three points with the intersection of the (x, y)-plane. The process repeats until a precise root is found. Estimates are on the values of a and b in $\bar{z} = a + b_i$ in the (x, y)-plane. Spiral behavior from an initial point close to the real root for the dry case is suggested. The estimate will be reasonably close if the density of the fluid is small with respect to the density of the structure.

After obtaining the complex eigenvalues k, the phase velocity c_p and attenuation α can be calculated in the usual manner:

$$c_p = \omega/k_{\mathrm{Re}}, \tag{12.43}$$

$$\alpha = k_{\mathrm{Im}} \tag{12.44}$$

(for decibels, use $20\log\alpha$). Once the complex eigensolution k of the dispersion equation (12.41) is obtained, corresponding wave structures of displacement and stress can be calculated using equations (12.34)–(12.38). It must be noted that such wave structures represent not absolute values but rather relative values with respect to one of the five unknown amplitudes in (12.41). This is because the system of equations in (12.41) does not become linearly independent when the complex wavenumber k is substituted. Consequently, for convenience we recommended that cross-sectional wave structures be normalized by the maximum of each distribution in order to keep the range of wave structure variation between 0 and 1. They could also be normalized against one of the displacement variables or the maximum of u_r or u_z, for example. Cross-sectional power flow can be expressed in terms of the inner product of displacement and stress as

$$p_i = \sigma_{ij}u_j \quad (i, j = r, \theta, z). \tag{12.45}$$

Table 12-2. *Material and geometric properties for sample problem of Figures 12-13–12-16*

Material	Density	c_l	c_t
Steel	7.8 g/cm^3	5.94 mm/μs	3.2 mm/μ
Water	1 g/cm^3	1.5 mm/μs	

Geometric properties of steel tubing
inner radius $a = 3, 3.5, 4$ mm
outer radius $b = 5$ mm
ratio $d/r = 0.500, 0.353, 0.222$

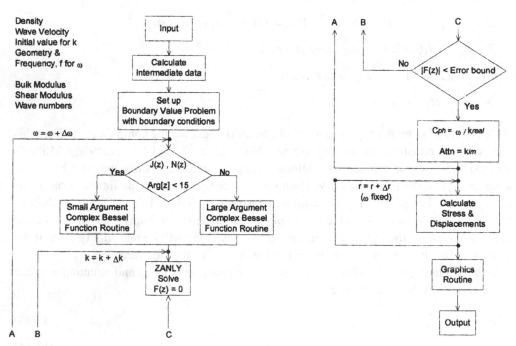

Figure 12-12. Flow chart for calculating dispersion and attenuation curves.

12.7 Attenuation Dispersion Curves

Equation (12.41) is called the "characteristic" or "dispersion" equation for the cylindrical guided wave under liquid loading. It can be evaluated as an implicit transcendental function relating applied frequency ω to complex wavenumber k under any given material and geometrical properties. Table 12-2 lists the material properties and geometrical parameters of a sample problem; Figure 12-12 shows a flow chart summarizing computational procedures.

In order to solve the complicated mathematical equations arising from both the loaded plate and loaded cylinder problems, we have developed a software routine integrating commercially available packages with additional software. The software is used to generate the dispersion and attenuation curves associated with the loaded plate and cylinder problems. Our approach is described in the flow chart of Figure 12-12, where the term

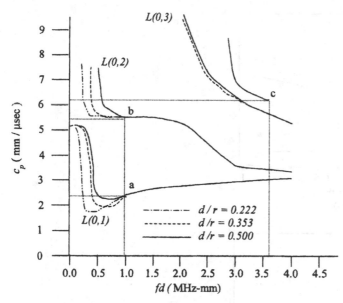

Figure 12-13. Phase velocity dispersion curves for a water-loaded steel tube (inner radius 4 mm, outer radius 5 mm).

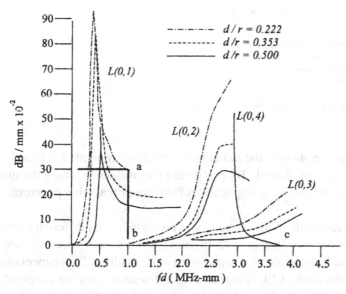

Figure 12-14. Attenuation dispersion curves for a water-loaded steel tube.

$F(z)$ represents the complex-valued equations that result from setting the dispersion relation determinants to zero. The software ZANLY (1992) is a commercial package that solves for the complex roots of the equation $F(z) = 0$. Sample results are presented in Figures 12-13 and 12-14, where $d = b - a$ and $r = (b + a)/2$.

12.8   Wave Structure Considerations

Three sample points are selected on the dispersion curves for phase velocity and attenuation; these points are labeled **a**, **b**, and **c** in Figures 12-13 and 12-14. Using these

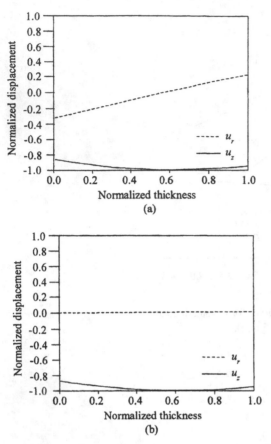

Figure 12-15. Displacements at point **b**: (a) the free tube; (b) the water-loaded tube.

points, which define propagation modes, the radial and axial displacements for a loaded and unloaded steel cylinder can be plotted. By comparing the manner in which the displacements vary across the water and tube, implications for wave propagation penetration become apparent.

It is evident that the displacements of particular modes are significantly different at various points on the dispersion curves. As Figure 12-14 shows, point **a** has a relatively large attenuation at 1 MHz-mm and is therefore very sensitive to water loading. Experimentally, it has been confirmed that the mode $L(0, 1)$ associated with point **a** does not propagate well in a water-loaded tube. This can be explained by noting the large radial displacement for this mode: ultrasonic energy radiates from the tube into the water through the radial displacement.

Point **b** exhibits much less attenuation than point **a** and is also less influenced by water loading, as seen in the overall shape of the displacement curves (see Figure 12-15). Therefore, the mode associated with point **b** would be a good candidate for defect detection. Additionally, this mode can travel long distances within the tube because of the low attenuation associated with it. Point **c** has attenuation that comes very close to zero and has displacements that are virtually no different from those of an unloaded tube – close to zero on the outside surface (Figure 12-16). This makes the mode associated with point **c** an excellent possibility for NDE applications because of strong penetration potential.

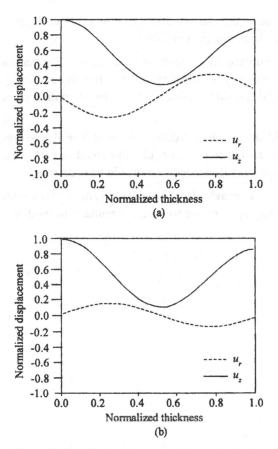

Figure 12-16. Displacements at point **c**: (a) the free tube; (b) the water-loaded tube.

12.9 Discussion

Ultrasonic wave propagation in tubing and piping has become a critically important subject for the oil, power generation, and chemical processing industries. Many new techniques of wave excitation and reception are in the process of being developed. For examples, see Alleyne, Lowe, and Cawley (1996), Kwun and Bartels (1997), Rose, Jiao, and Spanner (1996), and Rose, Pelts, and Quarry (1997b).

For additional related reading, see Chimenti and Nayfeh (1989), Chimenti and Rokhlin (1990), Cho and Rose (1996a), Ditri and Rose (1992), Meeker and Meitzler (1964), Simmons, Dreswcher-Krasicka, and Wadley (1992), and Yapura and Kinra (1995).

12.10 Exercises

1. From Navier's equation, develop the governing wave equation (12.3) for circumferential guided waves.

2. What is implied by attenuation of displacement in the wavevector direction?

3. How would you theoretically explore the possibilities of torsional waves in a hollow cylinder?

4. What order of determinant would lead to the characteristic equation of a two-layered hollow cylinder of two different materials immersed in a fluid?

5. Once the complex wavenumber is extracted from the dispersion equation for certain attenuation-oriented problems, how would you calculate and plot the phase velocity dispersion curves, the group velocity dispersion curves, and the attenuation dispersion curves?

6. It is quite often possible to study guided waves in a structure and yet still be able to predict energy leakage if the structure were immersed in a fluid. How could you tell what modes were susceptible to leakage and what fluid conditions would be necessary?

7. Describe the difference in the wave field behavior for a plate for the A0 and S0 modes when fd is large. Also consider the appropriate case for circumferential pipe modes.

13

Guided Waves in Multiple Layers

13.1 Introduction

Guided waves in multiple layers can be used in a number of different practical situations. Examples include coating problems of plasma spray on a turbine blade, painted structures, aircraft multiple layers (either bonded or with sealant layers), diffusion bonded or adhesively bonded structures in general, and even ice or contaminant detection via multiple layers of solids and fluids (e.g., water, glycol, etc.).

In this chapter we examine the theory of guided wave propagation in layered media by presenting a sample problem of guided waves in a four-layered medium. The resulting dispersion equation is in the form of the determinant of a 16×16 coefficient matrix, which can be reduced if there are fewer layers or if water layers are considered. The derivation is general in the sense that each layer can be either solid or liquid if appropriate boundary conditions are observed. This general result is then specialized to the case of four perfectly bonded layers, where the outermost two layers are assumed to be traction-free on their free surfaces. The interface conditions between layers is taken as either continuity of displacement and traction vectors or continuity of the normal component of displacement and traction vectors, depending on whether the layers adjacent to the interface are solid or liquid. For the four solid layers treated in detail, the results are 16 equations in 16 unknowns, leading to a matrix with $16 \times 16 = 256$ elements; we list these elements explicitly. In addition, this general matrix is simplified for the case where one layer is liquid. We also give explicit expressions for the coefficient matrix corresponding to these cases.

For more details on waves in multiple layers, see Abo-Zena (1979), Adler (1990), Auld (1990), Bragar and Herrmann (1992), Brekhovskikh (1960), Castaings and Hosten (1994), Chimenti, Nayfeh, and Butler (1982), Ewing et al. (1957), Haskell (1953), Kennett (1983), Knopoff (1964), Kundu and Mal (1985), Lowe (1995), Mal (1988a), Mal and Ting (1988), Nagy and Adler (1989), Nayfeh (1995), Nemat-Nasser and Minagawa (1977), Pilarski (1995), Rokhlin and Wang (1991a), Schwab (1970), and Thomson (1950).

13.2 *N*-Layered Plates

13.2.1 Notation and Definitions

The problem to be solved is depicted in Figure 13-1. We have a stratified elastic body composed of N layers rigidly joined to each other at $N - 1$ interfaces. The y-coordinate axis is affixed to the top layer and increases downward. The waves propagate in the z-coordinate direction and are infinitely extended in the x-coordinate direction (plane strain

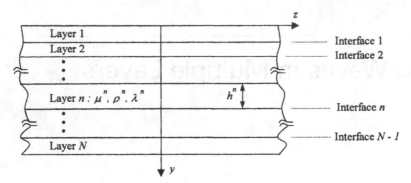

Figure 13-1. Geometry of N-layered plate.

problem), which is directed out of the plane of the figure. Each layer is labeled with an integer index n, starting at $n = 1$ for the topmost layer (i.e., the one containing the origin) and ending with $n = N$ (where N denotes the number of layers). The density and Lamé constants of layer n are denoted by ρ^n and λ^n, μ^n, respectively. (For elastic layers, λ and μ are real. Viscoelastic layers can be simulated by allowing the Lamé constants to be complex; see e.g. Section 18.5.) The thickness of layer n is denoted h_n, and $H \equiv \sum_{i=1}^{N} h_i$ denotes the total thickness of the body.

There are a total of $N - 1$ interfaces joining the N layers. The interfaces also are indexed by an integer n, which increases from $n = 1$ (the bottom of the first layer) to $n = N - 1$ (the top of the last layer). Traction and displacement vector fields are assumed to be continuous across each interface when both layers adjacent to the interface are solid; if either (or both) layers adjacent to an interface are fluid, then (a) only continuity of the normal component of these vectors is enforced, and (b) the shear traction component is required to vanish. Finally, all field variables within layer n are denoted with the superscript n. For instance, u^n denotes the displacement field in layer n.

13.2.2 Analysis

It is known that, within each layer, the displacement field must satisfy Navier's displacement equations of motion. These can be written as

$$\mu^n \nabla^2 u^n + (\lambda^n + \mu^n) \nabla(\nabla \cdot u^n) = \rho^n \frac{\partial^2 u^n}{\partial t^2}. \tag{13.1}$$

Equation (13.1) holds for each layer $n \in \{1, 2, \ldots, N\}$ and therefore represents N vector differential equations. It should be understood that all equations to follow, when written for general n, hold for every layer $n \in \{1, 2, \ldots, N\}$.

For fields that vary harmonically as $e^{i\omega t}$, (13.1) becomes

$$\mu^n \nabla^2 u^n + (\lambda^n + \mu^n) \nabla(\nabla \cdot u^n) = \rho^n \omega^2 u^n. \tag{13.2}$$

The harmonic time dependence will be dropped for brevity (else it would appear as a multiplying factor in every equation to come!), but its presence must be remembered when performing differentiations with respect to time t.

In order to determine the displacement fields in each layer, the generalized displacement is first decomposed using the Helmholtz decomposition,

$$u^n = \nabla\phi^n + \nabla \times \Psi^n, \tag{13.3}$$

into dilatational and equivoluminal parts. The potentials ϕ^n and Ψ^n are then the unknowns for each layer. For the case of Lamb-type waves, we require that the displacement vector component $u_x^n \equiv 0$ and that the other two components be functions only of y and z. This can be achieved by taking $\Psi_y^n = \Psi_z^n \equiv 0$ with $\Psi_x^n = \Psi_x^n(y, z)$ and $\phi^n = \phi^n(y, z) \equiv 0$. Hereafter, Ψ_x^n (the nonzero component of Ψ^n) will be denoted by ψ^n.

With these restrictions on ϕ and Ψ, equation (13.3) yields the displacement components

$$u_x^n \equiv 0,$$

$$u_y^n = \frac{\partial\phi^n}{\partial y} + \frac{\partial\psi^n}{\partial z}, \tag{13.4}$$

$$u_z^n = \frac{\partial\phi^n}{\partial z} - \frac{\partial\psi^n}{\partial y}.$$

If (13.3) is substituted into (13.1), the result is two uncoupled wave equations:

$$\left(\nabla^2 - \frac{1}{(c_L^n)^2}\frac{\partial^2}{\partial t^2}\right)\phi^n = 0,$$

$$\left(\nabla^2 - \frac{1}{(c_T^n)^2}\frac{\partial^2}{\partial t^2}\right)\psi^n = 0. \tag{13.5}$$

In (13.5), $\nabla^2 = \partial^2/\partial y^2 + \partial^2/\partial z^2$. The terms c_L^n and c_T^n denote the bulk longitudinal and transverse wave speeds in layer n. Solutions to (13.5) can be written in a most general form as

$$\phi^n = C_1^n e^{ik_L^n[z\sin(\theta_L^n)+y\cos(\theta_L^n)]} + C_2^n e^{ik_L^n[z\sin(\theta_L^n)-y\cos(\theta_L^n)]},$$

$$\psi^n = C_3^n e^{ik_T^n[z\sin(\theta_T^n)+y\cos(\theta_T^n)]} + C_4^n e^{ik_T^n[z\sin(\theta_T^n)-y\cos(\theta_T^n)]}, \tag{13.6}$$

where

$$k_L^n = \frac{\omega}{c_L^n}, \qquad k_T^n = \frac{\omega}{c_T^n}, \tag{13.7}$$

and the constants $C_1^n, C_2^n, C_3^n, C_4^n$ are now arbitrary.

Each of the potentials in (13.6) can be seen to be the sum of two terms, one representing a downward propagating plane wave (positive y in the exponential term) and one representing an upward propagating term (negative y in the exponential term). Physically, this corresponds to assuming four plane bulk waves in each layer, two longitudinal and two shear. Each of the longitudinal and shear modes propagates downward at the respective angles θ_L^n and θ_T^n to the normal; the other two modes (one longitudinal and one shear) propagate upward, with wavevectors making the same angles with respect to the normals. This description is interpreted physically as a partial wave.

We now have the general solutions for the potentials. After this analysis, what remains is to compute the displacement and traction components entering into the boundary conditions and to enforce those boundary conditions.

13.2.3   Displacement

The displacement field in layer n follows by substituting the general solutions for the potentials into the Helmholtz representation, equation (13.3). When this is done, the resulting displacement fields can be written as

$$u_y^n = ik_L^n \cos(\theta_L^n)\{C_1^n e^{ik_L^n[z\sin(\theta_L^n)+y\cos(\theta_L^n)]} - C_2^n e^{ik_L^n[z\sin(\theta_L^n)-y\cos(\theta_L^n)]}\}$$
$$+ ik_T^n \sin(\theta_T^n)\{C_3^n e^{ik_T^n[z\sin(\theta_T^n)+y\cos(\theta_T^n)]} + C_4^n e^{ik_T^n[z\sin(\theta_T^n)-y\cos(\theta_T^n)]}\}, \qquad (13.8)$$

$$u_z^n = ik_L^n \sin(\theta_L^n)\{C_1^n e^{ik_L^n[z\sin(\theta_L^n)+y\cos(\theta_L^n)]} + C_2^n e^{ik_L^n[z\sin(\theta_L^n)-y\cos(\theta_L^n)]}\}$$
$$+ ik_T^n \cos(\theta_T^n)\{C_3^n e^{ik_T^n[z\sin(\theta_T^n)+y\cos(\theta_T^n)]} - C_4^n e^{ik_T^n[z\sin(\theta_T^n)-y\cos(\theta_T^n)]}\}. \qquad (13.9)$$

13.2.4   Strain

The strain field in each layer is found from the displacement field by using the relations

$$\varepsilon_{ij}^n = \frac{1}{2}\left(\frac{\partial u_i^n}{\partial x_j} + \frac{\partial u_j^n}{\partial x_i}\right), \qquad (13.10)$$

where $x_1 \equiv x$, $x_2 \equiv y$, and $x_3 \equiv z$. Because $u_x \equiv 0$ and u_y, u_z are independent of x, we have $\varepsilon_{xx} = \varepsilon_{xy} = \varepsilon_{xz} \equiv 0$. This confirms our previous remark that we are solving a plane strain problem.

The other strain components follow by differentiating (13.8) and (13.9) according to (13.10). Once the results are algebraically simplified, we have

$$\varepsilon_{yy}^n = -(k_L^n)^2 \cos^2(\theta_L^n)\{C_1^n e^{ik_L^n[z\sin(\theta_L^n)+y\cos(\theta_L^n)]} + C_2^n e^{ik_L^n[z\sin(\theta_L^n)-y\cos(\theta_L^n)]}\}$$
$$- (k_T^n)^2 \sin(\theta_T^n)\cos(\theta_T^n)\{C_3^n e^{ik_T^n[z\sin(\theta_T^n)+y\cos(\theta_T^n)]}$$
$$- C_4^n e^{ik_T^n[z\sin(\theta_T^n)-y\cos(\theta_T^n)]}\}, \qquad (13.11)$$

$$\varepsilon_{zz}^n = -(k_L^n)^2 \sin^2(\theta_L^n)\{C_1^n e^{ik_L^n[z\sin(\theta_L^n)+y\cos(\theta_L^n)]} + C_2^n e^{ik_L^n[z\sin(\theta_L^n)-y\cos(\theta_L^n)]}\}$$
$$+ (k_T^n)^2 \sin(\theta_T^n)\cos(\theta_T^n)\{C_3^n e^{ik_T^n[z\sin(\theta_T^n)+y\cos(\theta_T^n)]}$$
$$- C_4^n e^{ik_T^n[z\sin(\theta_T^n)-y\cos(\theta_T^n)]}\}, \qquad (13.12)$$

and

$$2\varepsilon_{yz}^n = -(k_L^n)^2 \sin^2(2\theta_L^n)\{C_1^n e^{ik_L^n[z\sin(\theta_L^n)+y\cos(\theta_L^n)]} - C_2^n e^{ik_L^n[z\sin(\theta_L^n)-y\cos(\theta_L^n)]}\}$$
$$+ (k_T^n)^2 \cos(2\theta_T^n)\{C_3^n e^{ik_T^n[z\sin(\theta_T^n)+y\cos(\theta_T^n)]} + C_4^n e^{ik_T^n[z\sin(\theta_T^n)-y\cos(\theta_T^n)]}\}. \qquad (13.13)$$

In (13.13) we have used the trigonometric identities $\sin(2x) = 2\sin(x)\cos(x)$ and $\cos(2x) = \cos^2(x) - \sin^2(x)$.

In our calculation of the tractions, it will prove convenient to have an expression for Δ^n, the dilatation in layer n. Recall that the dilatation (sometimes denoted e) is given by

$$\Delta^n \equiv \nabla \cdot u^n. \qquad (13.14)$$

Although it can be calculated in this way (i.e., by taking the divergence of the displacement field), it is easier to recognize that, by the Helmholtz resolution, $u^n = \nabla\phi^n + \nabla \times \Psi^n$. Substituting this representation for u^n into (13.14) gives

$$\Delta^n \equiv \nabla \cdot (\nabla \phi^n + \nabla \times \Psi^n)$$
$$= \nabla^2 \phi^n, \tag{13.15}$$

where we have used that $\nabla \cdot \nabla \phi^n = \nabla^2 \phi^n$ and $\nabla \cdot (\nabla \times \Psi^n) = 0$. Finally, we make use of the fact that ϕ^n satisfies (13.5), that is, $\nabla^2 \phi^n = -(k_L^n)^2 \phi$. Therefore,

$$\Delta^n = -(k_L^n)^2 \phi^n$$
$$= -(k_L^n)^2 \{ C_1^n e^{ik_L^n [z \sin(\theta_L^n) + y \cos(\theta_L^n)]} + C_2^n e^{ik_L^n [z \sin(\theta_L^n) - y \cos(\theta_L^n)]} \}. \tag{13.16}$$

13.2.5 Traction

With general expressions for the strain components in hand, we now apply Hooke's law to calculate the traction components to be used in the boundary conditions. For isotropic media, Hooke's law is

$$\sigma_{ij}^n = \lambda^n \Delta^n \delta_{ij} + 2\mu^n \varepsilon_{ij}^n, \tag{13.17}$$

where λ^n, μ^n are the Lamé constants for layer n and δ_{ij} is the Kronecker delta.

Using (13.17), along with the strain components (13.11)–(13.13) and the dilatation (13.16), yields (after simplifying)

$$\sigma_{yy}^n = \lambda^n \Delta^n + 2\mu^n \varepsilon_{yy}^n$$
$$= -(k_L^n)^2 [\lambda^n + 2\mu^n \cos^2(\theta_L^n)] \{ C_1^n e^{ik_L^n [z \sin(\theta_L^n) + y \cos(\theta_L^n)]} + C_2^n e^{ik_L^n [z \sin(\theta_L^n) - y \cos(\theta_L^n)]} \}$$
$$- 2\mu^n (k_T^n)^2 \sin(\theta_T^n) \cos(\theta_T^n) \{ C_3^n e^{ik_T^n [z \sin(\theta_T^n) + y \cos(\theta_T^n)]}$$
$$- C_4^n e^{ik_T^n [z \sin(\theta_T^n) - y \cos(\theta_T^n)]} \}, \tag{13.18}$$

$$\sigma_{yz}^n = 2\mu^n \varepsilon_{yz}^n$$
$$= -(k_L^n)^2 2\mu^n \sin(2\theta_L^n) \{ C_1^n e^{ik_L^n [z \sin(\theta_L^n) + y \cos(\theta_L^n)]} - C_2^n e^{ik_L^n [z \sin(\theta_L^n) - y \cos(\theta_L^n)]} \}$$
$$+ \mu^n (k_T^n)^2 \cos(2\theta_T^n) \{ C_3^n e^{ik_T^n [z \sin(\theta_T^n) + y \cos(\theta_T^n)]}$$
$$+ C_4^n e^{ik_T^n [z \sin(\theta_T^n) - y \cos(\theta_T^n)]} \}. \tag{13.19}$$

13.2.6 Boundary Conditions

We now have general expressions for the displacement and traction fields; hence we are in a position to state and impose boundary conditions on our required solution. Boundary conditions are required at each interface and free surface of the medium. For an N-layered system, this means that there will be $N - 1$ interfaces, plus two free surfaces, upon which boundary conditions must be imposed. Strictly speaking, the conditions imposed at each interface are not boundary conditions but rather continuity conditions. The difference is that, at the interface, we do not specify the value of any field component. Rather, we specify that the field components of the layers above and below the interface be equal (slip conditions also fall into this category). In contrast, we actually do specify the value of the field component for a boundary condition. For example, we require traction to be zero (our imposed value) on the free surface(s) of the layered medium.

The number and type of conditions imposed at each surface depends on whether the layer(s) adjacent to that surface are solid or liquid. In the analytical derivations to follow, the most general interface conditions will be applied – that is, continuity of displacement and traction vector fields. In other words, both normal and transverse components of the displacement and traction fields will be required to be continuous across each interface. At the free surfaces, the traction vector will be required to vanish. Analytically, the simpler case where one or more layers are liquid can then be analyzed by making several changes to the general system of equations.

However, the approach of solving the general problem and then reducing it to a specific configuration is useful only when things are done analytically. There is no numerical or programming advantage to first formulating the general system, because it will not reduce to the appropriate specialized case merely by letting constants approach zero. In fact, completely bogus results will be obtained by doing so because the system of equations will, in general, become ill-conditioned. This is due to the fact that some of the interface conditions imposed for the general case may not be applicable to the specific case under consideration. With solid layers, for instance, we require continuity of both normal and transverse components of displacement (and traction), whereas for liquid layers we require continuity only of the normal components. It is not simply a matter of letting the shear modulus μ^n of the liquid layers become zero; instead, we must (a) physically remove the condition requiring continuity of transverse displacements and (b) replace the condition requiring continuity of transverse traction by one requiring zero transverse traction. Analytically, this can be done first only by reformulating the system of equations.

In this regard, once the analytical solution is obtained for the general case of all solid layers, the case of one or more liquid layers can be incorporated by means of the following steps.

(1) Since liquid layers cannot support shear waves, we must remove any shear waves present in the appropriate layer(s). The partial wave formalism allows us to do this in a very convenient manner: all we need do is formally set $C_3^n = C_4^n = 0$, in the system of equations developed for the general case, for each layer n that is liquid. (This follows because C_3^n and C_4^n are respectively the amplitudes of downward and upward propagating bulk shear waves in layer n.) In addition, the shear modulus μ^n of any liquid layer n should be set to zero.

(2) When one or both layers making up an interface are liquid, continuity of only the normal component of displacement can be enforced (since there can be tangential slip across the interface owing to the inviscous nature of the fluids being considered). Therefore, equations in the general system that enforce continuity of transverse components of displacement should be removed from the equation set if either of the layers adjacent to the interface is liquid.

(3) When one or both layers making up an interface are liquid, continuity of only the normal component of traction can be enforced (for the same reason as in the previous step). In addition, we must require that the shear traction vanish at the interface. If only one layer adjacent to an interface is liquid, and if the shear modulus of that layer has been set to zero, then the continuity condition of shear traction actually becomes a boundary condition on the shear traction of the solid layer adjacent to the interface. In this case (i.e., one liquid layer adjacent to an interface), the general equation remains valid. If both layers adjacent to a particular interface are liquid (and their shear moduli are set equal to zero), then the condition becomes trivial (i.e. $0 = 0$) and should be removed from the general equation set.

We may now proceed to obtain the general equation set. Referring to Figure 13-1, it can be seen that the $n = 1$ layer has at least one free surface (i.e., its top). At this free surface we require that the traction vector vanish, which can be written as

$$\sigma^1_{yy} = \sigma^1_{yz} \equiv 0 \quad (y = 0, \, -\infty < z < \infty). \tag{13.20}$$

In light of the general expressions (13.18) and (13.19) for traction, (13.20) becomes

$$(k^1_L)^2 [\lambda^1 + 2\mu^1 \cos^2(\theta^1_L)]\{C^1_1 + C^1_2\}e^{ik^1_L \sin(\theta^1_L)z}$$
$$+ 2\mu^1 (k^1_T)^2 \sin(\theta^1_T) \cos(\theta^1_T)\{C^1_3 - C^1_4\}e^{ik^1_T \sin(\theta^1_T)z} = 0 \tag{13.21a}$$

and

$$-(k^1_L)^2 \mu^1 \sin(2\theta^1_L)\{C^1_1 - C^1_2\}e^{ik^1_L \sin(\theta^1_L)z}$$
$$+ \mu^1 (k^1_T)^2 \cos(2\theta^1_T)\{C^1_3 + C^1_4\}e^{ik^1_T \sin(\theta^1_T)z} = 0. \tag{13.21b}$$

Equations (13.21a) and (13.21b) must be satisfied for all z along the entire upper surface of the layer. The only way this can happen is if the exponentials multiplying the individual terms in the equations are equal to each other. Otherwise, the two terms will vary as two different functions of z and hence there will be no way the equations can be satisfied simultaneously for all z. We thus require that

$$e^{ik^1_L \sin(\theta^1_L)z} = e^{ik^1_T \sin(\theta^1_T)z}, \tag{13.22}$$

which implies that

$$k^1_L \sin(\theta^1_L) = k^1_T \sin(\theta^1_T). \tag{13.23}$$

Equation (13.23) is, of course, Snell's law.

It will prove useful at this point to introduce additional notation so as to simplify the following equations. First, we note that (13.23) will remain valid if the index 1 is replaced by n. That is, in order for the boundary (and interface) conditions to be simultaneously satisfied for all z, we must have

$$k^n_L \sin(\theta^n_L) = k^n_T \sin(\theta^n_T). \tag{13.24}$$

Physically, the left- and right-hand sides of (13.24) represent the projection of the wavevector of the partial longitudinal and transverse waves in the z direction (see Figure 13-2). Since all terms of this form are equal, we denote all of them by k. Note that the guided waves in the layer have only a z wavevector component, which is also equal to k. We therefore have

$$k^n_L \sin(\theta^n_L) = k^n_T \sin(\theta^n_T) = k. \tag{13.25}$$

We will also introduce a symbol for the transverse component of any partial wave's wavevector:

$$K^n_T \equiv k^n_T \cos(\theta^n_T); \tag{13.26a}$$

$$K^n_L \equiv k^n_L \cos(\theta^n_L). \tag{13.26b}$$

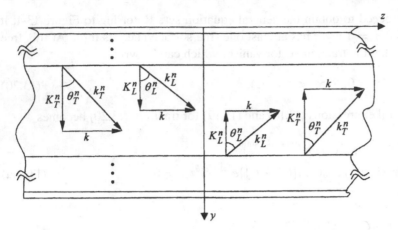

Figure 13-2. Four partial waves in layer n.

Finally, we introduce a symbol representing the y coordinate of the nth interface:

$$H_n \equiv \sum_{i=1}^{n} h_n, \tag{13.27}$$

where h_n represents the thickness of layer n. Of course, $H_1 = h_1$ (the thickness of the first layer) and $H_N = H$ (the total thickness of the stratified body).

The introduction of these definitions allows the boundary condition equations to be written in a more concise form. More importantly, it introduces the "Lamb"-wave wavenumber k into the equations explicitly. This will be useful when we undertake the task of numerically finding the roots of the dispersion equation.

With these notational definitions, equations (13.21a) and (13.21b) become

$$[\lambda^1(k_L^1)^2 + 2\mu^1(K_L^1)^2]\{C_1^1 + C_2^1\} + 2\mu^1 k K_T^1\{C_3^1 - C_4^1\} = 0 \tag{13.28a}$$

and

$$-2k\mu^1 K_L^1\{C_1^1 - C_2^1\} + \mu^1[(K_T^1)^2 - k^2]\{C_3^1 + C_4^1\} = 0, \tag{13.28b}$$

respectively. At each of the $N - 1$ interfaces, we require continuity of both normal and transverse components of traction and displacement. These boundary conditions result in the following equations.

Continuity of u_y. We require that

$$u_y^n = u_y^{n+1} \quad (y = H_n, \; -\infty < z < \infty, \; 1 \leq n \leq N - 1). \tag{13.29}$$

Using the general form for the displacement fields, equation (13.8), this becomes

$$K_L^n\{C_1^n e^{iK_L^n H_n} - C_2^n e^{-iK_L^n H_n}\} + k\{C_3^n e^{iK_T^n H_n} + C_4^n e^{-iK_T^n H_n}\}$$
$$- K_L^{n+1}\{C_1^{n+1} e^{iK_L^{n+1} H_n} - C_2^{n+1} e^{-iK_L^{n+1} H_n}\}$$
$$- k\{C_3^{n+1} e^{iK_T^{n+1} H_n} + C_4^{n+1} e^{-iK_T^{n+1} H_n}\} = 0. \tag{13.30}$$

Continuity of u_z. We require that

$$u_z^n = u_z^{n+1} \quad (y = H_n, \, -\infty < z < \infty, \, 1 \le n \le N - 1), \tag{13.31}$$

which leads to:

$$k\{C_1^n e^{iK_L^n H_n} + C_2^n e^{-iK_L^n H_n}\} - K_T^n\{C_3^n e^{iK_T^n H_n} - C_4^n e^{-iK_T^n H_n}\}$$
$$- k\{C_1^{n+1} e^{iK_L^{n+1} H_n} + C_2^{n+1} e^{-iK_L^{n+1} H_n}\}$$
$$+ K_T^{n+1}\{C_3^{n+1} e^{iK_T^{n+1} H_n} - C_4^{n+1} e^{-iK_T^{n+1} H_n}\} = 0. \tag{13.32}$$

Continuity of σ_{yy}. We require that

$$\sigma_{yy}^n = \sigma_{yy}^{n+1} \quad (y = H_n, \, -\infty < z < \infty, \, 1 \le n \le N - 1). \tag{13.33}$$

This condition can be written as

$$[\lambda^n (k_L^n)^2 + 2\mu^n (K_L^n)^2]\{C_1^n e^{iK_L^n H_n} + C_2^n e^{-iK_L^n H_n}\} + 2\mu^n k K_T^n\{C_3^n e^{iK_T^n H_n} - C_4^n e^{-iK_T^n H_n}\}$$
$$- [\lambda^{n+1} (k_L^{n+1})^2 + 2\mu^{n+1} (K_L^{n+1})^2]\{C_1^{n+1} e^{iK_L^{n+1} H_n} + C_2^{n+1} e^{-iK_L^{n+1} H_n}\}$$
$$- 2\mu^{n+1} k K_T^{n+1}\{C_3^{n+1} e^{iK_T^{n+1} H_n} - C_4^{n+1} e^{-iK_T^{n+1} H_n}\} = 0. \tag{13.34}$$

Continuity of σ_{yz}. We require that

$$\sigma_{yz}^n = \sigma_{yz}^{n+1} \quad (y = H_n, \, -\infty < z < \infty, \, 1 \le n \le N - 1), \tag{13.35}$$

which leads to

$$-2k\mu^n K_L^n\{C_1^n e^{iK_L^n H_n} - C_2^n e^{-iK_L^n H_n}\} + \mu^n[(K_T^n)^2 - k^2]\{C_3^n e^{iK_T^n H_n} + C_4^n e^{-iK_T^n H_n}\}$$
$$+ 2k\mu^{n+1} K_L^{n+1}\{C_1^{n+1} e^{iK_L^{n+1} H_n} - C_2^{n+1} e^{-iK_L^{n+1} H_n}\}$$
$$- \mu^{n+1}[(K_T^{n+1})^2 - k^2]\{C_3^{n+1} e^{iK_T^{n+1} H_n} + C_4^{n+1} e^{-iK_T^{n+1} H_n}\} = 0. \tag{13.36}$$

Finally, we impose traction-free boundary conditions on the bottom surface of the last layer (i.e., $n = N$). These conditions can be written as

$$\sigma_{yy}^N = 0 \quad (y = H, \, -\infty < z < \infty) \tag{13.37a}$$

and

$$\sigma_{yz}^N = 0 \quad (y = H, \, -\infty < z < \infty). \tag{13.37b}$$

Equations (13.37) then lead to

$$[\lambda^N (k_L^N)^2 + 2\mu^N (K_L^N)^2]\{C_1^N e^{iK_L^N H} + C_2^N e^{-iK_L^N H}\}$$
$$+ 2\mu^N k K_T^N\{C_3^N e^{iK_T^N H} - C_4^N e^{-iK_T^N H}\} = 0 \tag{13.38a}$$

and

$$-2k\mu^N K_L^N \{C_1^N e^{iK_L^N H} - C_2^N e^{-iK_L^N H}\}$$

$$-\mu^N[(K_T^N)^2 - k^2]\{C_3^N e^{iK_T^N H} + C_4^N e^{-iK_T^N H_n}\} = 0, \qquad (13.38b)$$

respectively.

13.2.7 Dispersion Equations

The four boundary conditions – equations (13.28a), (13.28b), (13.38a), and (13.38b) – must be imposed, regardless of the number N of layers. In the case of a single layer, the equations represent conditions on the upper and lower surfaces of the same layer. The four interface conditions – equations (13.29), (13.31), (13.33), and (13.35) – must be imposed at each of the $N - 1$ interfaces. This gives an additional $4 \times (N - 1) = 4N - 4$ equations for a total (including boundary conditions) of $4N$ equations. Each of the N layers contributes four unknowns (i.e., C_1^n, C_2^n, C_3^n, C_4^n) and hence there are $4N$ equations in $4N$ unknowns for an N-layered medium. As can be seen, the equations are homogeneous.

The system of $4N$ equations in $4N$ unknowns can be written in matrix form as

$$\begin{bmatrix} A_{11} & A_{12} & \cdots & A_{1(4N)} \\ A_{21} & A_{22} & \cdots & A_{2(4N)} \\ \vdots & \vdots & \ddots & \vdots \\ A_{(4N)1} & A_{(4N)2} & \cdots & A_{(4N)(4N)} \end{bmatrix} \begin{bmatrix} C^1 \\ C_2^1 \\ \vdots \\ C_4^1 \end{bmatrix} = \begin{bmatrix} 0 \\ 0 \\ \vdots \\ 0 \end{bmatrix}. \qquad (13.39)$$

The individual elements of the coefficient matrix $[A]$ can be "picked out" of the set of equations by identifying the factors of the respective unknowns.

The dispersion equation of the structure is obtained by setting the determinant of $[A]$ in (13.39) equal to zero. Mathematically, this is equivalent to requiring a solution (other than the trivial one) to exist for the $4N$ unknowns. The dispersion equation can thus be written as

$$|A(\omega, k, \lambda^n, \mu^n, h_n)| = 0; \qquad (13.40)$$

this can be considered to be an implicit equation relating frequency ω to wavenumber k for given material (λ^n, μ^n) and geometric (h_n) parameters. For a given frequency, there will be many (in fact, infinitely many) roots for the wavenumber, yet there will be a finite number of *real* roots.

Note that, for given material and geometric parameters, the determinant as written in (13.40) is a function only of ω and k. However, as seen from the boundary condition equations, the elements $[A]$ also contain the "transverse wavenumbers" K_L^n and K_T^n, which in turn depend on the angles θ_L^n and θ_T^n, respectively (see equations (13.26)). We can eliminate the angles as follows:

$$\begin{aligned} K_L^n &\equiv k_L^n \cos(\theta_L^n) \\ &= k_L^n \sqrt{1 - \sin^2(\theta_L^n)} \\ &= k_L^n \sqrt{1 - (k/k_L^n)^2} \\ &= \sqrt{(k_L^n)^2 - k^2}. \end{aligned} \qquad (13.41)$$

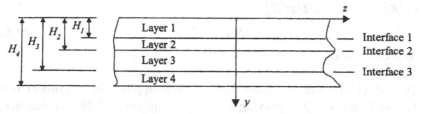

Figure 13-3. Four-layered medium.

In the third step of (13.41), Snell's law (equation (13.25)) has been invoked to eliminate the $\sin^2(\theta_L^n)$ term. Following steps analogous to (13.41), we also have

$$K_T^n = \sqrt{(k_T^n)^2 - k^2}. \tag{13.42}$$

The transverse wavenumbers (which can be real or purely imaginary) are therefore completely specified once the wavenumber k is given. If $k^2 > (k_L^n)^2$ then K_L^n will be purely imaginary and the positive square root can be taken. That is,

$$K_L^n = i\sqrt{k^2 - (k_L^n)^2} \quad \text{for } k^2 > (k_L^n)^2.$$

The same holds for K_T^n:

$$K_T^n = i\sqrt{k^2 - (k_T^n)^2} \quad \text{for } k^2 > (k_T^n)^2.$$

13.2.8 Specific Configurations

In this section we specialize the general dispersion equation (derived in the previous section) to specific layer configurations. Specifically, we analyze a four-layered configuration ($N = 4$). The general problem of four solid layers (Figure 13-3) is derived in detail, and procedures are given for reducing this system when one or all of the layers becomes a liquid.

Our goal is to determine the elements of the coefficient matrix $[A]$, the vanishing of whose determinant defines the dispersion relation of the structure. This will be accomplished by examining the terms in (13.28a), (13.28b), (13.30), (13.32), (13.34), (13.36), (13.38a), and (13.38b) to determine the coefficients of the respective constants C_j^i ($i \in \{1, 2, \ldots, N\}$, $j \in \{1, 2, 3, 4\}$). Because we are dealing with four layers, we have a total of 16 equations ($4 \times N$) in 16 unknowns. Our coefficient matrix will therefore be 16×16, giving a total of 256 elements! However, many of the elements will be zero.

For example, from (13.28a) we arrive at

$$A_{11} = [\lambda^1(k_L^1)^2 + 2\mu^1(K_L^1)^2], \qquad A_{12} = [\lambda^1(k_L^1)^2 + 2\mu^1(K_L^1)^2],$$

$$A_{13} = 2\mu^1 k K_T^1, \qquad A_{14} = -2\mu^1 k K_T^1, \tag{13.43}$$

$$A_{1n} = 0 \quad \text{for } 5 \leq n \leq 16,$$

which represents the coefficients of $C_1^1, C_2^1, C_3^1, C_4^1, C_1^2, C_2^2, \ldots, C_4^N$ in that equation. From (13.28b), we likewise find

$$A_{21} = -2k\mu^1 K_L^1, \qquad A_{22} = 2k\mu^1 K_L^1,$$

$$A_{23} = \mu^1[(K_T^1)^2 - k^2], \qquad A_{24} = \mu^1[(K_T^1)^2 - k^2], \qquad (13.44)$$

$$A_{2n} = 0 \quad \text{for } 5 \le n \le 16.$$

Equations (13.30), (13.32), (13.34), and (13.36) are now applied (in that order) to interfaces $n = 1$, $n = 2$, and $n = 3$. Applying $u_y^n = u_y^{n+1}$ (equation (13.29)) to interface 1 yields

$$A_{31} = K_L^1 e^{iK_L^1 h_1}, \qquad A_{32} = -K_L^1 e^{-iK_L^1 h_1},$$

$$A_{33} = k e^{iK_T^1 h_1}, \qquad A_{34} = k e^{-iK_T^1 h_1},$$

$$A_{35} = -K_L^2 e^{iK_L^2 h_1}, \qquad A_{36} = K_L^2 e^{-iK_L^2 h_1}, \qquad (13.45)$$

$$A_{37} = -k e^{iK_T^2 h_1}, \qquad A_{38} = -k e^{-iK_T^2 h_1},$$

$$A_{3n} = 0 \quad \text{for } 9 \le n \le 16.$$

Applying $u_z^n = u_z^{n+1}$ (equation (13.31)) to interface 1 yields

$$A_{41} = k e^{iK_L^1 h_1}, \qquad A_{42} = k e^{-iK_L^1 h_1},$$

$$A_{43} = -K_T^1 e^{iK_T^1 h_1}, \qquad A_{44} = K_T^1 e^{-iK_T^1 h_1},$$

$$A_{45} = -k e^{iK_L^2 h_1}, \qquad A_{46} = -k e^{-iK_L^2 h_1}, \qquad (13.46)$$

$$A_{47} = K_T^2 e^{-iK_T^2 h_1}, \qquad A_{48} = -K_T^2 e^{-iK_T^2 h_1},$$

$$A_{4n} = 0 \quad \text{for } 9 \le n \le 16.$$

Applying $\sigma_{yy}^n = \sigma_{yy}^{n+1}$ (equation (13.33)) to interface 1 yields

$$A_{51} = [\lambda^1(k_L^1)^2 + 2\mu^1(K_L^1)^2] e^{iK_L^1 h_1}, \qquad A_{52} = [\lambda^1(k_L^1)^2 + 2\mu^1(K_L^1)^2] e^{-iK_L^1 h_1},$$

$$A_{53} = 2\mu^1 k K_T^1 e^{iK_T^1 h_1}, \qquad A_{54} = -2\mu^1 k K_T^1 e^{-iK_T^1 h_1},$$

$$A_{55} = [\lambda^2(k_L^2)^2 + 2\mu^2(K_L^2)^2] e^{iK_L^2 h_1},$$

$$A_{56} = -[\lambda^2(k_L^2)^2 + 2\mu^2(K_L^2)^2] e^{-iK_L^2 h_1}, \qquad (13.47)$$

$$A_{57} = -2\mu^2 k K_T^2 e^{iK_T^2 h_1}, \qquad A_{58} = 2\mu^2 k K_T^2 e^{-iK_T^2 h_1},$$

$$A_{5n} = 0 \quad \text{for } 9 \le n \le 16.$$

Applying $\sigma_{yz}^n = \sigma_{yz}^{n+1}$ (equation (13.35)) to interface 1 yields

$$A_{61} = 2k\mu^1 K_L^1 e^{iK_L^1 h_1}, \qquad A_{62} = 2k\mu^1 K_L^1 e^{-iK_L^1 h_1},$$

$$A_{63} = \mu^1[(K_T^1)^2 - k^2] e^{iK_T^1 h_1}, \qquad A_{64} = \mu^1[(K_T^1)^2 - k^2] e^{-iK_T^1 h_1},$$

$$A_{65} = 2k\mu^2 K_L^2 e^{iK_L^2 h_1}, \qquad A_{66} = -2k\mu^2 K_L^2 e^{-iK_L^2 h_1}, \qquad (13.48)$$

$$A_{67} = -\mu^2[(K_T^2)^2 - k^2] e^{iK_T^2 h_1}, \qquad A_{68} = -\mu^2[(K_T^2)^2 - k^2] e^{-iK_T^2 h_1},$$

$$A_{6n} = 0 \quad \text{for } 9 \le n \le 16.$$

Applying $u_y^n = u_y^{n+1}$ to interface 2 (i.e., $n = 2$) yields

$A_{7n} = 0 \quad \text{for } 1 \le n \le 4,$

$A_{75} = K_L^2 e^{iK_L^2 H_2}, \qquad A_{76} = -K_L^2 e^{-iK_L^2 H_2},$

$A_{77} = k e^{iK_T^2 H_2}, \qquad A_{78} = k e^{-iK_T^2 H_2},$

$A_{79} = -K_L^3 e^{iK_L^3 H_2}, \qquad A_{7(10)} = K_L^3 e^{-iK_L^3 H_2}, \tag{13.49}$

$A_{7(11)} = -k e^{iK_T^3 H_2}, \qquad A_{7(12)} = -k e^{-iK_T^3 H_2},$

$A_{7n} = 0 \quad \text{for } 13 \le n \le 16.$

Note that H_2, which now appears in the exponential terms, is equal to $h_1 + h_2$ as in (13.27). Applying $u_z^n = u_z^{n+1}$ to interface 2 yields

$A_{8n} = 0 \quad \text{for } 1 \le n \le 4,$

$A_{85} = k e^{iK_L^2 H_2}, \qquad A_{86} = k e^{-iK_L^2 H_2},$

$A_{87} = K_L^2 e^{-iK_T^2 H_2}, \qquad A_{88} = K_L^2 e^{-iK_T^2 H_2},$

$A_{89} = -k_L^3 e^{iK_L^3 H_2}, \qquad A_{8(10)} = -k e^{-iK_L^3 H_2}, \tag{13.50}$

$A_{8(11)} = K_T^3 e^{iK_T^3 H_2}, \qquad A_{8(12)} = -K_T^3 e^{-iK_T^3 H_2},$

$A_{8n} = 0 \quad \text{for } 13 \le n \le 16.$

Applying $\sigma_{yy}^n = \sigma_{yy}^{n+1}$ to interface 2 yields

$A_{9n} = 0 \quad \text{for } 1 \le n \le 4,$

$A_{95} = [\lambda^2 (k_L^2)^2 + 2\mu^2 (K_L^2)^2] e^{iK_L^2 H_2}, \qquad A_{96} = [\lambda^2 (k_L^2)^2 + 2\mu^2 (K_L^2)^2] e^{-iK_L^2 H_2},$

$A_{97} = 2\mu^2 k K_T^2 e^{iK_T^2 H_2}, \qquad A_{98} = -2\mu^2 k K_T^2 e^{-iK_T^2 H_2},$

$A_{99} = -[\lambda^3 (k_L^3)^2 + 2\mu^3 (K_L^3)^2] e^{iK_L^3 H_2},$

$A_{9(10)} = -[\lambda^3 (k_L^3)^2 + 2\mu^3 (K_L^3)^2] e^{-iK_L^3 H_2}, \tag{13.51}$

$A_{9(11)} = -2\mu^3 k K_T^3 e^{iK_T^3 H_2}, \qquad A_{9(12)} = 2\mu^3 k K_T^3 e^{-iK_T^3 H_2},$

$A_{9n} = 0 \quad \text{for } 13 \le n \le 16;$

applying $\sigma_{yz}^n = \sigma_{yz}^{n+1}$ to interface 2 yields

$A_{(10)n} = 0 \quad \text{for } 1 \le n \le 4,$

$A_{(10)5} = -2k\mu^2 K_L^2 e^{iK_L^2 H_2}, \qquad A_{(10)6} = 2k\mu^2 K_L^2 e^{-iK_L^2 H_2},$

$A_{(10)7} = \mu^2 [(K_T^2)^2 - k^2] e^{iK_T^2 H_2}, \qquad A_{(10)8} = \mu^2 [(K_T^2)^2 - k^2] e^{-iK_T^2 H_2},$

$A_{(10)9} = 2k\mu^3 K_L^3 e^{iK_L^3 H_2}, \qquad A_{(10)(10)} = -2k\mu^3 K_L^3 e^{-iK_L^3 H_2}, \tag{13.52}$

$A_{(10)(11)} = -\mu^3 [(K_T^3)^2 - k^2] e^{iK_T^3 H_2}, \qquad A_{(10)(12)} = -\mu^3 [(K_T^3)^2 - k^2] e^{-iK_T^3 H_2},$

$A_{(10)n} = 0 \quad \text{for } 13 \le n \le 16.$

Applying $u_y^n = u_y^{n+1}$ to interface 3 (i.e., $n = 3$) yields

$A_{(11)n} = 0$ for $1 \le n \le 8$,

$A_{(11)9} = K_L^3 e^{iK_L^3 H_3}$, $A_{(11)(10)} = -K_L^3 e^{-iK_L^3 H_3}$,

$A_{(11)(11)} = k e^{iK_T^3 H_3}$, $A_{(11)(12)} = k e^{-iK_T^3 H_3}$,   (13.53)

$A_{(11)(13)} = -K_L^4 e^{iK_L^4 H_3}$, $A_{(11)(14)} = K_L^4 e^{-iK_L^4 H_3}$,

$A_{(11)(15)} = -k e^{iK_T^4 H_3}$, $A_{(11)(16)} = -k e^{-iK_T^4 H_3}$,

and applying $u_z^n = u_z^{n+1}$ to interface 3 yields

$A_{(12)n} = 0$ for $1 \le n \le 8$,

$A_{(12)9} = k e^{iK_L^3 H_3}$, $A_{(12)(10)} = k e^{-iK_L^3 H_3}$,

$A_{(12)(11)} = -K_T^3 e^{iK_T^3 H_3}$, $A_{(12)(12)} = K_T^3 e^{-iK_T^3 H_3}$,   (13.54)

$A_{(12)(13)} = -k e^{iK_L^4 H_3}$, $A_{(12)(14)} = -k e^{-iK_L^4 H_3}$,

$A_{(12)(15)} = K_T^4 e^{iK_T^4 H_3}$, $A_{(12)(16)} = -K_T^4 e^{-iK_T^4 H_3}$.

Applying $\sigma_{yy}^n = \sigma_{yy}^{n+1}$ to interface 3 yields

$A_{(13)n} = 0$ for $1 \le n \le 8$,

$A_{(13)9} = [\lambda^3 (k_L^3)^2 + 2\mu^3 (K_L^3)] e^{iK_L^3 H_3}$, $A_{(13)(10)} = [\lambda^3 (k_L^3)^2 + 2\mu^3 (K_L^3)] e^{-iK_L^3 H_3}$,

$A_{(13)(11)} = 2\mu^3 k K_T^3 e^{iK_T^3 H_3}$, $A_{(13)(12)} = -2\mu^3 k K_T^3 e^{-iK_T^3 H_3}$,

$A_{(13)(13)} = -[\lambda^4 (k_L^4)^2 + 2\mu^4 (K_L^4)^2] e^{iK_L^4 H_3}$,   (13.55)

$A_{(13)(14)} = -[\lambda^4 (k_L^4)^2 + 2\mu^4 (K_L^4)^2] e^{-iK_L^4 H_3}$,

$A_{(13)(15)} = -2\mu^4 k K_T^4 e^{iK_T^4 H_3}$, $A_{(13)(16)} = 2\mu^4 k K_T^4 e^{-iK_T^4 H_3}$;

applying $\sigma_{yz}^n = \sigma_{yz}^{n+1}$ to interface 3 yields

$A_{(14)n} = 0$ for $1 \le n \le 8$,

$A_{(14)9} = -2k\mu^3 K_L^3 e^{iK_L^3 H_3}$, $A_{(14)(10)} = 2k\mu^3 K_L^3 e^{-iK_L^3 H_3}$,

$A_{(14)(11)} = \mu^3 [(K_T^3)^2 - k^2] e^{iK_T^3 H_3}$,

$A_{(14)(12)} = \mu^3 [(K_T^3)^2 - k^2] e^{-iK_T^3 H_3}$,   (13.56)

$A_{(14)(13)} = 2k\mu^4 K_L^4 e^{iK_L^4 H_3}$, $A_{(14)(14)} = -2k\mu^4 K_L^4 e^{-iK_L^4 H_3}$,

$A_{(14)(15)} = -\mu^4 [(K_T^4)^2 - k^2] e^{iK_T^4 H_3}$, $A_{(14)(16)} = -\mu^4 [(K_T^4)^2 - k^2] e^{-iK_T^4 H_3}$.

This completes the interface conditions. All that remains is for us to apply (13.38a) and (13.38b) to the bottom of the last layer ($n = 4$, $y = H_4 \equiv H$). Equation (13.38a) yields

$$A_{(15)n} = 0 \quad \text{for } 1 \leq n \leq 12,$$

$$A_{(15)(13)} = [\lambda^4 (k_L^4)^2 + 2\mu^4 (K_L^4)^2] e^{iK_L^4 H},$$

$$A_{(15)(14)} = [\lambda^4 (k_L^4)^2 + 2\mu^4 (K_L^4)^2] e^{-iK_L^4 H}, \tag{13.57}$$

$$A_{(15)(15)} = 2\mu^4 k K_T^4 e^{iK_T^4 H}, \qquad A_{(15)(16)} = -2\mu^4 k K_T^4 e^{-iK_T^4 H},$$

and (13.38b) yields

$$A_{(16)n} = 0 \quad \text{for } 1 \leq n \leq 12,$$

$$A_{(16)(13)} = -2k\mu^4 K_L^4 e^{iK_L^4 H}, \qquad A_{(16)(14)} = 2k\mu^4 K_L^4 e^{-iK_L^4 H}, \tag{13.58}$$

$$A_{(16)(15)} = \mu^4 [(K_T^4)^2 - k^2] e^{iK_T^4 H}, \qquad A_{(16)(16)} = \mu^4 [(K_T^4)^2 - k^2] e^{-iK_T^4 H}.$$

This completes the specification of the elements of the matrix $[A]$, whose form is shown in Figure 13-4. The boundary conditions at the top of the first layer introduce the first two rows (with only four nonzero elements each, since only the four partial waves in the first layer contribute to the traction at its upper surface). Each interface then introduces four additional equations with eight elements each (i.e., four partial waves in each layer are needed to satisfy the interface conditions). Note that each set of four interface conditions is "indented" by four elements, since the partial waves of the $(n - 2)$th layer play no role at interface n.

As discussed previously, the cases where one or more layers become liquid can be obtained analytically from this general result by a straightforward procedure. The rules for reducing the general matrix for the case of one or more liquid layers are shown on the matrix in Figure 13-4. As can be verified, making a single layer liquid reduces the number of equations (and unknowns) by two, and making m layers liquid reduces the size by $2 \times m$. Of the many configurations possible, two are analyzed in detail in the next section: (i) the first layer liquid and all others solid and (ii) the second layer liquid and all others solid. Following the rules exhibited in Figure 13-4, the coefficient matrices of these two cases are as shown in Figures 13-5 and 13-6, respectively. As usual, the dispersion equation for all cases is obtained by setting the determinant of the appropriate $[A]$ matrix equal to zero.

13.3 Sample Problems

There exist a countless number of practical applications of waves in multiple layers that we might consider. Two examples are presented here: wing ice detection in aircraft; and titanium diffusion bond inspection, which is of potential value in testing structural components of high-speed civil transport.

13.3.1 Multilayer Model for Wing Ice Detection

If we were to develop an ice detection system for the wings of aircraft, we could consider guided wave propagation in a multilayer system. The four possible states are presented in Figure 13-7. Dispersion curves could be generated and compared for the different states,

			Remove if Layer 1 is liquid ↓	↓			Remove if Layer 2 is liquid ↓	↓			Remove if Layer 3 is liquid ↓	↓			Remove if Layer 4 is liquid ↓	↓	
	A_{11}	A_{12}	A_{13}	A_{14}	0	0	0	0	0	0	0	0	0	0	0	0	
	A_{21}	A_{22}	A_{23}	A_{24}	0	0	0	0	0	0	0	0	0	0	0	0	Remove if Layer 1 is liquid
	A_{31}	A_{32}	A_{33}	A_{34}	A_{35}	A_{36}	A_{37}	A_{38}	0	0	0	0	0	0	0	0	
	A_{41}	A_{42}	A_{43}	A_{44}	A_{45}	A_{46}	A_{47}	A_{48}	0	0	0	0	0	0	0	0	Remove if Layer 1 or 2 is liquid
	A_{51}	A_{52}	A_{53}	A_{54}	A_{55}	A_{56}	A_{57}	A_{58}	0	0	0	0	0	0	0	0	
	A_{61}	A_{62}	A_{63}	A_{64}	A_{65}	A_{66}	A_{67}	A_{68}	0	0	0	0	0	0	0	0	Remove if Layer 1 and 2 is liquid
	0	0	0	0	A_{75}	A_{76}	A_{77}	A_{78}	A_{79}	$A_{7(10)}$	$A_{7(11)}$	$A_{7(12)}$	0	0	0	0	
	0	0	0	0	A_{85}	A_{86}	A_{87}	A_{88}	A_{89}	$A_{8(10)}$	$A_{8(11)}$	$A_{8(12)}$	0	0	0	0	Remove if Layer 2 or 3 is liquid
[A] =	0	0	0	0	A_{95}	A_{96}	A_{97}	A_{98}	A_{99}	$A_{9(10)}$	$A_{9(11)}$	$A_{9(12)}$	0	0	0	0	
	0	0	0	0	$A_{(10)5}$	$A_{(10)6}$	$A_{(10)7}$	$A_{(10)8}$	$A_{(10)9}$	$A_{(10)(10)}$	$A_{(10)(11)}$	$A_{(10)(12)}$	0	0	0	0	Remove if Layer 2 and 3 is liquid
	0	0	0	0	0	0	0	0	$A_{(11)9}$	$A_{(11)(10)}$	$A_{(11)(11)}$	$A_{(11)(12)}$	$A_{(11)(13)}$	$A_{(11)(14)}$	$A_{(11)(15)}$	$A_{(11)(16)}$	
	0	0	0	0	0	0	0	0	$A_{(12)9}$	$A_{(12)(10)}$	$A_{(12)(11)}$	$A_{(12)(12)}$	$A_{(12)(13)}$	$A_{(12)(14)}$	$A_{(12)(15)}$	$A_{(12)(16)}$	Remove if Layer 3 or 4 is liquid
	0	0	0	0	0	0	0	0	$A_{(13)9}$	$A_{(13)(10)}$	$A_{(13)(11)}$	$A_{(13)(12)}$	$A_{(13)(13)}$	$A_{(13)(14)}$	$A_{(13)(15)}$	$A_{(13)(16)}$	
	0	0	0	0	0	0	0	0	$A_{(14)9}$	$A_{(14)(10)}$	$A_{(14)(11)}$	$A_{(14)(12)}$	$A_{(14)(13)}$	$A_{(14)(14)}$	$A_{(14)(15)}$	$A_{(14)(16)}$	Remove if Layer 3 and 4 is liquid
	0	0	0	0	0	0	0	0	0	0	0	0	$A_{(15)(13)}$	$A_{(15)(14)}$	$A_{(15)(15)}$	$A_{(15)(16)}$	
	0	0	0	0	0	0	0	0	0	0	0	0	$A_{(16)(13)}$	$A_{(16)(14)}$	$A_{(16)(15)}$	$A_{(16)(16)}$	Remove if Layer 4 is liquid

Figure 13-4. Coefficient matrix for four-layered structure where all layers are solid; the elements are given by equations (13.58). The modifications to be made when any layer(s) are liquid are shown above and to the right of the corresponding rows and columns.

$$
[A] = \begin{bmatrix}
\bar{A}_{11} & \bar{A}_{12} & 0 & 0 & 0 & 0 & 0 & 0 & 0 & 0 & 0 & 0 & 0 & 0 & 0 & 0 \\
0 & 0 & 0 & 0 & 0 & 0 & 0 & 0 & 0 & 0 & 0 & 0 & 0 & 0 & 0 & 0 \\
A_{31} & A_{32} & 0 & 0 & A_{35} & A_{36} & A_{37} & A_{38} & 0 & 0 & 0 & 0 & 0 & 0 & 0 & 0 \\
0 & 0 & 0 & 0 & 0 & 0 & 0 & 0 & 0 & 0 & 0 & 0 & 0 & 0 & 0 & 0 \\
\bar{A}_{51} & \bar{A}_{52} & 0 & 0 & A_{55} & A_{56} & A_{57} & A_{58} & 0 & 0 & 0 & 0 & 0 & 0 & 0 & 0 \\
0 & 0 & 0 & 0 & A_{65} & A_{66} & A_{67} & A_{68} & 0 & 0 & 0 & 0 & 0 & 0 & 0 & 0 \\
0 & 0 & 0 & 0 & A_{75} & A_{76} & A_{77} & A_{78} & A_{79} & A_{7(10)} & A_{7(11)} & A_{7(12)} & 0 & 0 & 0 & 0 \\
0 & 0 & 0 & 0 & A_{85} & A_{86} & A_{87} & A_{88} & A_{89} & A_{8(10)} & A_{8(11)} & A_{8(12)} & 0 & 0 & 0 & 0 \\
0 & 0 & 0 & 0 & A_{95} & A_{96} & A_{97} & A_{98} & A_{99} & A_{9(10)} & A_{9(11)} & A_{9(12)} & 0 & 0 & 0 & 0 \\
0 & 0 & 0 & 0 & A_{(10)5} & A_{(10)6} & A_{(10)7} & A_{(10)8} & A_{(10)9} & A_{(10)(10)} & A_{(10)(11)} & A_{(10)(12)} & 0 & 0 & 0 & 0 \\
0 & 0 & 0 & 0 & 0 & 0 & 0 & 0 & A_{(11)9} & A_{(11)(10)} & A_{(11)(11)} & A_{(11)(12)} & A_{(11)(13)} & A_{(11)(14)} & A_{(11)(15)} & A_{(11)(16)} \\
0 & 0 & 0 & 0 & 0 & 0 & 0 & 0 & A_{(12)9} & A_{(12)(10)} & A_{(12)(11)} & A_{(12)(12)} & A_{(12)(13)} & A_{(12)(14)} & A_{(12)(15)} & A_{(12)(16)} \\
0 & 0 & 0 & 0 & 0 & 0 & 0 & 0 & A_{(13)9} & A_{(13)(10)} & A_{(13)(11)} & A_{(13)(12)} & A_{(13)(13)} & A_{(13)(14)} & A_{(13)(15)} & A_{(13)(16)} \\
0 & 0 & 0 & 0 & 0 & 0 & 0 & 0 & A_{(14)9} & A_{(14)(10)} & A_{(14)(11)} & A_{(14)(12)} & A_{(14)(13)} & A_{(14)(14)} & A_{(14)(15)} & A_{(14)(16)} \\
0 & 0 & 0 & 0 & 0 & 0 & 0 & 0 & 0 & 0 & 0 & 0 & A_{(15)(13)} & A_{(15)(14)} & A_{(15)(15)} & A_{(15)(16)} \\
0 & 0 & 0 & 0 & 0 & 0 & 0 & 0 & 0 & 0 & 0 & 0 & A_{(16)(13)} & A_{(16)(14)} & A_{(16)(15)} & A_{(16)(16)}
\end{bmatrix}
$$

Figure 13-5. Coefficient matrix for four-layered structure when first layer is liquid, where $\bar{A}_{ij} \equiv \lim_{\mu^1 \to 0} A_{ij}$ and $\bar{A}_{61} = \bar{A}_{62} = 0$.

$$
[A] =
\begin{bmatrix}
A_{11} & A_{12} & A_{13} & A_{14} & 0 & 0 & 0 & 0 & 0 & 0 & 0 & 0 & 0 & 0 & 0 & 0 \\
A_{21} & A_{22} & A_{23} & A_{24} & 0 & 0 & 0 & 0 & 0 & 0 & 0 & 0 & 0 & 0 & 0 & 0 \\
A_{31} & A_{32} & A_{33} & A_{34} & A_{35} & A_{36} & 0 & 0 & 0 & 0 & 0 & 0 & 0 & 0 & 0 & 0 \\
A_{51} & A_{52} & A_{53} & A_{54} & \bar{A}_{55} & \bar{A}_{56} & 0 & 0 & 0 & 0 & 0 & 0 & 0 & 0 & 0 & 0 \\
A_{61} & A_{62} & A_{63} & A_{64} & 0 & 0 & 0 & 0 & 0 & 0 & 0 & 0 & 0 & 0 & 0 & 0 \\
0 & 0 & 0 & 0 & A_{75} & A_{76} & 0 & 0 & A_{79} & A_{7(10)} & A_{7(11)} & A_{7(12)} & 0 & 0 & 0 & 0 \\
0 & 0 & 0 & 0 & \bar{A}_{95} & \bar{A}_{96} & 0 & 0 & A_{99} & A_{9(10)} & A_{9(11)} & A_{9(12)} & 0 & 0 & 0 & 0 \\
0 & 0 & 0 & 0 & 0 & 0 & 0 & 0 & A_{(10)9} & A_{(10)(10)} & A_{(10)(11)} & A_{(10)(12)} & A_{(10)(13)} & A_{(10)(14)} & A_{(10)(15)} & A_{(10)(16)} \\
0 & 0 & 0 & 0 & 0 & 0 & 0 & 0 & A_{(11)9} & A_{(11)(10)} & A_{(11)(11)} & A_{(11)(12)} & A_{(11)(13)} & A_{(11)(14)} & A_{(11)(15)} & A_{(11)(16)} \\
0 & 0 & 0 & 0 & 0 & 0 & 0 & 0 & A_{(12)9} & A_{(12)(10)} & A_{(12)(11)} & A_{(12)(12)} & A_{(12)(13)} & A_{(12)(14)} & A_{(12)(15)} & A_{(12)(16)} \\
0 & 0 & 0 & 0 & 0 & 0 & 0 & 0 & A_{(13)9} & A_{(13)(10)} & A_{(13)(11)} & A_{(13)(12)} & A_{(13)(13)} & A_{(13)(14)} & A_{(13)(15)} & A_{(13)(16)} \\
0 & 0 & 0 & 0 & 0 & 0 & 0 & 0 & A_{(14)9} & A_{(14)(10)} & A_{(14)(11)} & A_{(14)(12)} & A_{(14)(13)} & A_{(14)(14)} & A_{(14)(15)} & A_{(14)(16)} \\
0 & 0 & 0 & 0 & 0 & 0 & 0 & 0 & 0 & 0 & 0 & 0 & A_{(15)(13)} & A_{(15)(14)} & A_{(15)(15)} & A_{(15)(16)} \\
0 & 0 & 0 & 0 & 0 & 0 & 0 & 0 & 0 & 0 & 0 & 0 & A_{(16)(13)} & A_{(16)(14)} & A_{(16)(15)} & A_{(16)(16)}
\end{bmatrix}
$$

Figure 13-6. Coefficient matrix for four-layered structure when second layer is liquid, where $\bar{A}_{ij} \equiv \lim_{\mu^2 \to 0} A_{ij}$ and $\bar{A}_{65} = \bar{A}_{66} = \bar{A}_{10\,5} = \bar{A}_{10\,6} = 0$ has been used.

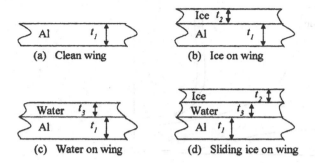

(a) Clean wing   (b) Ice on wing

(c) Water on wing   (d) Sliding ice on wing

$c_{L\,Al} = 6.3\text{ km}/\text{sec}$ $c_{L\,Ice} = 3.98\text{ km}/\text{sec}$ $c_{L\,Water} = 1.48\text{ km}/\text{sec}$

$c_{T\,Al} = 6.3\text{ km}/\text{sec}$ $c_{T\,Ice} = 1.99\text{ km}/\text{sec}$

Figure 13-7. Models used in the ice detection study ($t_1 = 40$ mil, $t_2 = t_3 = 3$ mil).

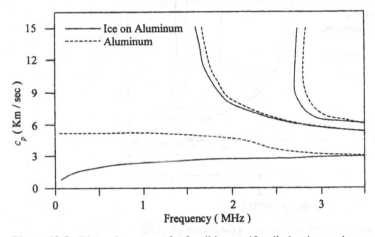

Figure 13-8. Dispersion curves for 3-mil ice on 40-mil aluminum plate.

as illustrated in Figures 13-8, 13-9, and 13-10. We could obtain an amplitude-versus-frequency profile for a particular phase velocity with an angle beam transducer. Then, using for reference the values associated with a clean wing, comparisons could be made with the frequency shifts that characterize water, ice, or sliding-ice conditions.

Obviously, a larger library of curves might be needed. Better yet, if we could evaluate wave structure via elasticity considerations once dispersion curves are available, we could initiate a mode with dominant in-plane displacement on the wing surface that would attenuate only if ice were present (as opposed to water). In-plane displacement, equivalent to shear loading of the water, would not be possible, so severe attenuation of this special mode would serve as an ice detector. Other situations, wing contaminants, and thicknesses could be incorporated, in which case multimode inspection might prove valuable.

13.3.2 Multilayer Model for Titanium Diffusion Bond Inspection

The fine diffusion bonded interface can be viewed as a thin layer; this is shown in Figure 13-11. From the perspective of ultrasonic wave propagation, poor contact can be

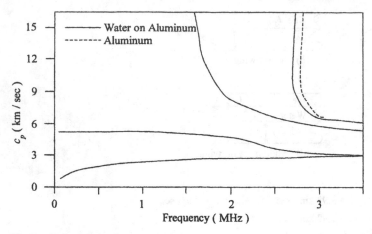

Figure 13-9. Dispersion curves for 3-mil water on 40-mil aluminum plate.

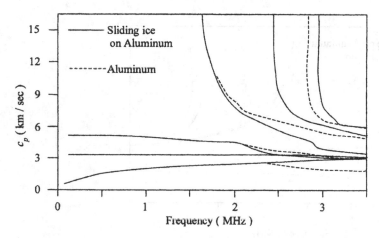

Figure 13-10. Dispersion curves for sliding ice on the aluminum plate (aluminum, 40 mil; ice, 3 mil; water, 3 mil).

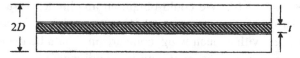

Figure 13-11. Three-layer titanium diffusion bond model.

modeled as a distinct layer with modulus and density degradation values; thus we can depict poor, intermediate, or excellent diffusion bonding (see Rose, Zhu, and Zaidi 1998 for more details). By doing this, dispersion curves for a perfect bond can be compared with degradation models. Figure 13-12 shows phase velocity dispersion curves; in general, degradation model dispersion curves shift to the left. Wave structure is illustrated in Figure 13-13 (note that there are resonable amounts of energy along the interface for the mode studied). An experiment was conducted to see if the frequency shift could be observed; the experimental arrangement is shown in Figure 13-14. A frequency shift experimental result is shown in Figure 13-15. Additional mode features are described in Rose et al. (1998).

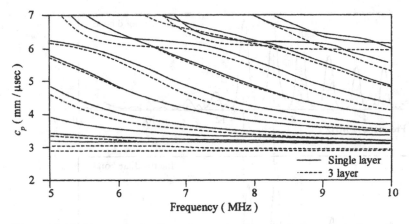

Figure 13-12. Phase velocity dispersion curves for a single-layer and a three-layer titanium model, showing the mode frequency shifts.

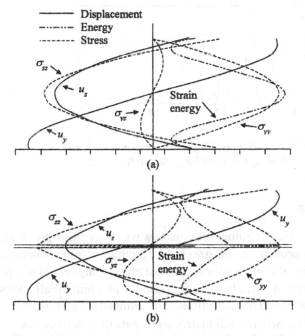

Figure 13-13. Wave structure in titanium plates of (a) the single-layer model and (b) the three-layer model.

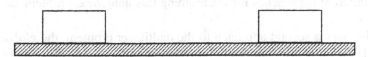

Figure 13-14. Guided wave experimental arrangement.

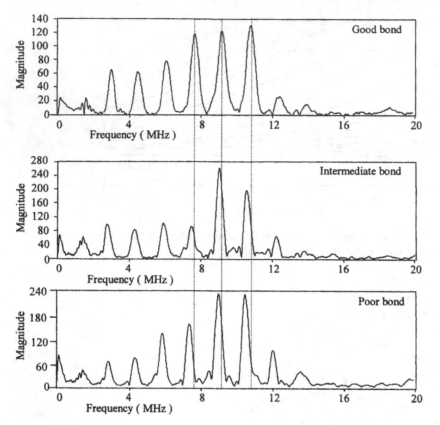

Figure 13-15. Frequency spectrum of the 15-MHz broadband pulse response for 2-mm diffusion bond panels at 60° incidence, showing the mode frequency shifts.

13.4 Concluding Remarks

There are two popular approaches for obtaining solutions to wave propagation problems in multilayer structures. The first approach, the *transfer matrix* method, was introduced in Thomson (1950) and Haskell (1953). The basic idea consists of eliminating all unknowns introduced by the intermediate layers; as a result, the solution to the problem for all layers is expressed in terms of the external boundary conditions. The elimination algorithm can be described by matrix multiplication, where each matrix represents the interface boundary conditions for the intermediate layer. For large fd values, when the waves on the top layers have little influence on those of the bottom layers, the transfer matrix becomes ill-conditioned and so a numerical solution of the characteristic equations becomes unstable. References to the different approaches for overcoming this numerical problem can be found in Lowe (1995).

In this chapter we discussed a second approach to the multilayer problem, the *global matrix* method. A single global matrix can easily be constructed for any N numbers of layers (elastic, viscoelastic, isotropic, anisotropic) by following the technique described here. This method remains numerically robust for any range of fd values, but often involves root extraction from a large-order determinant. Numerical implementation of the transfer matrix may thus be faster, but we prefer the robustness of the global matrix method for a larger class of problem parameters.

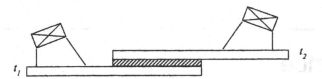

Figure 13-16. Exercise 6.

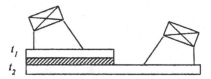

Figure 13-17. Exercise 7.

13.5 Exercises

1. In general, what order determinant must be solved for guided waves in a ten-layered traction-free structure (all solid layers)? Twenty layers?

2. What order determinant must be solved for guided waves in a ten-layer fluid-loaded structure (all solid layers)?

3. What order determinant must be solved for guided waves in a four-layered structure if one layer has a slip or smooth boundary condition?

4. What order determinant must be solved for guided waves in a five-layer structure consisting of four solid layers and one fluid layer?

5. Even for an n-layer structure, the order of the determinant can be reduced. Specify what order might be possible for smooth interfaces, liquid layers, and so on.

6. Explain how it might be possible to generate strong guided wave modes across a three-layered bonded joint structure if entry and reception were only on the single layers having different thicknesses (see Figure 13-16).

7. How might it be possible to generate guided waves across the three-layer structure of Figure 13-17 in such a way that strong modes can be received on a single layer?

8. How would you calculate the wave structure of in-plane and out-of-plane displacement in a four-layer structure?

9. How would you model wave propagation in an airplane wing to differentiate ice from water on the surface?

10. How would you use wave structure to locate a defect in a specific interface layer of a four-layer structure?

11. What would the coefficient matrix look like if the first and third layers were fluids in a four-layer structure?

14

Source Influence

14.1 Introduction

Theoretical analysis is usually based on ideal conditions (e.g., uniform loading and infinite plane wave excitation), and a precise transfer to experimental work is often difficult. In this chapter we consider the practical problem of loading with a finite-sized transducer or source. In particular, we shall address two problems: guided wave application in a traction-free plate, and wave propagation in bulk three-dimensional anisotropic media.

14.2 Guided Waves in a Traction-Free Structure

14.2.1 Background

All the dispersion curves developed for a variety of structural configurations are based on the assumption of an infinite plane wave excitation producing a particular phase velocity value at a specific frequency. In order to determine the utility of these curves (from an experimental point of view) when using transducers of some finite size, we must first investigate a series of experimental, instrumentation, and transducer parameters. In this chapter, we establish guidelines for evaluating the effect of such parameters – or source influence – on the dispersion curves.

In previous chapters we investigated the types of waves that could exist in various structures, from the simple taut string to more complex systems such as a thick layer or cylinder. One prevalent feature of all the wave motions evaluated is that there were various ways in which the structures could vibrate. The string, for example, could have maximal vibration (i.e., an anti-node) at its center and zero displacement on its ends, or it could have zero displacement at the center (i.e, a node) and anti-nodes located halfway between the center and (say) its fixed ends. The membrane stretched across the head of a drum can also vibrate in any one (or more) of a number of sometimes fascinating and amazing ways. Figure 14-1, for example, shows some of the ways in which a square membrane, which is fixed (zero displacement) on all four sides, can vibrate. The vibration modes of a square membrane with fixed ends can be expressed in the form $w_{mn} = \sin(n\pi x)\sin(m\pi y)$.

These different types of vibrations (and the infinitely many more which are not shown) have historically been called "modes" or sometimes "normal modes" of vibration of the membrane. Recall that the normal modes of a structure come about as the solutions of homogeneous boundary value problems. That is, the equations governing the motion of the structure (whether a string, layer, cylinder, membrane, etc.) is given, but the loading terms in the equations were assumed to be zero. In addition, the boundary conditions were also

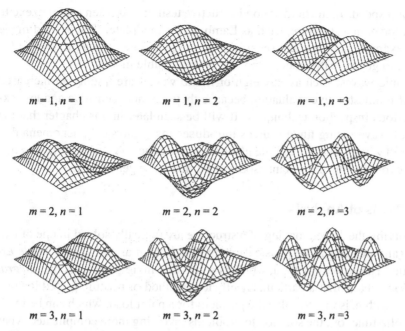

$m = 1, n = 1$ $m = 1, n = 2$ $m = 1, n = 3$

$m = 2, n = 1$ $m = 2, n = 2$ $m = 2, n = 3$

$m = 3, n = 1$ $m = 3, n = 2$ $m = 3, n = 3$

Figure 14-1. First nine vibration modes of a square membrane, $w(x, y) = \sin(n\pi x)\sin(m\pi y)$.

taken as zero (i.e., zero displacement or slope at the ends of a string, zero tractions on the surfaces of a layer or cylinder, zero displacement around the periphery of the membrane, etc.). The normal modes therefore represent the possible vibration characteristics of the structure, and they are independent of any load that may be applied to the structure. For this reason, the normal modes are also often called "natural" modes of the vibration of the structure.

The fact that all structures can vibrate in an infinite number of ways (modes) leads to some interesting questions. In what manner will the structure vibrate in any given situation? How can we get it to vibrate in the manner we desire? That a square membrane can vibrate in any of the modes shown in Figure 14-1 does not mean that it necessarily will vibrate in one of these ways. We might wonder just how to induce the membrane to vibrate in one of the different manners shown – for example, what would cause the membrane to vibrate in the relatively simple (1, 1) pattern in the upper left frame as opposed to the intricate (3, 3) pattern in the lower right frame?

In this chapter we will see that the manner in which a structure vibrates depends essentially upon the type of loading applied to the structure. (Indeed, an unloaded structure will not move at all, regardless of its potential vibrations.) We will examine the question of how to excite the desired types of waves in a structure. Our attention will be focused primarily on the selective excitation of guided Lamb waves in layers – using realistic, finite-sized transducers.

The results of this chapter are extremely useful in many areas. In the design of structures, for example, it is critical to determine the types of loads a structure may be subjected to and thereby avoid situations where the structure vibrates uncontrollably in response to those loads. A good example of this was a bridge that began to oscillate in one of its natural modes of vibration (and eventually collapsed) because of transverse winds blowing across it at a certain speed. In the field of nondestructive testing, it is often useful to excite particular wave modes in a structure, such as Lamb waves in a plate. The desired mode would have, for example, an excellent sensitivity to the defect being sought, and so for testing we'd want to generate that mode – not the other possible modes.

Guided ultrasonic waves, such as Rayleigh or Lamb waves, are finding an increased use in the field of nondestructive evaluation because they are more sensitive and less expensive than previous inspection techniques. It will be seen later in this chapter that the excitation of these waves using finite sources introduces many physical phenomena that are not encountered when assuming infinite plane wave incidence. A major portion of this chapter will be devoted to this important aspect of guided wave generation.

14.2.2 Methods of Analysis

Problems involving the forced loading of a structure are typically solved in one of two ways. The first (and perhaps more physical) method is known as the *normal mode expansion* technique. The second method, which is somewhat more general, is the *integral transform* technique. Note that a particularly efficient method of treating forced loading problems, when possible, is first to solve the problem for a point load, which can be modeled by a Dirac delta function; the solution to problems involving more complicated types of loading can then be found by superposition of the point-load solution, assuming the governing equation is linear. The solution to a forced loading problem involving a point source is often termed the *Green's function* (after the nineteenth-century English mathematician, George Green) of the associated governing equation. However, this technique does not warrant a separate treatment here because the solution for the Green's function is most often accomplished by the use of integral transform methods. In this section we will concentrate on the normal mode expansion technique, although the integral transform technique will be briefly discussed also.

Many integral transforms have been developed over the years to aid in solving a variety of ordinary and partial differential equations. Some of the more popular transforms are Laplace, Fourier, Hankel, and Mellin. As the term suggests, the various integral transforms are used to transform a given function into another function. This transformation is done via an integration (over some domain) of the original function multiplied by some known "kernel" function. The transforms mentioned have different kernels and, in some cases, different integration domains.

Each of the integral transforms can be written in the symbolic form as

$$F(\alpha) = \int_a^b f(x)K(x,\alpha)\,dx, \tag{14.1}$$

where $F(\alpha)$ is called the transform of the function $f(x)$, $K(x,\alpha)$ is the kernel of the integral transform, and the domain is $[a, b]$. In most cases, either or both of a and b is infinite. Of course, the functions $f(x)$ and $K(x,\alpha)$ and the interval $[a, b]$ must be such that the integral converges to some finite value for every choice of a in its domain, which may be different from the domain of the original variable x and may even be complex. The kernels, domains, and conditions for existence of the particular transforms mentioned here can be found in math books treating integral transforms (see e.g. Davies 1985; Sneddon 1995).

We will not discuss the intricacies of any particular transform (this is best left to math texts), but we will discuss why they are useful in solving boundary value problems. The main reason for their utility is that, by proper choice of integral transform, differential equations involving any number of independent variables can be reduced to differential equations involving fewer independent variables. The use of the Laplace transform to convert ordinary differential equations (i.e., in one independent variable) to algebraic equations should be familiar. Similarly, a partial differential equation (PDE) in two independent variables (say, x and t) can be reduced to an ordinary differential equation (ODE) in either x or t by performing an integral transform on the other variable, thus effectively reducing it to a constant parameter in the resulting ODE. The reason the simplification occurs is that, for most of the transforms that find frequent use, there is a simple relation between the transform of a function's derivative and the transform of the function itself. This relation usually amounts to a multiplication of the transform by some power of the transform variable, the power depending upon the order of the derivative. The resulting equation, with fewer number of independent variables, is often easier to solve than the original equation. Of course, it is not the originally sought function that is solved for after transformation; instead, it is the transform of the original function that is obtained as the solution of the simpler differential equations.

Having obtained the transform of the desired function – say, $F(\alpha)$ – from the solution of a simpler problem, the task remaining is to recover from it the original (nontransformed) function $f(x)$. This task is invariably the most difficult part of applying integral transforms to the solution of boundary value problems. Of course, a method must be specified for obtaining the original function from its transform, a process known as *inverse transformation*. Any transform used in analysis has as a counterpart an inverse transformation, which specifies how the original function is to be recovered from the transformed function. This inverse transform can be written in the general form of

$$f(x) = \int_c^d F(\alpha)G(x, \alpha)\,d\alpha, \tag{14.2}$$

where $G(x, \alpha)$ represents the kernel of the inverse transform and the inversion is over the domain $[c, d]$. Again, either c or d (or both) can be infinite.

In most situations, the inversion integrals cannot be evaluated exactly in closed form. In such cases, the general approach is as follows.

(1) Expand the integrand into a series of terms and integrate (if possible) the series term by term, retaining only the first several terms.
(2) Resort to some type of asymptotic expansion techniques (Bleistein and Handelsman 1986) to estimate the values of the original function for specific (usually very large or small) values of the independent variable x.
(3) Evaluate the integral numerically.

In cases where the inversion integrals can be evaluated exactly, this is usually achieved through integration in the complex plane, treating the transform parameter α as a complex variable. Application of the residue theorem from complex analysis often enables the inversion integral to be written as the sum of terms, each one coming about from a particular singularity (e.g., a pole or branch point) of the integrand. In wave propagation problems,

each of the resulting terms usually has a direct physical meaning – for instance, as a reflected wave or a particular mode of a waveguide. The example of shear horizontal wave
generation in a flat layer (discussed in Chapter 15) illustrates all of these points.

14.2.3 Normal Mode Expansion Technique

We first discuss some preliminaries. The normal mode expansion technique of solving
forced loading problems is analogous to the so-called eigenfunction expansion methods
discussed in most mathematics texts (see Friedman 1990). The normal modes of the structure serve as the eigenfunctions. The main idea of the method is to assume, a priori, that
the sought function (or functions) can be written in the form of a series of other, known
functions (i.e., the eigenfunctions or normal modes), each with an unknown amplitude.
The goal is then to find either a general expression for the unknown amplitudes, or a numerical estimate of them. In the former case, where a general expression can be found
for the amplitudes, the resulting series should represent the exact solution to the original
equation(s). In the latter case, where numerical evaluation is necessary, an approximate
solution is obtained since (i) the expansion coefficients are only approximately evaluated and (ii) only a finite number of such amplitudes can be calculated, resulting in a
finite-series approximation to the true infinite-series solution. Note that the efficacy of
normal mode expansion depends directly on two main considerations: completeness and
orthogonality of the eigenfunctions.

Completeness of Eigenfunctions

We must emphasize that the normal modes of the structure being analyzed (Lamb wave
modes of a flat layer, vibration modes of a membrane or of a stretched string, etc.) are
assumed to be complete in the space of solutions of the governing equation(s). A set of
functions (such as the normal modes) are said to be *complete* in a particular space of functions if any of the functions in that space can be exactly represented in terms of a finite
or infinite number of functions in the set. For example, in the space of twice differen-
tiable functions $u(x)$ satisfying the boundary conditions $u(0) = u(1) = 0$, it is known
that the set of functions $\sin(n\pi x)$ for $n \in \{1, 2, \ldots, \infty\}$ are complete (Friedman 1990).
This means that any function which can be differentiated twice (without having singular
points) and which vanishes at $x = 0$ and $x = 1$ can be expressed, for any x in the interval
$[0, 1]$, as a series in the form

$$u(x) = \sum_{n=0}^{\infty} A_n \sin(n\pi x)$$

for some choice of expansion coefficients A_n. That is, there is no twice-differentiable
function vanishing at $x = 0$ and $x = 1$ that cannot be expressed in the form of the equality
just displayed. A more rigorous definition of completeness, stating essentially the same
thing, can be found in many math texts (see e.g. Hochstadt 1989).

If the set of normal modes were *not* complete then the solution of the governing equation might not be expressible in the form of a series of the normal modes. On the other
hand, if the normal modes are complete then we are assured our expansion is valid, since
all solutions to the governing equations are thereby expressible in that form. Of course,
the assumed series is useful only if we can somehow determine the appropriate expansion
coefficients.

Orthogonality of Eigenfunctions

The concept of orthogonality of a set of functions is an abstraction of the familiar notion of the orthogonality of vectors in space, that is, vectors whose dot products with each other vanish.

In function spaces, the dot product is generalized as the integral of the product of two functions over some definite domain. That is, given any set of functions $\phi_m(x)$ for $m \in \{0, 1, \ldots, \infty\}$ defined over the domain $x \in [a, b]$, we say that the set is *orthogonal* (or the functions are orthogonal to each other) over the interval $[a, b]$ if and only if

$$\int_a^b \phi_m(x)\phi_n(x)\,dx = 0 \quad \text{for } m \neq n. \tag{14.3}$$

An example of functions that are orthogonal over $[0, 1]$ is the familiar $\sin(m\pi x)$ for $m \in \{1, \ldots, \infty\}$; these functions are hence both complete and orthogonal in the interval $[0, 1]$.

The orthogonality of a set of functions used to expand other functions permits a straightforward method of determining the appropriate expansion coefficients. All one need do is multiply both sides of the expansion by any member of the set and integrate both sides of the resulting equality over the appropriate domain, integrating the series term by term. Owing to orthogonality, the only term remaining in the summation is that which contains the coefficient of the function that multiplied both sides. Formally, the procedure may be described as follows:

$$u(x) = \sum_{n=0}^{\infty} A_n \phi_n(x); \tag{14.4}$$

therefore, multiplying both sides by (say) $\phi_m(x)$ and integrating over $[a, b]$, we have

$$\int_a^b u(x)\phi_m(x)\,dx = \int_a^b \left\{ \phi_m(x) \sum_{n=0}^{\infty} A_n \phi_n(x) \right\} dx. \tag{14.5}$$

Taking the integral inside the summation yields

$$\int_a^b u(x)\phi_m(x)\,dx = \sum_{n=0}^{\infty} A_n \left\{ \int_a^b \phi_m(x)\phi_n(x)\,dx \right\}. \tag{14.6}$$

Finally, from the orthogonality of the set of functions $\{\phi_n(x)\}$, the only nonzero term in the summation is that for which $n = m$. Therefore, we immediately have the following result:

$$A_m = \frac{\int_a^b u(x)\phi_m(x)\,dx}{\int_a^b \phi_m^2(x)\,dx}. \tag{14.7}$$

Orthogonality of the normal modes (or eigenfunctions) is therefore seen to be a useful tool that enables the expansion coefficients to be determined.

In the application of these concepts to the case of wave excitation in waveguides, it will be assumed that the normal modes used in the expansion are complete. It should be noted, however, that no general proof of such completeness exists for the normal modes of even an isotropic flat layer (i.e., the Lamb wave modes). Nonetheless, their completeness is almost always assumed. In contrast, we must establish orthogonality of the normal modes because that property constitutes the tool by which the expansion amplitudes are determined. It will be seen in the example to follow that the orthogonality property of the normal modes is slightly more complex than the simple case discussed here. This should not confuse the issue, however; regardless of the form taken by the orthogonality condition, it remains useful because it allows us to determine the expansion coefficients.

14.2.4 Sample Problem

As an example of the use of the normal mode expansion technique to solve problems involving the forced loading of waveguides, in this section we will solve the problem of the excitation of an arbitrarily anisotropic layer due to arbitrarily applied, time-harmonic surface tractions. This is essentially the same presentation as in Ditri and Rose (1994). Other examples of the use of the normal mode expansion technique to solve edge load problems for plates can be found in Folk and Herczynski (1986) and, for solid circular cylinders, in Herczynski and Folk (1989); excitation problems for hollow circular cylinders are examined in Ditri and Rose (1992). The key to the technique's success is the establishment and use of an orthogonality condition between the free waveguide modes, analogous to the orthogonality of the trigonometric functions used in Fourier series expansions.

Note that this example has been chosen not only because it illustrates the application of the normal mode expansion technique but also because it sheds much light on the technically important problem of using finite sources to generate guided waves. A simpler example of the application of the normal mode expansion technique is left to the exercises.

Starting with the expansion amplitudes for the case of harmonic excitation, the more general problem of arbitrary time-dependent loading can be treated by linear superposition of harmonic solutions. This type of approach leads to solutions (expansion amplitudes) that are identical in form for isotropic or generally anisotropic layers. The difference is accounted for when evaluating the quantities appearing in the solutions. Because of this, the true physical nature of the excitation process becomes clear, unobstructed by the peculiarities of a particular material. This is the beauty of the normal mode expansion technique as compared with integral transform methods. When treating the same problem with integral transforms, the amount of algebra quickly becomes untractable because the explicit details of the solution must be carried along at every step. Compare, for instance, the solution of the current problem for generally anisotropic layers with that for an isotropic layer using integral transform techniques (Viktorov 1967).

Consider the normal modes. The governing equations for wave propagation in flat, linearly elastic, non-piezoelectric, and generally anisotropic layers, for zero body forces, can be written as

$$\frac{\partial T_{ij}}{\partial x_j} = \rho \frac{\partial^2 u_i}{\partial t^2}, \tag{14.8a}$$

$$T_{ij} = C_{ijkl} \frac{\partial u_k}{\partial x_l}, \tag{14.8b}$$

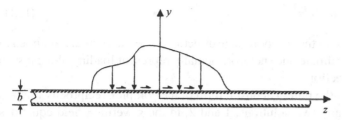

Figure 14-2. Loading of an anisotropic layer by harmonic surface traction.

where T denotes the Cauchy stress tensor, u the particle displacement vector, C the stiffness tensor, and ρ the mass density. The usual summation convention over repeated subscripts is assumed.

Along with the governing equations (14.8a,b), the normal modes of the layer are required to satisfy the traction-free boundary conditions

$$t_i \equiv T_{ij} n_j = 0 \tag{14.9}$$

at both the upper and lower surfaces of the layer, where the n_j are components of a unit vector normal to the surfaces of the layer. Solutions to the boundary value problem – via equations (14.8) and (14.9) – can be found in the literature for layers of various classes of anisotropy (Li and Thompson 1990; Nayfeh, Taylor, and Chimenti 1988; Solie and Auld 1973).

When propagating in an anisotropic layer, the characteristics of the normal modes (i.e., phase and group velocities, field distributions, etc.) are generally very dependent on the orientation between the wavevector of the mode and the natural or crystallographic coordinate system of the media. To see when such preferred axes exist, consult for instance Datta et al. (1988b,c). In all the derivations that follow, it is assumed that the z-coordinate is aligned with the wavevector of the generated mode and that the y-axis is transverse to the layer (i.e., in the thickness direction). For anisotropic layers, this will generally necessitate a coordinate transformation of the elastic stiffness tensor since, in general, the principal axes of the anisotropic medium will not coincide with the chosen coordinate axes. With this choice of coordinate system (and the subsequent transformation of the elastic stiffness tensor), the fields associated with any given mode are uniform in the x-coordinate direction. This serves to make the problem one of plane strain in the (y, z)-plane (i.e., $\varepsilon_{xx} = \varepsilon_{xy} = \varepsilon_{xz} \equiv 0$ and $\varepsilon_{ij} = \varepsilon_{ij}(y, z, t)$ where $i, j \in \{y, z\}$).

The fields caused by the waveguide modes can now be written in the general form of

$$\xi(y, z, t) = \bar{\xi}(y) e^{i(\omega t - kz)}, \tag{14.10}$$

where $\bar{\xi}(y)$ represents the *modal distribution* or y-variation of the field ξ, and k is the wavenumber of the mode which, for a given frequency ω, is determined by a generalized Rayleigh–Lamb type of equation whose form depends on the type of anisotropy exhibited by the layer (Li and Thompson 1990). Just as in the isotropic case, the wavenumber k can be real or complex for a given real and positive frequency ω.

The specific problem to be addressed is shown in Figure 14-2. A linearly elastic, generally anisotropic layer (i.e., having up to 21 independent elastic moduli) is loaded over a finite portion of its top surface by a traction force of

$$\bar{t}(z)e^{i\omega t} = [t_z(z)\hat{e}_z + t_y(z)\hat{e}_y]e^{i\omega t}, \tag{14.11}$$

where $\hat{e}_i$ denotes a unit vector in the x_i-coordinate direction. The tractions are assumed to be independent of the x-coordinate and therefore actually represent loading along a strip of infinite width in the x direction.

We start the analysis from the purely acoustic form of the complex reciprocity relation (Auld 1990), which relates any two solutions, 1 and 2, to the governing field equations (14.8). This relation can be written in differential form as

$$\nabla \cdot (\tilde{\bar{v}}_2 \cdot \bar{T}_1 + \bar{v}_1 \cdot \tilde{\bar{T}}_2) = 0, \tag{14.12}$$

where $\bar{v}$ represents the particle velocity field and the tilde denotes complex conjugation. In (14.12), both solutions (1 and 2) are assumed to have the same frequency ω; the $e^{i\omega t}$ dependence of all field variables will be dropped for brevity.

The fields in the loaded waveguide are then expressed as a summation over all of the normal modes of the free layer, specified by index μ and multiplied by unknown, generally complex, and z-dependent amplitudes $A_\mu(z)$. For the field variables appearing in (14.12), these expansions are in the following form:

$$\bar{v}(y, z) = \sum_\mu A_\mu(z)\bar{v}_\mu(y), \tag{14.13a}$$

$$\bar{T}(y, z) = \sum_\mu A_\mu(z)\bar{T}_\mu(y). \tag{14.13b}$$

At this point, we make two important notes. (1) The same set of amplitude coefficients, $A_\mu(z)$, is used for expansion of both the particle velocity and stress fields. This may seem inconsistent in that, starting with the expansion for the particle velocity field, the expansion for the stress field can be obtained by (i) integrating with respect to time to obtain the particle displacement field, (ii) differentiating spatially to obtain the strains, and (iii) applying Hooke's law to discover the stresses. In the process of doing this, multipliers to the amplitude coefficients A_μ will be generated; for example, the elastic constants of the medium arising from the application of Hooke's law show this phenomenon. It is assumed, when using the same coefficients for both expansions, that these extra multiplying factors have been absorbed into the modal stress field $\bar{T}_\mu$. (2) Unlike the simple example discussed previously, the expansion amplitudes are now considered to be functions of the position along the plate z. This means that the relative importance of each individual mode, specified by the index μ, is a function of position along the plate.

In order to determine the unknown mode amplitudes $A_\mu(z)$ in (14.13a,b), we invoke the reciprocity relation (14.12) with solution 1 being the $\bar{v}$ and $\bar{T}$ of (14.13a,b) and solution 2 being the νth mode of the free layer:

$$\bar{v}_2 = \bar{v}_\nu(y)e^{-ik_\nu z}, \tag{14.14a}$$

$$\bar{T}_2 = \bar{T}_\nu(y)e^{-ik_\nu z}. \tag{14.14b}$$

Substituting (14.13a,b) and (14.14a,b) into the reciprocity equation (14.12), integrating the resulting equation across the waveguide thickness $-b/2 \le y \le b/2$, and making use of the fact that the normal modes satisfy traction-free boundary conditions, we derive a first-order ODE that governs the amplitudes of the generated modes. The wavenumber k_ν is real for propagating modes, and it can be shown by specializing a general equation in Auld (1990) that, for the case considered here, the equation governing the amplitude of the generated modes reduces to the form

$$4P_{\nu\nu}\left(\frac{d}{dz} + ik_\nu\right)A_\nu(z) = \bar{\tilde{v}}_\nu\left(\frac{b}{2}\right) \cdot \bar{t}(z), \tag{14.15}$$

where the quantity $P_{\nu\nu}$ has been defined as

$$P_{\nu\nu} = \text{Re}\left[-\frac{1}{2}\int_{-b/2}^{b/2}\bar{\tilde{v}}_\nu \cdot \bar{\tilde{T}}_\nu \cdot \hat{e}_z\,dy\right] \tag{14.16}$$

with Re[·] denoting the real part of the quantity in brackets. From the definition of the acoustic Poynting vector (Auld 1990), $P_{\nu\nu}$ is recognized as the time-averaged power flow carried by the νth mode in the z direction per unit waveguide width (in the x direction), with typical units of watts per meter. It should be remembered that, since the amplitudes of the modal fields ($\bar{v}_\nu$ and $\bar{T}_\nu$) can themselves be arbitrarily assigned (within the restrictions of infinitesimal elasticity theory), the power carried by the modes, equation (14.16), is also arbitrary and depends on the assigned amplitude. We will see that this indeterminacy will disappear in the final result, which will be independent of the arbitrary amplitude that may be assigned to $\bar{v}_\nu$ and $\bar{T}_\nu$.

We note that the orthogonality of the normal modes has been used in reducing equation (14.12) to (14.15). The orthogonality of Lamb wave modes is considerably more complicated than the cases discussed here (see Auld 1990), but it remains a useful feature for enabling extraction of a single amplitude coefficient from the normal mode summation. In contrast to the simple case discussed previously (where the amplitude coefficients were solved for directly by using the orthogonality relation), we have now obtained an ordinary differential equation governing the coefficients.

Equation (14.15) is a first-order ODE and can therefore be integrated using standard techniques. The solution can be written as

$$A_\nu(z) = \frac{e^{-ik_\nu z}}{4P_{\nu\nu}}\int_c^z e^{ik_\nu\eta}\bar{\tilde{v}}_\nu\left(\frac{b}{2}\right)\cdot\bar{t}(\eta)\,d\eta, \tag{14.17}$$

where c is a constant used to satisfy the boundary condition on $A_\nu(z)$. Suppose that the external tractions $\bar{t}$ are nonzero only in the interval $-L \leq z \leq L$. Then, using the boundary condition

$$A_\nu(z) = 0 \quad (z \leq -L), \tag{14.18}$$

we have

$$A_{+\nu}(z) = \frac{e^{-ik_\nu z}}{4P_{\nu\nu}}\int_{-L}^{L} e^{ik_\nu\eta}\bar{\tilde{v}}_\nu\left(\frac{b}{2}\right)\cdot\bar{t}(\eta)\,d\eta \quad (z \geq L) \tag{14.19}$$

for the amplitude of the rightward traveling modes. The corresponding expression for the leftward propagating modes can be obtained from (14.19) by replacing $P_{\nu\nu}$ with $-P_{\nu\nu}$, replacing $\tilde{v}_\nu$ with $\tilde{v}_{-\nu}$, replacing the wavenumber k_ν with $-k_\nu$, and finally using the boundary condition $A_{-\nu} = 0$ for $z \leq -L$. The result is

$$A_{-\nu}(z) = \frac{e^{-ik_\nu z}}{4P_{\nu\nu}} \int_{-L}^{L} e^{-ik_\nu \eta} \tilde{\bar{v}}_{-\nu} \left(\frac{b}{2}\right) \cdot \bar{t}(\eta)\, d\eta \quad (z \leq -L). \tag{14.20}$$

Because the applied tractions vanish outside the interval $|z| > L$, the integration limits in (14.19) and (14.20) can be extended indefinitely (i.e., to $+\infty$ and $-\infty$). The two equations can also be combined:

$$A_{\pm\nu}(z) = \frac{e^{\pm ik_\nu z}}{4P_{\nu\nu}} \tilde{\bar{v}}_{\pm\nu} \left(\frac{b}{2}\right) \cdot \int_{-\infty}^{\infty} e^{\pm ik_\nu \eta} \bar{t}(\eta)\, d\eta. \tag{14.21}$$

Now that we have determined the expansion amplitudes $A_{\pm\nu}(z)$ of the propagating modes (real k_ν), the fields in the loaded layer due to the propagating modes follow from the original expansions (14.13a,b). For instance, the particle velocity field in the loaded layer can be written as

$$\bar{v}(y, z) = \sum_\nu \left\{ \frac{e^{\mp ik_\nu z}}{4P_{\nu\nu}} \tilde{\bar{v}}_{\pm\nu} \left(\frac{b}{2}\right) \cdot \int_{-\infty}^{\infty} e^{\pm ik_\nu \eta} \bar{t}(\eta)\, d\eta \right\} \bar{v}_{\pm\nu}(y)$$
$$+ \sum_\nu A^e_{\pm\nu}(z) \bar{v}_{\pm\nu}(y). \tag{14.22}$$

For $z \geq L$, only the summation over the positive traveling (or decaying) modes $(+)$ contribute, whereas for $z \leq L$ only the summation over the negative traveling (or decaying) modes $(-)$ contribute. The first summation represents the contribution of the propagating modes (real wavenumber k_ν); the second represents the contribution of the evanescent modes (imaginary and/or complex wavenumbers) with expansion amplitudes $A^e_{\pm\nu}(z)$. Because the evanescent modes decay exponentially away from the source that excites them, the contribution from the second summation becomes smaller as z becomes larger. The actual distance from the source at which the amplitude of an evanescent mode becomes negligible depends on the magnitude of the imaginary part of its wavenumber. In fact, the dependence of the expansion amplitudes of the evanescent modes on z will be a factor of the form $e^{-\alpha z}$ for rightward traveling modes and $e^{\alpha z}$ for leftward traveling modes.

Once the applied tractions $\bar{t}(z)$ are specified, (14.21) can be used to determine the amplitudes of the forward and backward traveling modes, which – when substituted into the original normal mode expansion (14.13a,b) – allow us to attain that part of the total field in the loaded layer that is due to the propagating modes. The part of the total field due to the nonpropagating modes can also be determined by this procedure, but their contribution is of less importance for nondestructive evaluation since they decay exponentially from the source that excites them. Their amplitudes are not, however, given by equation (14.21), which is valid only for propagating modes. For the evanescent modes, the differential equation (14.15) governing the amplitudes must be modified.

14.2.5 Specific Cases

In this section, we will use the general results just derived to investigate the effects of using a finite source to excite Lamb waves in a layer. It will be seen that we can obtain

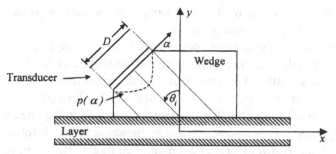

Figure 14-3. Model of the wedge method of generating guided waves.

several interesting and practically useful results that have no counterpart when dealing with the hypothetical case of infinite plane wave generation of Lamb waves.

It is now assumed that the tractions produced at the upper surface of the layer are due to loading by an ultrasonic transducer mounted to an angle beam wedge, as shown in Figure 14-3. The ultrasonic transducer is assumed to produce a time-harmonic plane stress wave that travels within the wedge and strikes the layer at a given angle θ_i. We ignore the effects of beam spreading and beam shifting upon striking the layer, and the wedge is assumed to be coupled to the layer by a thin layer of nonviscous liquid. As a result, only normal tractions $t_y(z)$ are assumed to be transferred; the shear tractions $t_z(z)$ are assumed to vanish. Finally, the transducer is allowed to have an arbitrary pressure distribution, $p(\alpha)$, across its face. A similar problem for isotropic layers and a pistonlike transducer was solved using Fourier integrals by Viktorov (1967).

The transducer, of width D, is mounted to a wedge and produces a collimated plane wave with arbitrary pressure distribution, $p(\alpha)$. Projecting $p(\alpha)$ onto the layer surface (and neglecting an unnecessary phase factor and the $e^{i\omega t}$ dependence), the traction produced on the top of the layer is assumed to be of the form

$$\bar{t}(b/2, z) = \begin{cases} p(z\cos\theta_i)|R(\theta_i)|e^{-ik_w \sin\theta_i z}\hat{e}_y & \text{if } |z| \leq L, \\ 0 & \text{if } |z| > L, \end{cases} \qquad (14.23)$$

where $|R(\theta_i)|$ is a numerical factor introduced to account for the fact that the actual traction at the wedge–layer interface will in general be different from that due solely to the incident wave. We assume, however, that this factor is independent of frequency (a discussion of this result, along with limited experimental verification, can be found in Ditri 1992). In (14.23), k_w represents the wavenumber of the incident wave in the wedge and is numerically equal to ω/c_w, where c_w is the velocity of longitudinal waves in the wedge material.

Substituting this assumed form of surface traction into the general result (equation (14.21)) yields, after some manipulation,

$$A_{\pm\nu}(z) = \frac{|R(\theta_i)|}{4} \frac{\tilde{\upsilon}_{\pm\nu y}(b/2)}{P_{\nu\nu}} \frac{e^{\mp ik_\nu z}}{\cos(\theta_i)} \int_{-\infty}^{\infty} p(\alpha)e^{\pm i\chi_\pm\alpha}\, d\alpha, \qquad (14.24)$$

where

$$\chi_+ \equiv \frac{k_\nu - k_w \sin\theta_i}{\cos\theta_i}, \qquad (14.25a)$$

$$\chi_- \equiv \frac{k_\nu + k_w \sin\theta_i}{\cos\theta_i}, \qquad (14.25b)$$

and $v_{\pm vy}$ represents the complex conjugate of the y component of the particle velocity field of positive $(+)$ and negative $(-)$ propagating modes.

The appearance of the factor P_{vv} in the denominator of (14.24) for the amplitudes ensures that the final result is independent of the arbitrarily chosen amplitude of $\bar{v}_v$. This is so because, when A_v of (14.24) is substituted into the normal mode expansions (14.13a,b), the latter contain products of the form $v_{vy}(b/2) \cdot \bar{v}_v(y)/P_{vv}$ and $v_{vy}(b/2) \cdot \bar{T}_v(y)/P_{vv}$. If $\bar{v}_v$ is replaced by $\alpha \bar{v}_v$ (where α is an arbitrary complex constant) then, owing to the linearity of the governing elasticity equations, $\bar{T}_v$ is also multiplied by the same factor. It follows that the terms $v_{vy}(b/2) \cdot \bar{v}_v(y)$ and $v_{vy}(b/2) \cdot \bar{T}_v(y)$ are multiplied by α^2. However, reference to equation (14.16) shows that P_{vv} is also proportional to the product $\bar{v}_v(y) \cdot \bar{T}_v(y)$ and is therefore also multiplied by α^2. This ensures that the ratios $v_{vy}(b/2) \cdot \bar{v}_v(y)/P_{vv}$ and $v_{vy}(b/2) \cdot \bar{T}_v(y)/P_{vv}$ remain the same. Therefore, the final fields caused in the layer by the individual modes – equations (14.13a,b) – are independent of the arbitrarily assigned amplitudes of the modal fields appearing in their definitions. One could have started the normal mode expansion technique by expanding the fields in the loaded layer in terms of the power normalized normal modes, $\bar{v}_v/\sqrt{P_{vv}}$ and $\bar{T}_v/\sqrt{P_{vv}}$, which are already independent of the arbitrarily chosen multiplicative amplitudes for $\bar{v}_v$ and $\bar{T}_v$. In this case, $A_{\pm v}$ would have a $\sqrt{P_{vv}}$ factor instead of P_{vv} in its denominator.

The integral appearing in (14.24) can be interpreted as the Fourier transform of the pressure distribution function $p(\alpha)$ with transform parameter χ given by (14.25a) or (14.25b). It clearly exhibits the influence of the transducer's pressure distribution on its guided wave–generation characteristics.

In order to clearly manifest the physics of the excitation phenomenon, we rewrite equation (14.24) as

$$A_{\pm v}(z) = GF^{\pm}E_{\pm v}e^{\mp ik_v z}, \tag{14.26}$$

where

$$G \equiv \frac{|R(\theta_i)|}{4}, \tag{14.27a}$$

$$E_{\pm v} \equiv \frac{\tilde{v}_{\pm vy}(b/2)}{P_{vv}}, \tag{14.27b}$$

$$F^{(\pm)} \equiv \frac{1}{\cos(\theta_i)} \int_{-\infty}^{\infty} p(\alpha)e^{\pm i\chi \pm \alpha} \, d\alpha. \tag{14.27c}$$

The function G represents a numerical factor depending solely on the output power of the transducer. The function $E_{\pm v}$, termed the *excitability function* of mode v, depends only on properties of the mode being excited and not on the properties of the source used for excitation. The function $F^{(\pm)}$ depends only on properties of the transducer and wedge used to excite the layer. The product function, FG, which is also dependent only on transducer parameters, is termed the *excitation function* of the source. We can see that the amplitude with which guided waves are generated is proportional to the product of two factors:

(1) the excitation function FG, determined by transducer and wedge parameters; and
(2) the excitability function $E_{\pm v}$, which depends on the mode that is excited and also on where (on its dispersion curve) it is excited.

The actual pressure distribution $p(\alpha)$ of a transducer depends essentially on how the transducer is manufactured. The simplest approximation to an actual pressure variation is the piston source, defined as

$$
p(\alpha) = \begin{cases} \sigma_0 & \text{if } |\alpha| \leq D/2, \\ 0 & \text{if } |\alpha| > D/2. \end{cases} \tag{14.28}
$$

If this pressure variation is substituted into (14.24), the resulting expression for the excitation amplitudes is found to be:

$$
A_{\pm\nu}(z) = \frac{\sigma_0 |R(\theta_i)|}{4} \frac{\tilde{v}_{\pm\nu y}(b/2)}{P_{\nu\nu}} 2 \frac{\sin\left(\dfrac{(k_\nu \mp k_w \sin\theta_i)D}{2\cos\theta_i}\right)}{(k_\nu \mp k_w \sin\theta_i)} e^{\mp i k_\nu z}, \tag{14.29}
$$

where the positive amplitudes are valid for $z \geq L$ and the negative amplitudes for $z \leq -L$. Comparing (14.29) with (14.26) shows that, for a piston-source transducer,

$$
G = \frac{\sigma_0 |R(\theta_i)|}{4}, \qquad E_{\pm\nu} = \frac{\tilde{v}_{\pm\nu y}(b/2)}{P_{\nu\nu}},
$$

and

$$
F_{\text{piston}}^{(\pm)} = 2\frac{\sin\left(\dfrac{(k_\nu \mp k_w \sin\theta_i)D}{2\cos\theta_i}\right)}{(k_\nu \mp k_w \sin\theta_i)}; \tag{14.30}
$$

except for notation, this is identical to the result found by Viktorov (1967) in his Fourier integral treatment of piston-source excitation of isotropic layers.

It is interesting to note from (14.29) that, unlike the case of infinite plane wave excitation, Snell's law is not rigorously applicable in describing the generation process of Lamb waves when using finite sources, whose amplitude function is nonzero for a *range* of wavenumbers (or phase velocities) and not just for the single wavenumber determined by Snell's law (equation (14.32)). Even so, the most common method found in the literature of determining the appropriate shoe angle to use when trying to generate a Lamb wave mode with phase velocity c_p is to use Snell's law,

$$
\theta_i = \sin^{-1}(c_w/c_p), \tag{14.31}
$$

where c_w is the longitudinal wave velocity in the shoe material (or coupling fluid if immersion is used). In terms of wavenumbers, this relation can be written as

$$
k = k_w \sin\theta_i. \tag{14.32}
$$

Moreover, it is usually yet erroneously assumed that, given an incident angle θ_i and incident wavenumber k_w, only a single wavenumber (given by Snell's law) can be generated – or at least will be the dominant wavenumber of the generated waves.

We shall proceed to show the error of these assumptions. Note that, although the function $F_{\text{piston}}^{(\pm)}$ is strictly not defined when (14.32) is satisfied, it can be defined as the limit of

$$
F_{\text{piston}}^{(+)}|_{k=k_w \sin\theta_i} \equiv \lim_{k \to k_w \sin\theta_i} F_{\text{piston}}^{(+)} = \frac{D}{\cos\theta_i}, \tag{14.33}
$$

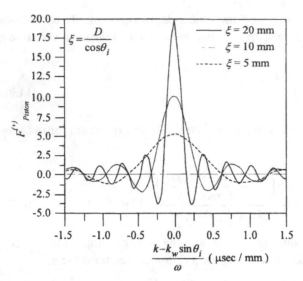

Figure 14-4. Parametric plots of $F_{\text{piston}}^{(+)}$ with the size of the loading region ξ as a parameter ($\omega = 2$ rad/μs).

in which case $F_{\text{piston}}^{(+)}$ will be continuous for all k and have a maximum at the Snell's law wavenumber, $k = k_s \sin(\theta_i)$. Equation (14.29) then shows that, for $k_v \neq k_w \sin(\theta_i)$, $F_{\text{piston}}^{(+)}$ and $F_{\text{piston}}^{(-)}$ do not vanish identically, even though they decrease as k_v becomes different from $k_w \sin(\theta_i)$. Therefore, even if the "Snell's law" angle is not used, the mode can still be generated (albeit less efficiently).

We can define a *wavenumber bandwidth* σ_k associated with the piston transducer as the value of $k - k_s \sin(\theta_i)$, where the function $F_{\text{piston}}^{(+)}$ decreases to $1/e$ (≈ -9 dB), its maximum: $F_{\text{piston}}^{(+)}(k - k_s \sin(\theta_i) = \pm \sigma_k) = F_{\text{piston}}^{(+)}(0)/e$. The approximate value is

$$\sigma_k \approx \frac{4.398 \cos \theta_i}{D}. \tag{14.34}$$

From (14.34) we can see that $\sigma_k \to 0$ as $D \to \infty$. That is, for plane wave incidence (infinite-diameter transducer), the wavenumber bandwidth is zero and so only one value of wavenumber can be excited, as given by Snell's law. For a (finite-sized) piston-source transducer, however, we see that the wavenumber bandwidth is inversely proportional to the diameter D. The bandwidth is directly proportional to $\cos \theta_i$, which (in the interval $0° \leq \theta_i < 90°$) decreases monotonically from 1.0 to 0.0. Recognizing $D/(\cos \theta_i)$ as the length of the loaded region (see Figure 14-3), we can conclude that the wavenumber bandwidth decreases as the size of the loaded (or *insonified*) region increases.

Figure 14-4 is a parametric plot of $F_{\text{piston}}^{(+)}$ with the size of the loading region, $D/(\cos \theta_i)$, as the parameter (in all plots, $\omega = 2$ rad/μs). We can see that, as the insonified region becomes larger, the transducer–wedge source becomes more and more selective to the Snell's law wavenumber $k_w \sin(\theta_i)$.

A somewhat more realistic model of an actual ultrasonic transducer is a parabolic pressure distribution. In this case, the pressure distribution function can be written as

$$p(\alpha) = \begin{cases} \sigma_0 (1 - \alpha^2/(D/2)^2) & \text{if } |\alpha| \leq D/2, \\ 0 & \text{if } |\alpha| > D/2, \end{cases} \tag{14.35}$$

where σ_0 now represents the maximum pressure that occurs at the center of the transducer face when $\alpha = 0$.

Substituting (14.35) into (14.24) and comparing the result with equation (14.26), we find that

$$F_{\text{parabolic}}^{(+)}(\chi_+) = \frac{8}{D\chi_+^2 \cos \theta_i} \left[\frac{2 \sin \left(\chi_+ \dfrac{D}{2} \right)}{D\chi_+} - \cos \left(\chi_+ \frac{D}{2} \right) \right],$$ (14.36)

where χ_+ is defined in (14.25a).

Again, we can define the wavenumber bandwidth σ_k of the parabolic source transducer to be the value of $k - k_s \sin \theta_i$ at which $F_{\text{parabolic}}^{(+)}$ drops to $1/e$ of its maximum value (at $\chi_+ = 0$). Performing these calculations yields

$$\sigma_k \approx \frac{5.852 \cos \theta_i}{D}.$$ (14.37)

A comparison of this bandwidth with that obtained for the piston source, (14.34), shows that the parabolic source has a somewhat wider wavenumber bandwidth. The dependence of σ_k on the transducer diameter D and the incident angle θ_i is the same as for the piston transducer.

It is important to know how the transducer parameters (or, equivalently, the wavenumber bandwidth) influence the range of phase velocities that may be generated by the applied source. Because of the finite size of the loading region, there is actually a range of phase velocities within which the magnitude of $F^{(+)}$ (for both piston and parabolic sources) remains approximately 9 dB below its maximum value. As a result, it should be expected that any mode whose dispersion curve passes through this phase velocity region (for the given frequency ω) may be excited by the source, while modes whose dispersion curves are far from this region should contribute less to the total generated field. How strongly each mode is generated also depends on the value of its excitability function E_ν at this frequency.

To arrive at an estimate of the range of phase velocities within which significant excitation may occur, we can use the definition of the wavenumber bandwidth σ_k. Recalling that σ_k is the value of $k - k_s \sin \theta_i$ where the function $F^{(+)}$ drops to $1/e$ (≈ -9 dB) of its maximum value (at $k = k_w \sin \theta_i$), we have

$$F^{(+)}(k - k_s \sin \theta_i) = \frac{F^{(+)}(0)}{e} \Rightarrow k - k_w \sin \theta_i = \pm \sigma_k$$

$$\Rightarrow \omega \left(\frac{1}{c_p^*} - \frac{1}{c_p^0} \right) = \pm \sigma_k,$$ (14.38)

where c_p^* is the phase velocity at which $F^{(+)}$ drops by 9 dB of its maximum value and $c_p^0 = c_w/(\sin \theta_i)$. Considering the plus-or-minus signs in (14.38), we can solve for two values of c_p^* as follows:

$$c_{p(-)}^* = \frac{\omega c_p^0}{\omega + \sigma_k c_p^0},$$ (14.39)

$$c_{p(+)}^* = \frac{\omega c_p^0}{\omega - \sigma_k c_p^0}.$$ (14.40)

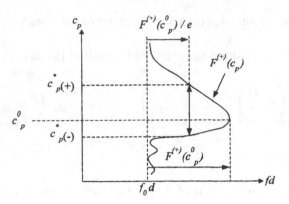

Figure 14-5. Definitions of the lower $(c^*_{p(-)})$ and upper $(c^*_{p(+)})$ phase velocities where $F^{(+)}$ drops by 9 dB.

Under certain conditions (to be developed shortly), these two terms signify the phase velocity at which the function $F^{(+)}$ (considered as a function of phase velocity) drops by 9 dB of its maximum value, which occurs at $c_p = c_p^0$. Thus, $c^*_{p(-)}$ represents the velocity smaller than c_p^0 at which $F^{(+)}$ drops by 9 dB, and $c^*_{p(+)}$ represents the velocity greater than c_p^0 at which $F^{(+)}$ drops by 9 dB. See Figure 14-5 showing a phase velocity spectrum.

Having determined the values of $(c^*_{p(-)})$ and $(c^*_{p(+)})$, we can now calculate the range of phase velocities over which significant excitation may occur. As a percentage of the velocity c_p^0 obtained via Snell's law (at which $F^{(+)}$ is maximum), this range is given by

$$\sigma_\upsilon \equiv \frac{c^*_{p(+)} - c^*_{p(-)}}{c_p^0} = \frac{2\sigma(c_p^0/\omega)}{1 - \sigma_k^2(c_p^0/\omega)^2}. \tag{14.41}$$

The quantity σ_υ will hereafter be called the "-9-dB phase velocity bandwidth" (or "-9-dB PVB") of the transducer–wedge combination. Equation (14.41) for the -9-dB PVB is valid only in the range $0 \leq \sigma_k < \omega/c_p^0$. This is due to the fact that whereas $c^*_{p(-)}$ remains valid for $0 \leq \sigma_k < \infty$ (i.e., for any value of σ_k it represents the value of phase velocity smaller than c_p^0 at which $F^{(+)}$ drops by 9 dB), $c^*_{p(+)}$ is defined only for $0 \leq \sigma_k < \omega/c_p^0$. As σ_k approaches ω/c_p^0, the upper velocity at which $F^{(+)}$ drops by 9 dB (i.e. $c^*_{p(+)}$) quickly approaches infinity. For $\sigma_k > \omega/c_p^0$, the term $c^*_{p(+)}$ becomes negative and hence there is no longer an upper value of phase velocity at which $F^{(+)}$ will drop by 9 dB. Instead, it asymptotically approaches some finite amount above the -9-dB value as $c_p \rightarrow \infty$. Therefore, the -9-dB PVB approaches infinity as $\sigma_k \rightarrow \omega/c_p^0$, and we will say that it is infinite for $\sigma_k > \omega/c_p^0$ with the understanding that this means there is no upper velocity at which $F^{(+)}$ drops by 9 dB of its maximum value.

Making use of the relation $c_p^0/\omega = \lambda^0/2\pi$ and using that the wavenumber bandwidths of both the piston and parabolic sources can be written as $\sigma_k = K \cos(\theta_i)/D$ (see (14.34) and (14.37)), the -9-dB PVB can be written as

$$\sigma_\upsilon = \frac{\dfrac{K}{\pi}\left(\dfrac{\lambda^0}{D}\right)}{1 - \left(\dfrac{K}{2\pi}\right)^2\left(\dfrac{\lambda^0}{D}\right)^2}, \tag{14.42}$$

where $\bar{D} \equiv D/\cos(\theta_i)$ represents the length of the insonified region, $\lambda^0 \equiv 2\pi c_p^0/\omega$, and $K = 4.398$ for piston sources or 5.852 for parabolic sources.

Equation (14.42) shows that, for a given source: (i) the -9-dB PVB depends only on the ratio of the length of the insonified region to the wavelength at the chosen phase velocity and frequency; and (ii) the -9-dB PVB approaches zero as this ratio approaches infinity.

Physically, this is a manifestation of the fact that if the wavelength is very small when compared with the size of the insonified region, then the waveguide modes generated in the layer can properly "phase match" to the incident wave field and hence (owing to destructive interference) will have a narrow phase velocity spectrum. On the other hand, for wavelengths that are large when compared with the size of the insonified region, the generated wave modes cannot properly phase match to the incident field, and the generated modes may be excited over a significant range of phase velocities. Because the guided wave modes are more or less excitable at different points of their dispersion curves (determined by their excitability functions), this means that the modes may actually be generated more strongly at phase velocities other than that given by Snell's law, even though the excitation function $-F^{(+)}$ is maximum at the Snell's law phase velocity.

14.2.6 Transient Loading

The results presented thus far concern harmonic excitation at a single but arbitrary frequency ω. The excitation amplitudes, $A_{\pm v}$, are therefore frequency-dependent. If the transducer mounted on the wedge actually has a frequency spectrum, say $\hat{g}(\omega)$, then by linear superposition the total fields in the loaded waveguide can be expressed as integrals over frequency (times thickness product), $\Omega \equiv fd$. For instance, the velocity field in front of $(+)$ or behind $(-)$ the transducer due to the propagating modes, v_p, can be written as

$$\bar{v}_p^{(\pm)}(y, z, t) = \int \left(\sum_\mu A_{\pm\mu}(z, \Omega, c_p) \bar{v}_{\pm\mu}(y, \Omega, c_p) \exp\left[i\frac{2\pi\Omega}{d}t \right] \right) d\Omega \qquad (14.43)$$

(Ewing et al. 1957). The term G, which now appears in the definition of $A_{\pm v}$, is given by $G = \hat{g}(\Omega)|R(\theta_i)|/4$; since we are considering only propagating modes, the limits of the integration extend from the cutoff fd of the given mode to infinity. Also, the index μ is taken to include only the propagating modes of the structure at a given frequency thickness product Ω.

Because there are only a finite number of propagating modes at any frequency thickness product, the summation in (14.43) contains a finite number of terms. We can therefore interchange the summation and integration, whereafter (14.43) can be written as

$$\bar{v}_p^{(\pm)}(y, z, t)$$
$$= \sum_\mu \int A_{\pm\mu}(z, \Omega, c_p) \bar{v}_{\pm\mu}(y, \Omega, c_p) \exp\left[i\frac{2\pi\Omega}{d}t \right] d\Omega$$
$$= \sum_\mu \int G(\Omega) F^{(\pm)}(c_p, \Omega) \bar{v}_{\pm\mu}(y, \Omega, c_p) \exp\left[i\left(\frac{2\pi\Omega}{d}t \mp k_\mu z \right) \right] d\Omega. \qquad (14.44)$$

The phase velocity c_p and frequency thickness Ω of a given mode μ are restricted to lie on that mode's dispersion curve, so the integrals in (14.44) are actually along the dispersion curves of the individual modes. It is interesting to note that, even when integral

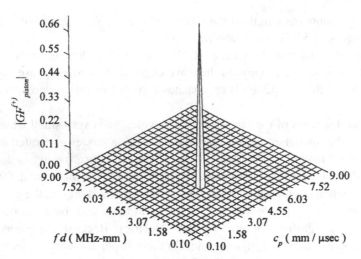

Figure 14-6. Plot of the magnitude of the excitation function $|GF^{(+)}_{piston}|$ for a transducer whose length is 1×10^5 mm (3,937") with a central frequency of 3.5 MHz and −9-dB frequency bandwidth of 0.5 MHz, mounted to a wedge at a 45°.

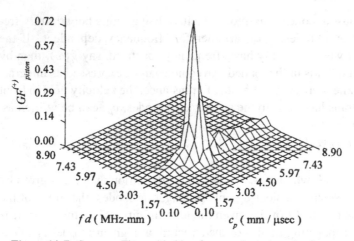

Figure 14-7. Same as Figure 14-6 but for transducer length of 6.35 mm (1/4").

transform techniques are used to solve waveguide loading problems, a similar form (i.e., summation of dispersion curve integrals) results for the excited fields (see Achenbach 1984; Miklowitz 1978). However, the present technique gives a direct physical interpretation to the integrand whereas the integral transform technique does not. Some limited verification of equation (14.44), comparing experimentally obtained RF signals with the numerical evaluation of (14.44) for specific y and z, can be found in Ditri (1992).

Two functions, $G(\Omega)$ and $F(c_p, \Omega)$, determine the region of the phase velocity frequency thickness plane in which the transducer–wedge combination is most capable of exciting waveguide modes. Figures 14-6 through 14-8 illustrate the effect of the transducer diameter D on the excitation function $G(\Omega) F(c_p, \Omega)$ – and hence on the selectivity

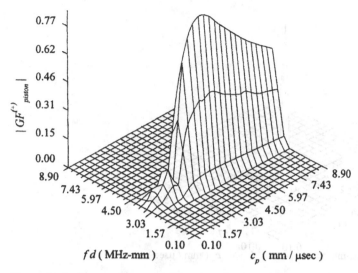

Figure 14-8. Same as Figure 14-6 but for transducer length of 0.77 mm (1/33").

Table 14-1. *Parameters used in Figures 14-6–14-8*

Figure	D (mm)	$\bar{D}/\lambda^0$	σ_k (1/mm)	σ_v
14-6	1.0×10^5 [3,937"]	3.29×10^6	1.22×10^{-6}	4.233×10^{-5}
14-7	6.35 [1/4"]	8.24	0.490	17.13
14-8	0.77 [1/33"]	1.00	4.035	274.57

of the transducer–wedge combination to a particular phase velocity and frequency. The transducer is assumed to have a Gaussian frequency spectrum,

$$\hat{g}(\Omega) = A \exp\left[-\left(\frac{\Omega - \Omega_0}{\sigma_t}\right)^2\right], \tag{14.45}$$

where the central frequency is taken as 3.5 MHz and the frequency bandwidth σ_τ is taken as 0.5 MHz (for a 1-mm-thick layer). The transducer is also assumed to generate a piston pressure profile, so $F^{(+)}$ represents $F_{piston}^{(+)}$ given by (14.30). The figures are surface plots of the magnitude of the excitation function $|G(\Omega)F(c_p, \Omega)|$ over the (c_p, fd)-plane. The figures show the effect of varying the diameter of the transducer for fixed incident angle, frequency, and frequency bandwidth – and hence of varying the ratio $\bar{D}/\lambda^0$. The constant parameters in all the plots are a center frequency of 3.5 MHz, a frequency bandwidth σ_τ of 0.5 MHz, an incident angle of 45°, and a (Plexiglas) shoe with longitudinal velocity c_w of 2.7 mm/μs. This gives a wavelength $\lambda^0 = 1.09$ mm at the peak frequency of 3.5 MHz. The parameter that is varied is the diameter D of the transducer and hence the size of the loading region on the layer, $\bar{D} = D/(\cos\theta_i)$. Since the wavelength λ^0 remains the same for each plot, decreasing $\bar{D}$ decreases the ratio $\bar{D}/\lambda^0$. The diameters used in Figures 14-6–14-8, together with calculated wavenumber and phase velocity bandwidths, are given in Table 14-1.

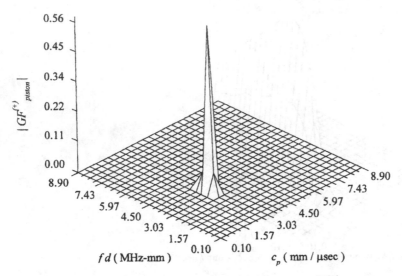

Figure 14-9. Plot of the magnitude of the excitation function $|GF_{\text{piston}}^{(+)}|$ for a transducer whose length is 25.4 mm (1") with a central frequency of 3.5 MHz and −9-dB frequency bandwidth of 0.5 MHz, mounted to a wedge at an 80° angle.

In each of the plots, one should imagine the dispersion curves of the particular layer under consideration as being superimposed on the underlying grid of phase velocity versus fd. Then, the amplitude with which each mode, μ, is generated at each particular point of its dispersion curve is given by the product of (a) the excitability $E_\mu(\Omega, c_p)$ of the mode at that point and (b) the amplitude of the GF surface at that point. Therefore, the sharper the peak of the GF surface, the more selective the transducer–shoe combination is to a particular mode at a particular point on its dispersion curve.

As can be seen from Figures 14-6–14-8 and Table 14-1, the selectivity of the transducer–wedge combination to a particular phase velocity becomes worse with decreasing $\bar{D}/\lambda^0$, the ratio of the insonified region to the wavelength at the central frequency. Figure 14-6 is used to illustrate that if a plane wave is incident (extremely large D), then there is essentially only one value of phase velocity that can be generated with any appreciable intensity; that value is given by Snell's law, once the material properties of the wedge and the incident angle are known. For $\bar{D}/\lambda^0 \approx 10$, Figure 14-7 illustrates that (i) the selectivity of the transducer–wedge combination is still highly peaked at the Snell's law phase velocity, but (ii) the width of the peak (in the phase velocity direction) at the −9-dB point is now around 17% of the Snell's law velocity. Finally, Figure 14-8 shows that, when $\bar{D}/\lambda^0 \approx$ 1, the transducer–wedge combination has virtually no selectivity to phase velocity, since the excitation function exceeds −9 dB of its maximum over a 275% velocity range.

Figures 14-9 and 14-10 illustrate that it is the ratio $\bar{D}/\lambda^0$ and not merely the size of the transducer that determines its selectivity to a particular phase velocity. The figures are parametric plots, where the diameter and all frequency parameters are kept the same as in the three previous figures while the incident angle is varied. This has the effect of varying the phase velocity c_p^0 determined by Snell's law and hence the wavelength $\lambda^0 \equiv 2\pi c_p^0/\omega$. For a given diameter transducer, the insonified region $\bar{D}$ increases as the incident angle does. The net result is that – although the transducer diameter D is kept constant in all

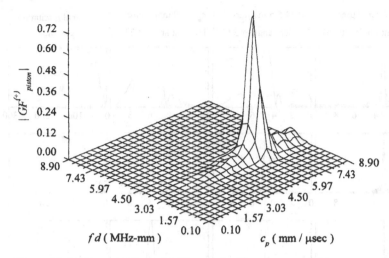

Figure 14-10. Same as Figure 14-9 but for incidence angle of 25°.

Table 14-2. *Parameters used in Figures 14-9 and 14-10*

Figure	D (mm)	θ_i (%)	c_p^0	$\bar{D}/\lambda^0$	σ_k (1/mm)	σ_v
14-9	25.40	80	2.74	186.73	0.024	0.59
14-10	25.40	25	6.38	15.35	0.286	16.76

plots – the ratio $\bar{D}/\lambda^0$ of loading length to wavelength is being varied. The incident angles used in Figures 14-9 and 14-10 are given in Table 14-2 along with the phase velocities calculated by Snell's law, as well as wavenumber and phase velocity bandwidths.

It can be seen from Figures 14-9 and 14-10 that a given diameter transducer (with a given central frequency and frequency bandwidth) can be more or less selective in phase velocity depending on the incident angle θ_i. This is actually a manifestation of the fact that the ratio $\bar{D}/\lambda^0$ of loading region to wavelength increases with increasing incident angle.

Figures 14-11 and 14-12 present additional graphs of the source influence results. Several interesting observations can be made from Figure 14-11. As an example, for a specified center frequency of 4.3 MHz and bandwidth of 0.6 MHz for all of the curves, note the improvement in the phase velocity spectrum as the transducer diameter is increased – particularly for low phase velocity values. Phase velocity value is listed with respect to an incident angle using the Cremer hypothesis for wave propagation (oblique incidence) through a Plexiglas wedge. The last column in the diagram is for normal excitation corresponding to a very high phase velocity value, theoretically of infinity. This is the case for normal beam incidence or an incident angle of 0°. Large transducer size helps in most cases, but please note that, with normal beam excitation, the phase velocity spectrum is quite broad (extending from zero to infinity). It is therefore quite difficult to isolate a particular mode and frequency when using a normal beam excitation.

Another interesting presentation is shown in Figure 14-12. A few particular phase velocity values and transducer diameter sizes were selected to illustrate the phase velocity

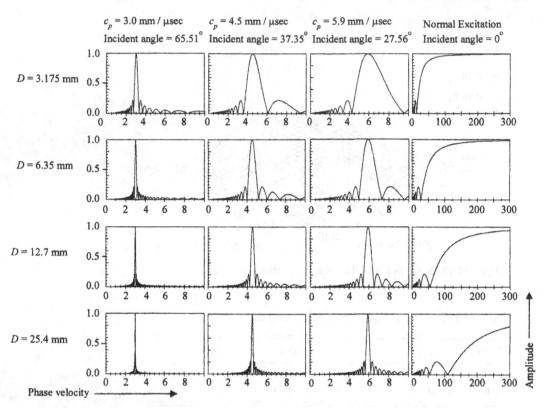

Figure 14-11. Phase velocity spectra showing excitation amplitude versus phase velocity value (frequency $= 4.3$ MHz, bandwidth $= 0.6$ MHz).

spectrum variations as a function of these input parameters. The results are shown in a gray scale, where the darkest regions indicate the largest amplitude on the phase velocity spectrum. Note that mode and frequency isolation become possible for the smallest excitation zones, which are possible at large transducer diameters and high frequency values. The normal incidence excitation shows side lobes and excitation energy arising from the higher phase velocity values. As the modes become closer together, mode and frequency isolation become more difficult, which summons the need for an even stronger understanding of source influence parameters on the phase velocity spectrum used in guided wave excitation.

14.2.7 Concluding Remarks

The excitation of generally isotropic and anisotropic layers by applied surface tractions has been analyzed using the normal mode expansion technique. We have examined in detail two loading configurations that are typical of those used in nondestructive evaluation for generating waveguide modes. We have shown how the relevant parameter for determining the selectivity of the source to a specific phase velocity is the ratio of the length $\bar{D}$ of the loaded region to the wavelength λ^0 in the phase velocity–frequency thickness plane. It was also demonstrated that, for $\bar{D}/\lambda^0$ greater than about 10, the significant excitation

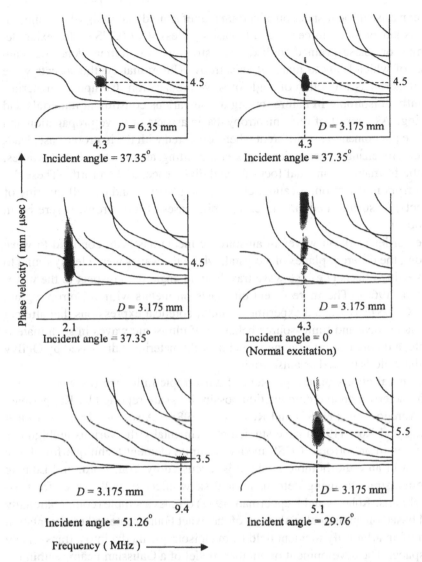

Figure 14-12. Phase velocity spectra for various phase velocities, transducer sizes, and frequencies for a frequency bandwidth of 0.6 MHz.

(above -9 dB of the maximum) may occur over approximately a 17% range, whereas for $\bar{D}/\lambda^0 \approx 1$ there is virtually no selectivity to a particular phase velocity.

14.3 Waves in a Bulk Anisotropic Media

14.3.1 Introduction

Wave propagation in anisotropic media for plane wave or point source excitation was previously discussed in Chapter 3. Here we examine wave propagation characteristics of a finite-sized source.

Difficulties can arise in the inspection of certain manufactured materials whose unusual grain structure is anisotropic. In centrifugally cast stainless steel (CCSS), for example, there may be fine- or coarse-grained equiaxe grain structure or columnar dendritic grain structure, either of which leads to material anisotropy. Unfortunately, this affects wave velocity in the material as a function of angle of wave propagation. Composite materials are also inherently anisotropic in nature, owing to the various constituent materials and layers or weaving. As a result of this anisotropy, basic aspects of wave propagation and wave interference phenomenon are different when compared with the isotropic case, leading to energy velocity changes, beam skewing, beam splitting, asymmetrical field profiles, unusual side lobe formations, unusual focusing and divergence, and so forth. These deviations cause errors in detection, location, classification, sizing, and overall imaging of reflectors (defects) in such a medium. Even for mild cases of anisotropy, severe beam distortion can occur.

Much of the earlier work on waves in anisotropic media has been confined to wave propagation along the different planes of a crystal, where the waves were clearly found to exhibit wave skewing effects (i.e., the wave travels at an angle with respect to the wavefront; see Fedorov 1968). The more recent emphasis on metals with anisotropic grain structures (e.g., CCSS) has yielded experiments and theoretical expressions that attempt to explain the plane wave and finite-source behavior of ultrasonic waves in such materials. For example, a theoretical ray-tracing model in weld material is discussed by Ogilvy (1986), where the weld is treated as anisotropic.

Some of the initial efforts on the problem of anisotropic influence (as well as various material characterization and computation possibilities) are reported by Kupperman, Reimann, and Abrego-Lopez (1987) and Rose et al. (1988). Also, Good and Van Fleet (1986) and Jeong and Rose (1986) have studied experimentally the various ambiguities of ultrasonic waves in anisotropic CCSS media. A finite element technique for elastic wave propagation in an orthorhombic medium is discussed by You, Lord, and Ludwig (1987), who demonstrate focusing/defocusing and skew effect of bulk waves for two-dimensional problems. Roberts and Kupperman (1987) discuss a simple (computationally quicker) model based on numerical evaluation of the exact Fourier integral representation of transmission of an arbitrarily incident field from an isotropic media into a transversely isotropic half-space. The development of another model of a Gaussian beam, within the Fresnel approximation region and generalized to the case of anisotropic media, is reported by Thompson and Newberry (1987).

Most of these models suffer one or more of the following drawbacks: excessive computational requirements; lack of flexibility to handle an ultrasonic pulse RF signal; arbitrarily shaped transducers and defects; approximation due to plane wave assumption or pre-defined beam functions. In contrast, the numerical integration (Green's function) model previewed in Section 14.2.2 provides simple computation and a flexible approach to the study of ultrasonic bulk wave propagation in anisotropic media.

In order to satisfy our goal of advancing the state of the art in ultrasonic nondestructive evaluation (NDE), we hereby launch a two-phase numerical integration program. The first phase is associated with efforts to characterize the material properties. Once this is accomplished, phase two becomes possible: optimal penetration and inspection through a known material for reliable reflector detection and analysis. A complete understanding of wave propagation and wave scattering within an anisotropic material is necessary for

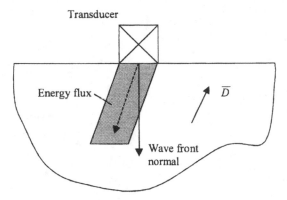

Figure 14-13. Beam-skewing phenomenon in an anisotropic material.

the process of detecting, locating, and identifying anomalies. Hence we are faced with the necessity of developing a computational model that can be used to evaluate ultrasonic field profiles in anisotropic media. As a result of this work, we can establish guidelines and transducer selection rules for the optimal inspection of both isotropic and anisotropic materials.

Consider some general statements on the problem of anisotropic material. The anisotropic influence on ultrasonic NDE can be summarized as follows.

(1) Errors can arise in reflector location analysis owing to beam skewing and to energy velocity variations throughout the material.

(2) Unusual beam focusing, diverging, splitting, and nonsymmetrical wave propagation can occur and so produce imaging artifacts and distortions.

(3) Confusion in feature selection and analysis for reflector classification can come about because of the unusual superposition characteristics associated with waves traveling in anisotropic media. For example, planar reflectors could appear to be volumetric and vice versa.

(4) Confusion is likely in defect sizing because the key variable of ultrasonic wave velocity now changes with angle; it is not constant (as is assumed for isotropic media).

As a result of these four factors, standard imaging techniques developed for isotropic material are bound to produce errors in anisotropic media. Unless the proper anisotropic corrections are made, ultrasonic imaging can produce serious errors with respect to the items that are seen inside an anisotropic material – from producing multiple images to distorted ones.

We will now provide a detailed demonstration of one of the key problems of ultrasonic wave propagation in anisotropic material. Consider the beam-skewing phenomenon shown in Figure 14-13. In this figure, a transducer is placed on a test surface that produces ultrasonic waves traveling in the direction directly ahead of the transducer; this is labeled as "wavefront normal." Unfortunately, because of the anisotropic character of the material and its alignment with respect to the transducer, the actual energy flux direction can be at some (skew) angle other than normal to the test probe, as illustrated in the

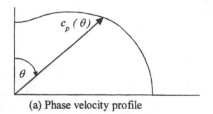

(a) Phase velocity profile

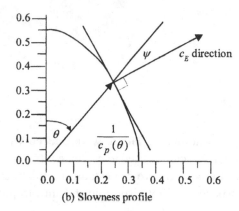

(b) Slowness profile

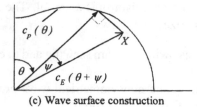

(c) Wave surface construction

Figure 14-14. Determining the energy velocity wave surface for a given phase velocity profile: (a) plot c_p in a particular θ direction; (b) plot the direction $c_E(\theta + \psi)$ normal to the tangent to the slowness profile; (c) determine $|c_E|$ such that $c_E(\theta + \psi)\cos\psi = c_p$. (Note that X is now a point on the wave surface.)

figure. Further examples and actual quantitative results will be presented that illustrate beam skewing, energy velocity variations, lateral resolution changes with focusing and divergence, and unusual side-lobe and beam-splitting possibilities for longitudinal, shear vertical, and shear horizontal modes.

For now, let us consider some basic elements of wave propagation in anisotropic media. If we go through the details of a computation process (reported in many references, including Auld 1990), we may start with a typical phase velocity surface $c_p(\theta)$ as shown in part (a) of Figure 14-14, where θ is the direction of phase velocity or of the wavevector. To find the direction of energy propagation, the direction of group (energy) velocity c_E, or the skew angle between phase velocity direction and energy propagation direction, we should construct a slowness surface $1/c_p(\theta)$ as illustrated in part (b) of the figure. The energy velocity vector must now be normal to the slowness surface, as shown in the diagram;

hence, the direction of energy propagation or skew angle (ψ) is found from the construction of tangent and normal lines. We can then calculate (or graph) the value of energy velocity by using the simple trigonometric equation $c_E(\theta + \psi)\cos(\psi) = c_p(\theta)$, as shown in part (c) of Figure 14-14. Hence, we can establish the wave surface by: (1) plotting the phase velocity in the θ direction, as shown in the diagram; (2) plotting the deviation angle; (3) plotting the actual energy velocity component; and (4) continuing to plot the locus of all points as we move around the diagram in determining the total wave surface.

These essential concepts apply to plane wave propagation, which in most cases gives us a reasonable approach to understanding the basic elements of wave propagation in anisotropic media. Moreover, the numerical integration (Green's function) model discussed next goes beyond plane wave analysis and can consider wave propagation from even a finite-size transducer source. These concepts are important, but they represent only the beginning with respect to the understanding of waves in anisotropic media that is required for ultrasonic NDE applications.

14.3.2 Theoretical Considerations

One of the most commonly used techniques for solving the boundary value problem is the Green's function method. The idea is as follows. Suppose we have an elastic field generated by the source distribution, like the summation of fields from elementary forces. The field distribution – at an appropriate point r of the media – of the force with a unit value applied at point ξ will be called $G(r, \xi)$. This is the form of the Green's function; the field resulting from the force $f(\xi)$ is $G(r, \xi)f(\xi)$. The field for all ξ takes the form of an integral,

$$u(r) = \int_{R_n} G(r, \xi)f(\xi)\,d\xi.$$

Now consider the inhomogeneous wave equation with body forces $f(r, t)$ distributed over a finite region V:

$$\nabla^2 u - \frac{1}{c^2}\frac{\partial^2 u}{\partial t^2} = f(r, t). \tag{14.46}$$

Using a Green's function representation, the solution of (14.46) with the homogeneous initial conditions can be obtained in the form

$$u(r_{\text{obs}}, t) = \int_0^t ds \int_V f(r_{\text{sou}}, S)G(r_{\text{obs}}, t, r_{\text{sou}}, S)\,dV \tag{14.47}$$

(see Achenbach 1984), where $(G_{\text{obs}}, t, r_{\text{sou}}, S)$ is the Green's function. The terms r_{obs} and r_{sou} denote the radius vectors of the observation point and the source point (resp.), $R = |r_{\text{obs}} - r_{\text{sou}}|$. Here, $u(z, t)$ is the amplitude of the specific mode (longitudinal or transverse wave) characterized by the corresponding sound velocity c – if this specific mode of interest has been generated by the appropriate external volume force of $f(r, t)$. For unbounded domain, the Green's function is:

$$G(x, t, \xi, s) = \frac{-1}{4\pi|x - \xi|}\delta\left(t - s - \frac{|x - \xi|}{C}\right), \tag{14.48}$$

where $\delta(x)$ is the Dirac delta function. Substituting (14.48) into (14.47) and using features of the delta function, we finally discover that

$$u(r_{\mathrm{obs}}, t) = \int_S \frac{f(r_{\mathrm{sou}}, t - R/C)}{R} \, dV, \tag{14.49}$$

where integration is over the sphere S with radius ct.

Equation (14.47) leads to a numerical model for mildly anisotropic media (see Rose, Balasubramaniam, and Tverdokhlebov 1989a; Tverdokhlebov and Rose 1989). The wave equation for a generally anisotropic media is written as

$$C_{jklm} \nabla_k \nabla_m u_1 - \rho d^2 u_j / dt^2 = 0, \tag{14.50}$$

where ∇ is the gradient vector, C_{jklm} are the stiffness coefficients of the anisotropic media, and u_i are the Cartesian components of the displacement vector. The displacement function of the plane harmonic wave solution for each mode M is

$$u_k^M = A_k^M \exp[i\omega(p_m r_m)/c_m], \tag{14.51}$$

where A_k^M is the corresponding displacement vector, ω is the angular frequency, p_m is the unit vector in the direction of the plane wave propagation, and r_m is the radius vector.

Using Christoffel's equation – as described, for example, in Auld (1990) – we can find plane wave solutions (for all three modes) that include the phase velocity profiles and polarization vectors as functions of propagation direction. As we shall demonstrate, the phase velocities for the three modes can be determined from a pair of decoupled expressions found when we solve for the eigenvalues of the secular representation.

For a transversely isotropic material (chosen here as an example), the quasi-longitudinal (c_L) and quasi-transverse (c_T) phase velocities are the two different positive roots of

$$0 = c^4 - c^2(c_l^2 + S\cos^2\theta + c_t^2) + c_t^2(c_l^2 + S\cos^2\theta) - E\cos^2\theta\sin^2\theta; \tag{14.52}$$

phase velocity (c_h) of the horizontally polarized shear wave is given by

$$c_h^2 = c_t^2 + B\sin^2\theta, \tag{14.53}$$

where θ is the angle between the phase velocity direction and the normal to the plane of transverse isotropy. The terms B, S, E are the three modified constants, which – along with c_l and c_t (resp. the longitudinal and transverse phase velocities along the isotropic (2, 3)-plane) – completely describe the given transversely isotropic material. The relationships between the conventional stiffness coefficients C_{ij} and the constants c_l, c_t, S, E, B used in the numerical integration model of a transversely isotropic material are as follows:

$$
\begin{aligned}
&C_{11} = \rho(c_l^2 + S), \qquad c_{22} = \rho c_l^2, \qquad C_{33} = \rho c_l^2, \\
&C_{12} = \rho\left[\sqrt{(c_l^2 - c_t^2)^2 + S(c_l^2 - c_t^2) + E} - c_t^2\right], \\
&C_{13} = C_{12}, \qquad C_{23} = \rho(c_l^2 - 2c_t^2 - 2B), \\
&C_{44} = \rho(c_t^2 + B), \qquad C_{55} = C_{66} = \rho c_t^2,
\end{aligned}
\tag{14.54}
$$

where the subscript l denotes direction along the axis of symmetry.

The exact solution for the Green's function is obtained by Radon's method in the form of an integral over all the propagation directions, with the plane wave representation of the Green's functions as an integrand (see John 1955). This method is a modification of the spatial Fourier transform that reduces the three-dimensional domain of the Fourier integral to the unit sphere, which is more suitable for this problem than the straightforward Fourier technique. Projection operators in k-space, which consist of the plane wave polarization vectors, are used to break the Green's function into a generalized Helmholtz decomposition of quasi-longitudinal, quasi-transverse, and shear horizontal modes (see Tverdokhlebov and Rose 1989).

The convolution form for the field solution is

$$u(R, \omega) = G(R - r, \omega) * f(r, \omega),\tag{14.55}$$

where u is the sought field, G is Green's function, and f is the distributed body force associated with the sender. With this form we have switched the projection so that operators on f function in $\bar{r}$-space. This transforms G into three scalar Green's functions, corresponding to three different modes of wave propagation. They are all the same function of $c_M(\theta)$, where c_M is the phase velocity of the corresponding plane waves. (Further on, we drop the subscript M whenever the formulas are valid for any mode.)

The next step is valid in the weak anisotropy approximation that allows us to express the integral representation of the scalar Green's function in the form

$$G(r) = (1/r) \exp[i\omega r/c_E(\phi)];\tag{14.56}$$

c_E is the energy velocity, which is dependent on the direction of the energy propagation (angle $\phi = \theta + \psi$; see Figure 14-14).

The exact expression for the Green's function is derived in Tverdokhlebov and Rose (1989) as an integral over the finite integration area – the unit sphere in three-dimensional space. It allows an approximation of the Green's function with a uniformly convergent series, which characterizes mildly anisotropic media. Equation (14.56) represents the first term of the approximation series. For materials with mild anisotropy, the quasi-longitudinal mode is approximated by purely longitudinal waves, whose only difference from the isotropic case is that the velocity of propagation waves changes with direction.

In the frequency domain, (14.56) leads to the generalized retarded potential of (14.49) with the sound velocity c replaced by the group velocity function c_E. This relationship may be utilized in a numerical computation routine to compute the ultrasonic field in any given anisotropic media.

The natural limitation on use of equation (14.56) is that the energy velocity must be a single-valued function in any direction. The accuracy of the model is roughly the same as the deviation of the velocity from its averaged value. For more on application domains of the model, see Tverdokhlebov and Rose (1989).

14.3.3 The Numerical Integration Model

The numerical model makes use of the energy flux deviation angle and the resulting group velocity in summing the point source responses over a transmitter or a reflector surface. The Green's function scalar retarded potential serves as a point source inside a surface integral for calculating the superimposed ultrasonic scattering function of waves

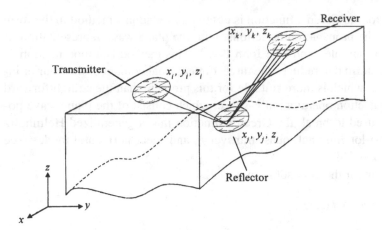

Figure 14-15. Schematic representation of the numerical integration model, where the transmitter, reflector, and receiver are approximated using a mesh of point sources; sources are centered within each mesh element.

impinging onto a reflector and/or reflecting from it. Thus, the model is capable of handling arbitrarily shaped transmitters and reflectors located in three-dimensional space for any generalized phase velocity profile. The numerical integration model is represented pictorially in Figure 14-15, where a sender and a receiver, as well as the defect, are approximated as distributed point sources/receivers. Each point on the sender acts as a point source transmitting an ultrasonic pulse function into the anisotropic media and is received by each and every point on the defect. Subsequently, the defects become point sources themselves and the final response reaches the summation at the receiver. For the beam profiles illustrated here, the computations are carried out only to the first stage; that is, the defect and the receiver are not considered.

As compared with a finite element approach, programs for computing the retarded potential have the obvious and tremendous advantages of short running time, accuracy, and insight. Some prior work in this direction (for isotropic media) may be found in Rose and Meyer (1975) and Singh and Rose (1982). We shall detail the development and application of the numerical integration process in the balance of this section.

In an anisotropic medium, the sound velocity c is no longer a constant (to be exact, there are different constants for longitudinal and any of the two degenerate transverse wave modes). Here, we must deal with three different sound velocity profiles for so-called quasi-longitudinal and the two quasi-transverse wave modes. In the case of transversely isotropic solids (or unidirectionally anisotropic media), one of the three modes is of a pure transverse wave – also known as horizontally polarized shear mode, meaning that the displacement amplitude is along the horizontal isotropic plane. The three velocities are functions of the direction of propagation. In addition, the phase and energy (group) velocities of the plane waves differ from one another.

We now seek the energy velocity profile $c_E(\bar{n})$ for use as model input data that effectively characterizes the anisotropic material. There are several ways to acquire this information, including a direct measurement of the energy velocity. However, in this investigation (and as a recommendation for similar future efforts), the phase velocity profile $c_p(\bar{n})$ is chosen as the primary description of the anisotropic material; the energy velocity $c_E(\bar{n})$ appears as secondary data that is derived from the phase velocity.

Table 14-3. *Elastic constants for transversely isotropic materials* (GPa)

Constants	Case I	Case II	Case III	Case IV
C_{11}	269	238	242	262
C_{22}	269	246	282	282
C_{33}	269	246	282	282
C_{12}	103	85	140	76
C_{13}	103	85	140	76
C_{23}	103	120	100	56
C_{44}	83	63	91	113
C_{55}	83	135	135	135
C_{66}	83	135	135	135

Thus, we can see that the anisotropic medium can be represented – and the phase velocity profiles inversely obtained – by using five independent measurements of the material and by simultaneously solving equations (14.52) and (14.53) for the five independent modified elastic constants. Four of the phase velocity measurements may be along the 0° and 90° directions for both longitudinal and shear waves, and the fifth one may be measured at any other angle. Here, the experimental work of Jeong (1987) is used for comparison of the elastic constants. This is of value in real-time applications where there is no prior knowledge about the material and speed of inspection is a factor.

With phase velocity as an input, we can calculate the group velocity as well as the theoretically predicted skew angles for a plane wave. These values have been verified on many different CCSS columnar specimens, all of which transversely isotropic in nature. Only five independent measurements of phase velocities are required for reconstruction of the phase velocity profile and for subsequent computation of the group velocity profile as a function of the incident angle and the corresponding skew angles. Table 14-3 lists elastic constants of the various CCSS materials; case I is isotropic equiaxial grain structure, while the other cases are columnar varieties. Table 14-4 displays the high degree of correspondence between theoretical and experimental results (Jeong 1987) for a CCSS case-II specimen of a type used in nuclear power plants.

As soon as all five parameters are evaluated for any given transversely isotropic material (usually through limited experimentation), the group velocity vector may be constructed as outlined previously (see Figure 14-14). Alternatively, the material's module $c_E(\phi)$ may be obtained as a function of its own directivity angle by the purely algebraic procedure of simultaneously solving the following equations:

$$c_E^2(\phi) = c_p^2(\theta)\cos^2\theta + (c')^2\sin\theta, \tag{14.57}$$

$$c_E^2(\phi)\cos^2\phi = c_p^2(\theta)\cos^2\theta + 2c_p(\theta)c'\cos\theta\sin^2\theta + (c')^2\sin^4\theta, \tag{14.58}$$

where c' is the derivative of $c_p(\theta)$ with respect to $\cos(\theta)$. In Figure 14-16, the group velocity profile $c_E(\phi)$ is compared with the phase velocity profile $c_p(\theta)$ and the group function profile $c_E(\theta)$; the latter is the value of group velocity corresponding with phase velocity at angle θ. Note that the group velocity $c_E(\phi)$ is always less than the phase velocity $c_p(\theta)$ at any given propagation angle, although the group function $c_E(\theta)$ is always larger than

Table 14-4. *Verifying the transversely isotropic model* (cm/μs)

Angle of entry	Phase velocities		Group velocities		Beam skew angles	
	Theory	Experiment	Theory	Experiment	Theory	Experiment
0°	0.545	0.545	0.545	0.552	0.000	0.000
5°	0.548	0.546	0.552	0.558	7.000	8.314
10°	0.556	0.552	0.567	0.558	12.000	8.314
15°	0.567	0.558	0.582	0.573	13.000	12.815
20°	0.578	0.577	0.592	0.594	13.000	13.694
25°	0.589	0.575	0.600	0.586	11.000	11.031
30°	0.598	0.591	0.605	0.597	9.000	9.229
35°	0.605	0.591	0.609	0.595	6.000	6.487
40°	0.609	0.605	0.611	0.607	3.000	4.645
45°	0.611	0.611	0.611	0.611	1.000	0.000
50°	0.611	0.608	0.611	0.608	−2.000	0.000
55°	0.607	0.601	0.610	0.605	−5.000	−6.487
60°	0.602	0.597	0.607	0.600	−7.000	−5.569
65°	0.594	0.591	0.602	0.600	−10.000	−7.405
70°	0.584	0.591	0.595	0.597	−11.000	−8.319
75°	0.574	0.581	0.586	0.592	−11.000	−11.000
80°	0.565	0.564	0.573	0.574	−10.000	−11.000
85°	0.557	0.559	0.561	0.583	−6.000	−16.301
90°	0.555	0.555	0.555	0.555	0.000	−0.932

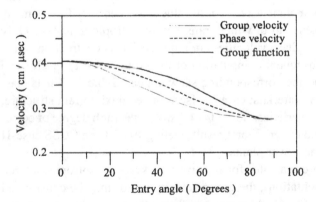

Figure 14-16. Phase, group, and group function velocity profiles versus the angle of the direction of phase velocity (note that group velocity is always less than phase velocity, and the latter is always less than the group function).

the phase velocity $c_p(\theta)$. This is true for any mode, not just for the horizontal shear mode presented in Figure 14-16.

The solution $c_E(\phi)$ of (14.57) and (14.58) is then incorporated into the scalar potential function, equation (14.47), and numerically integrated. This provides the displacement function at any particular location and time in an infinite elastic anisotropic medium.

In our theoretical calculations, the transmitter was approximated by a mesh of point sources that were approximately one quarter of a wavelength (of the input signal) apart

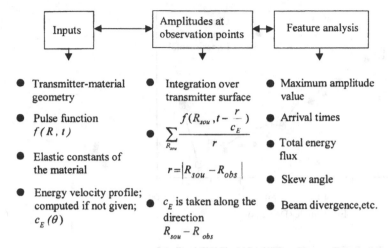

Figure 14-17. Schematic representation of the numerical model computation.

and equally spaced. The radio frequency (RF) impulse function, used as a point-source time function, was acquired by digitizing actual reflected signals of different ultrasonic transducers off a flat steel surface that was completely immersed in water. The computational scheme is shown in Figure 14-17.

The numerical integration computer program utilizes the phase velocity profiles (or five independent velocity measurements) to first compute the group velocity function, which then is fed – along with the transmitted pulse time function – into the scalar potential function, which is then integrated over the surface of the source. For all the computations in this section, a flat, circular-shaped transducer was considered as the source.

The skew or deviation angles for each type of CCSS material are obtained by picking the peak energy location from transverse pressure profiles parallel to the face of the upper surface for different entry angles (also referred to as anisotropy angles). The skew angle response to different phase velocity profiles, which is derived from five independent experimental phase velocity measurements (obtained in Jeong 1987), was used as a benchmark to confirm the validity of the numerical integration model by comparing it with the classical plane wave closed-form technique and experimental values, as shown in Figure 14-18 for the longitudinal and horizontal shear wave modes. A 2.25-MHz transducer of half-inch diameter was used in the computations; the material used has the velocity profile summarized in Table 14-4. The various field profiles were obtained via considering the "director" vector (in CCSS, the columnar grain direction) to be at specific angles to the plane of the piston-type source transducer. The excellent correlation of the numerically obtained values to both the standard plane wave solution and the experimental data inspires confidence in the Green's function model.

14.3.4 Ultrasonic Transducer Field Profiles

The study of beam patterns provides useful details of the wave interaction within an anisotropic medium. Analyzing the influence of such external factors as input signal frequency and transducer diameter can increase the odds of choosing appropriate experimental parameters. Such parametric studies also help explain several unusual phenomena (e.g., beam skewing, beam splitting and bending, asymmetrical beam patterns,

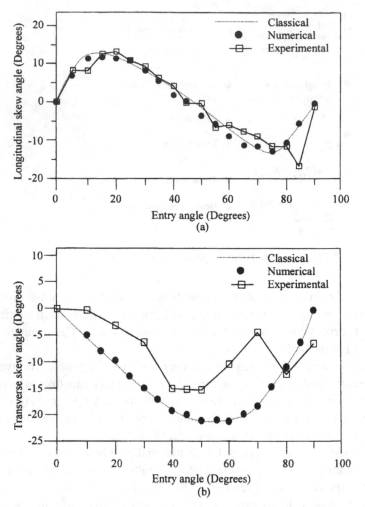

Figure 14-18. Skew angles derived using numerical models as compared with the classical and experimental approaches for (a) longitudinal waves and (b) horizontal shear waves.

beam convergence, etc.) observed in anisotropic materials. Most of these characteristics, both quantitative and qualitative, may be better understood with the help of Figure 14-19, which depicts a typical beam profile. The transducer location, director vector direction, beamwidth definition, skew angle, and so forth are explicitly shown.

We calculated field profiles for the three different wave modes: quasi-longitudinal, quasi-transverse, and pure transverse. Four different transducers (0.5 cm, 1.0 cm, 2.0 cm, and 5.0 cm in diameter) are considered for the three input RF broadband signals (of 0.5 MHz, 1.0 MHz, and 3.0 MHz central frequency) used in these studies, where a 15-cm × 25-cm area of the acoustic field is displayed. The digitized RF waveform (time domain) is fed into the program during the numerical computation.

In Figure 14-20, the beam pattern for a longitudinal wave mode is shown for different angles of entry. The 3-MHz, 1-cm transducer choice is employed for simulating the field profiles. The 0° incident beam shows a diverging pattern with a pair of side lobes symmetrically positioned. As the entry angle is varied to 15° with respect to the anisotropic

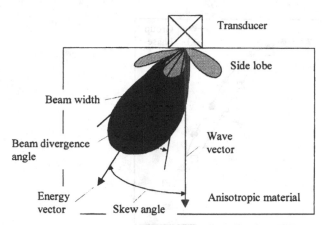

Figure 14-19. Schematic representation of typical beam profile characteristics.

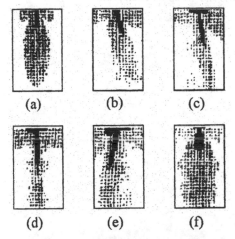

Figure 14-20. Sample ultrasonic field patterns for longitudinal wave modes in an anisotropic medium (case II) for six different angles of propagation with respect to the anisotropic direction: (a) 0°, (b) 15°, (c) 30°, (d) 45°, (e) 65°, and (f) 90°.

vector (in our case, the columnar grain direction), the beam skews to the direction where phase velocity is larger. Skew angles are also observed (although in opposite directions) with entry angles of 30° and 65°; however, at 45° and 90° the waves do not skew at all. A glance at Figure 14-21 shows that the group and phase velocities at these angles are identical, which explains the lack of skewing. The beam-focusing effect at 45° (observed experimentally by Jeong 1987) is pronounced. In many of these profiles, an asymmetrical beam pattern is also seen.

Figure 14-22 shows the quasi-transverse mode (shear vertical) for the same 3-MHz, 1-cm transducer. Here, the beam deviations are found to be more severe; we notice the beam splitting at 45° incidence, and the side lobes are clearly visible. Figure 14-23 shows the pure transverse (shear horizontal) mode used for a 0.5-MHz, 1-cm transducer; again, the skew angles and asymmetrical beam shape are easily recognized.

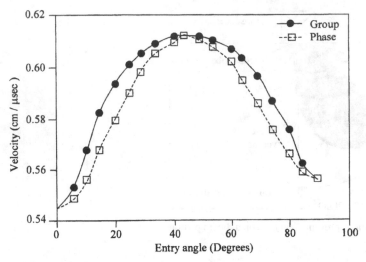

Figure 14-21. Comparison of the longitudinal phase and group function profiles.

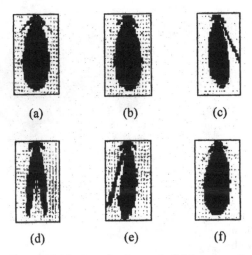

Figure 14-22. Sample ultrasonic field patterns for quasi-transverse (shear vertical) wave modes in an anisotropic medium (case II) for six different angles of propagation with respect to the anisotropic direction: (a) 0°, (b) 5°, (c) 30°, (d) 45°, (e) 80°, and (f) 90°.

Parametric studies on the effect of transducer diameter on the field profiles reveal an increasingly parallel beam at higher diameter sizes; this can be seen in parts (a)–(c) of Figure 14-24. Here, the input frequency is kept constant at 3 MHz, and the longitudinal wave mode is used at a fixed incident angle of 15° to the anisotropic vector. Parts (d)–(f) of the figure display the effect of changes in the input signal frequency for the horizontal shear mode. This study (for an incident angle of 65°) brings out the beam-bending and beam-splitting characteristics of the anisotropic material. Such characteristics have often been observed experimentally (Good and Van Fleet 1986; Kupperman et al. 1987).

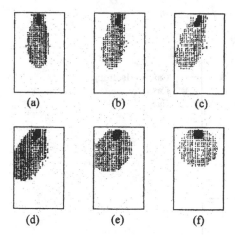

Figure 14-23. Sample ultrasonic field patterns for transverse (shear horizontal) wave modes in an anisotropic medium (case II) for six different angles of propagation with respect to the anisotropic direction: (a) 0°, (b) 15°, (c) 30°, (d) 45°, (e) 65°, and (f) 90°.

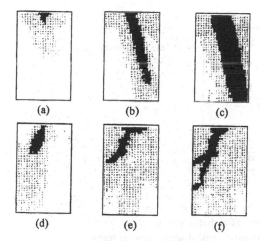

Figure 14-24. Parametric studies (for case II) of changing input frequency and transducer size. For 3-MHz longitudinal wave: transducer diameter of (a) 0.5 cm, (b) 1 cm, and (c) 2 cm. For 1-cm transducer: horizontal shear wave of (d) 1 MHz, (e) 3 MHz, and (f) 5 MHz.

14.3.5 Discussion

As a demonstration of the unusual wave interference patterns leading to pulse modifications and the anisotropic filter influence on wave propagation in angular dispersive anisotropic media, Figure 14-25 shows a set of longitudinal phase velocities for three different materials – one equiaxial (case I) and two different columnar CCSS types (cases III and IV) – obtained through computation. These curves clearly illustrate the effect of

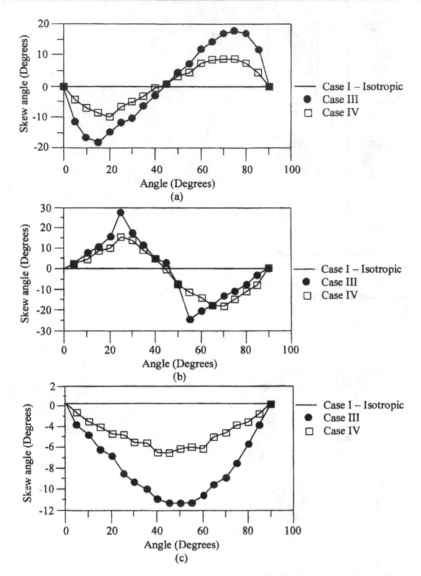

Figure 14-25. Beam skew angles of three materials with different levels of anisotropy for (a) longitudinal wave, (b) quasi-transverse vertical wave, and (c) transverse horizontal wave.

the increased degree of skewing with higher severity of anisotropy. It can also be inferred
from this diagram that the best angle of incidence to the anisotropy is around 45°, where
beam-skewing effects are negligible. The presence of beam skewing and beam splitting
complicates the ultrasonic evaluation of these anisotropic materials.

Defect artifacts, improper defect location, and poor identification are byproducts of the
anisotropic filter influence on ultrasonic waves. However, the basic models described here
enable analysis of commonly used anisotropic materials; a received signal that can be interpreted more accurately will improve anomaly imaging. It is also possible to use these
results to characterize the anisotropic property of any material through a series of skew
angle measurements. When matched with parametric curves generated by the numerical
integration model, the comparison will quantitatively identify the material constants.

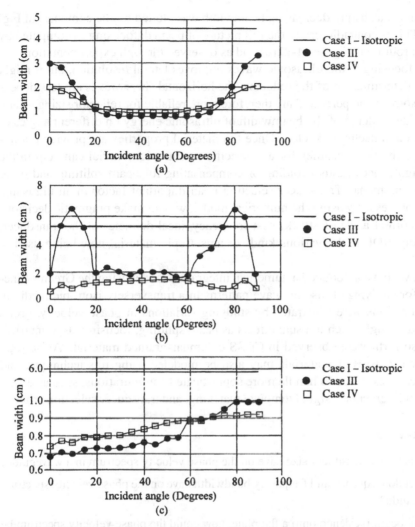

Figure 14-26. Lateral resolution transducer characterization at a distance of three times the transducer size for (a) longitudinal wave, (b) quasi-transverse vertical wave, and (c) transverse horizontal wave, showing better lateral resolution for certain anisotropic cases.

One of the important benefits of this study is the capability to characterize the ultrasonic transducers that could be useful on anisotropic materials. We have observed that, within a given medium, the ultrasonic beam either diverges or converges depending on the wave type and the orientation of the wavefront. Parts (a)–(c) of Figure 14-26 illustrate an analytical (numerical integration) study of these effects, observing the beamwidth (defined here as the half-amplitude beam) for two different anisotropic materials. The longitudinal waves exhibit a natural focusing effect between 30° and 60° of incidence angle as compared with the 0° or the 90° divergence.

With respect to the focusing effect, the results shown in Figure 14-26(b) for the shear vertical mode are misleading. Figure 14-22(d) exhibits the rather complicated shape of the actual beam profile. The computer program routinely calculated the numerical value of the beamwidth for the visible portion on Figure 14-22(d) across the narrow central part of the beam; hence, a sharply decreasing beamwidth was plotted by the computer in Figure 14-26(b). This is contradictory to the full picture of beam defocusing as computed by the field profile routine in Figure 14-22(d) and as observed through experimentation.

The natural focusing of the ultrasonic waves improves lateral resolution, but it might also affect size determination of the reflector: the horizontal shear waves are less dependent on the anisotropic property and are therefore more reliable for reflector sizing. Here, we observe the dependence of the beamwidth of ultrasonic energy on different angles of incidence with the anisotropic vector. Once the material properties are provided, a numerical study of the beam profiles for any specific anisotropic material can help in the transducer design process, thus avoiding or compensating for beam splitting and other such unusual phenomena. Transducer selection is an important factor in an ultrasonic nondestructive process. By using the numerical model, we can make pragmatic decisions (regarding the correct angle of attack, best frequency, ideal diameter of the transducer, etc.) for applying NDE to the various kinds of anisotropic material now being used in industry.

In summary, we have introduced a numerical integration model using the Green's function approach for studying ultrasonic wave patterns in a transversely isotropic medium. Its current applications were illustrated by studying variations in phase velocity, group velocity, and skew angle; such unusual effects as beam splitting, focusing, asymmetrical behavior, and so forth were observed in CCSS columnar-grained material. Anisotropy of many different classes of materials may also be modeled using this technique, and the analysis may be extended to handle more sophisticated configurations, such as defect modeling, defect interactions, signal template matching, and transducer selection criteria.

14.4 Exercises

1. What impact does transducer size have on the phase velocity spectrum in a waveguide?

2. What effect do frequency and frequency bandwidth have on the phase velocity spectrum in a waveguide?

3. For normal beam incidence onto a flat plate, how could the phase velocity spectrum be changed?

4. Why does beamwidth increase or decrease in size for waves traveling in anisotropic media as compared with isotropic media?

5. What influence does transducer diameter have on beamwidth in anisotropic media?

6. Calculate the phase velocity spectrum for parabolic and piston pressure distribution for the following cases.

    (a) Incident angle $\alpha = 32.68°$; $c_p = 5$ mm/μs; $D = 2$ mm; bandwidth = 0.5 MHz for the center frequencies $f = 1.725$ MHz, $f = 5.765$ MHz, $f = 9.725$ MHz, and $f = 13.685$ MHz.

    (b) Incident angle $\alpha = 22.69°$; $c_p = 7$ mm/μs; $D = 4$ mm; bandwidth = 0.5 MHz for the center frequencies $f = 2.405$ MHz, $f = 8.525$ MHz, $f = 5.365$ MHz, and $f = 11.545$ MHz.

    (c) Incident angle $\alpha = 15.66°$; $c_p = 10$ mm/μs; $D = 8$ mm; bandwidth = 0.5 MHz for the center frequencies $f = 1.885$ MHz, $f = 4.825$ MHz, $f = 7.445$ MHz, and $f = 8.825$ MHz.

15

Horizontal Shear

15.1 Introduction

Many aspects of horizontal shear wave propagation are intriguing and quite valuable for applications involving wave propagation, including ultrasonic NDE. Traditionally, the longitudinal and vertical shear modes of wave propagation have been the most commonly used – probably because they are simple to understand and to generate. Yet horizontal shear waves can also be generated quite easily through a variety of different transducers. This chapter covers the fundamental concepts of such propagation. A basic experiment on horizontal shear wave propagation is outlined in Section E.7.

15.2 Dispersion Curves

In addition to the Lamb wave modes that exist in flat layers, there also exists a set of time-harmonic wave motions known as shear horizontal (SH) modes. The term "horizontal shear" means that the particle vibrations (displacements and velocities) caused by any of the SH modes are in a plane that is parallel to the surfaces of the layer. This is depicted in Figure 15-1, where the wave propagates in the x_1 direction and the particle displacements are in the x_3 direction.

Physically, any mode in the SH family can be considered as the superposition of up- and down-reflecting bulk shear waves, polarized along x_3, with wavevectors lying in the (x_1, x_2)-plane and inclined at such an angle that the system of waves satisfies traction-free boundary conditions on the surfaces of the layer.

The dispersion equation governing the SH modes can be derived in several ways, including the use of Helmholtz potentials, partial wave analysis, or transverse resonance (Auld 1990). Because of the simple physical nature of the SH modes, the most straightforward way to solve the problem is to deal directly with the displacement equations of motion. This is the approach taken here; for more discussion of this technique, see Achenbach (1984).

For any isotropic medium, the particle displacement field $u(x, t)$ must satisfy Navier's displacement equations of motion:

$$\mu\nabla^2 u(x, t) + (\lambda + \mu)\nabla\nabla \bullet u(x, t) = \rho\frac{\partial^2 u(x, t)}{\partial t^2} \tag{15.1}$$

(Malvern 1969). For the SH modes, we consider particle displacement vectors that have only an x_3 component, that is, $u_1(x, t) = u_2(x, t) = 0$. Furthermore, for the x_3 displacement component, we specify a variation of the form at the outset:

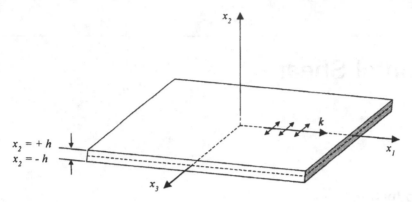

Figure 15-1. SH wave mode propagation, where the propagation is along x_1 and particle displacements are along x_3.

$$u_3(x_1, x_2, t) = f(x_2)e^{i(kx_1 - \omega t)}, \tag{15.2}$$

where k is the wavenumber of the mode ($k = \omega/c_p = 2\pi/\lambda$) and ω represents circular frequency. Notice that u_3 is independent of x_3 and hence the wavefronts are infinitely extended in the x_3 direction.

This form of the solution is chosen because it represents a wave motion that propagates along the x_1-coordinate direction (due to the exponential term) and has a fixed distribution in the x_2 direction given by $f(x_2)$. As usual, the actual physical displacement vector field is the real part of the expression on the right-hand side of equation (15.2).

If only the u_3 component of the particle displacement field is nonzero and if u_3 is independent of x_3, then (15.1) reduces to

$$\frac{\partial^2 u_3}{\partial x_1^2} + \frac{\partial^2 u_3}{\partial x_2^2} = \frac{1}{c_T^2}\frac{\partial^2 u_3}{\partial t^2}, \tag{15.3}$$

where $c_T^2 = \mu/\rho$.

Substituting the assumed solution (equation (15.2)) into (15.3) results in

$$\frac{\partial^2 f(x_2)}{\partial x_2^2} + \left(\frac{\omega^2}{c_T^2} - k^2\right)f(x_2) = 0. \tag{15.4}$$

This equation has the general solution

$$f(x_2) = A\sin(qx_2) + B\cos(qx_2), \tag{15.5}$$

where q is defined as

$$q = \sqrt{\frac{\omega^2}{c_T^2} - k^2} \tag{15.6}$$

and A and B are arbitrary constants. The general form of the displacement field is therefore

$$u_3(x_1, x_2, t) = [A\sin(qx_2) + B\cos(qx_2)]e^{i(kx_1 - \omega t)}. \tag{15.7}$$

At this point it is advantageous (though hardly necessary) to separate the total displacement field, equation (15.7), into symmetric and antisymmetric components (with respect to x_2). The parts of the total displacement field that represent symmetric and antisymmetric motions are the $\cos(qx_2)$ and $\sin(qx_2)$ terms, respectively. We thus consider two separate displacement fields:

$$
\begin{aligned}
u_3^s(x_1, x_2, t) &= B \cos(qx_2) e^{i(kx_1 - \omega t)}, \\
u_3^a(x_1, x_2, t) &= A \sin(qx_2) e^{i(kx_1 - \omega t)}.
\end{aligned}
\tag{15.8}
$$

The superscript s denotes a symmetric mode and a denotes an antisymmetric mode.

The boundary conditions imposed on either type of mode are that the surfaces of the layer ($x_2 = \pm h$) be free of tractions:

$$
\sigma_{22}(x_1, x_2, t)|_{x_2 = \pm h} = \tau_{12}(x_1, x_2, t)|_{x_2 = \pm h} = \tau_{23}(x_1, x_2, t)|_{x_2 = \pm h} = 0.
$$

However, for a displacement field of either form in (15.8), the stresses σ_{22} and τ_{12} vanish identically. Hence, the only remaining nontrivial boundary conditions are

$$
\tau_{23}(x_1, x_2, t)|_{x_2 = \pm h} = 0.
\tag{15.9}
$$

The strain field associated with either displacement field in (15.8) has only two nonzero components, $\varepsilon_{13} = (\partial u_3 / \partial x_1)/2$ and $\varepsilon_{23} = (\partial u_3 / \partial x_2)/2$. The traction component τ_{23} is therefore given by

$$
\tau_{23} = 2\mu\varepsilon_{23} = \mu \frac{\partial u_3}{\partial x_2},
\tag{15.10}
$$

the form of which will depend on whether the symmetric or antisymmetric displacement field of equation (15.8) is used. Calculating τ_{23} yields

$$
\tau_{23}(x_1, x_2, t) = \begin{cases} -\mu B q \sin(qx_2) e^{i(kx_1 - \omega t)} & \text{for symmetric modes,} \\ \mu A q \cos(qx_2) e^{i(kx_1 - \omega t)} & \text{for antisymmetric modes.} \end{cases}
\tag{15.11}
$$

Finally, imposing the boundary conditions (15.9) yields the dispersion equations

$$
\sin(qh) = 0
\tag{15.12a}
$$

and

$$
\cos(qh) = 0,
\tag{15.12b}
$$

for the traction due to symmetric and antisymmetric displacements, respectively. Owing to the simplicity of these equations, we can obtain explicit solutions. Since $\sin(x) = 0$ when $x = n\pi$ ($n \in \{0, 1, 2, \ldots\}$) and $\cos(x) = 0$ when $x = n\pi/2$ ($n \in \{1, 3, 5, \ldots\}$), the solutions to (15.12a) and (15.12b) can be written as

$$
qh = n\pi/2,
\tag{15.13}
$$

where $n \in \{0, 2, 4, \ldots\}$ for symmetric SH modes and $n \in \{1, 3, 5, \ldots\}$ for antisymmetric ones. Thus, the dispersion equation has an infinite number of solutions for either SH mode. The individual SH modes are specified by the integer n, which should be even for symmetric modes and odd for antisymmetric modes.

Using the form of q given in (15.13), denoting the thickness of the plate by $d = 2h$, and taking the real parts of equation (15.8), the displacement fields of the symmetric and antisymmetric SH modes can be written as

$$u_3^s(x_1, x_2, t) = B \cos(n\pi x_2/d) \cos(kx_1 - \omega t),$$

$$u_3^a(x_1, x_2, t) = A \sin(n\pi x_2/d) \cos(kx_1 - \omega t).$$

Note that, since the arguments of the sine and cosine terms are independent of frequency and wavenumber, the variation of the fields of the SH modes across the thickness of the plate do not vary along any mode's dispersion curve. This is in sharp contrast to Lamb wave behavior, where the fields are actually functions of the position of the dispersion curves.

15.3   Phase Velocities and Cutoff Frequencies

We have found explicit solutions to the dispersion equation. Hence we can also construct explicit solutions for the phase velocity-versus-fd curves of any mode specified by the integer n.

With q as defined in (15.6) and using (15.13) and the definition of wavenumber, $k = \omega/c_p$, the dispersion equation can be written as

$$\frac{\omega^2}{c_T^2} - \frac{\omega^2}{c_p^2} = \left(\frac{n\pi}{2h}\right)^2. \tag{15.14a}$$

This equation can easily be solved for the phase velocity c_p in terms of the frequency thickness product fd (where $d = 2h$ and $\omega = 2\pi f$). The result is

$$c_p(fd) = \pm 2c_T \left\{ \frac{fd}{\sqrt{4(fd)^2 - n^2 c_T^2}} \right\}. \tag{15.14b}$$

When $n = 0$ (corresponding to the zeroth-order symmetric SH mode) we have $c_p = c_T$, a dispersionless wave propagating at the shear wave speed c_T. All other SH modes (i.e., for all $n \neq 0$) are dispersive. Figure 15-2 plots the phase velocity dispersion curves for the first eight SH modes over a frequency thickness range of 0–15 MHz-mm. The (solid) even-numbered curves represent symmetric modes and the (dashed) odd-numbered curves represent antisymmetric modes.

The cutoff frequencies of the SH modes can be found by setting the denominator in (15.14b) equal to zero. This corresponds to infinite phase velocities and (as will be seen later) zero group velocities. The nth cutoff frequency is therefore given by:

$$(fd)_n = \frac{nc_T}{2}; \tag{15.15}$$

once again, even integer n represents symmetric modes and odd integer n represents antisymmetric modes.

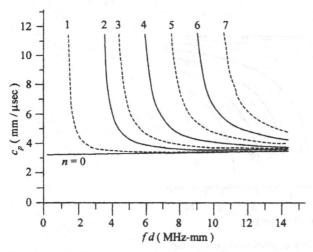

Figure 15-2. SH mode phase velocity dispersion curves for an aluminum layer ($c_T = 3.1$ mm/μs): solid curves denote symmetric modes; dashed curves denote antisymmetric modes.

15.4 Group Velocity

Having explicit solutions for the dispersion equation also enables us to determine explicit expressions for the group velocity of any given SH mode. From the dispersion equation,

$$\frac{\omega^2}{c_T^2} - k^2 = \left(\frac{n\pi}{2h}\right)^2, \tag{15.16}$$

we take the differential of both sides (using that the RHS is a constant for any n). The result is

$$\frac{2\omega \, d\omega}{c_T^2} - 2k \, dk = 0. \tag{15.17}$$

Solving this equation for the quantity $d\omega/dk$ (by definition, the group velocity) yields

$$\frac{d\omega}{dk} = \frac{kc_T^2}{\omega^2}. \tag{15.18}$$

Solving (15.16) for k, substituting the result into (15.18), and simplifying, we have

$$c_g(fd) = c_T \sqrt{1 - \frac{(n/2)^2}{(fd/c_T)^2}} \quad (fd \geq (fd)_n). \tag{15.19}$$

From this expression for the group velocity, we can see that – at the cutoff frequencies (given by (15.15)) – the group velocity of any mode is zero. As fd approaches infinity for any given fixed n, the group velocity of any SH mode approaches that of bulk shear waves, c_T.

Figure 15-3 plots the SH mode group velocity curves for the first eight SH modes in an aluminum layer with shear wave speed $c_T = 3.1$ mm/μs. As in Figure 15-2, the solid

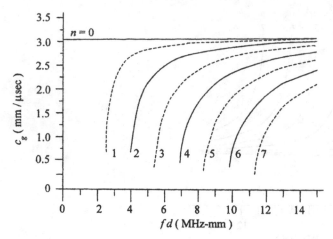

Figure 15-3. SH mode group velocity dispersion curves for the aluminum layer of Figure 15-2.

curves (even n) represent symmetric modes and the dashed curves (odd n) represent antisymmetric modes.

15.5 Summary

The main results concerning the propagation of SH modes in flat layers are summarized here for easy reference. First, recall the definitions

$$q^2 = \frac{\omega^2}{c_T^2} - k^2, \quad d = 2h = \text{plate thickness}, \quad c_T = \text{bulk shear wave speed},$$

and that the wave is assumed to propagate in the x_1 direction with polarization entirely in the x_3 direction. We then have the following equations, where $n = 0, 2, 4, \ldots$ for symmetric modes and $n = 1, 3, 5, \ldots$ for antisymmetric modes.

Dispersion

$\sin(qh) = 0$ (symmetric modes)

$\cos(qh) = 0$ (antisymmetric modes)

Cutoff frequency

$$(fd)_n = \frac{n c_T}{2}$$

Phase velocity

$$c_p(fd) = 2c_T \left\{ \frac{fd}{\sqrt{4(fd)^2 - n^2 c_T^2}} \right\}$$

Group velocity

$$c_g(fd) = c_T \sqrt{1 - \frac{(n/2)^2}{(fd/c_T)^2}}$$

Displacement

$$u_3(x_1, x_2, t) = \begin{cases} B\cos(n\pi x_2/d)\,e^{i(kx_1 - \omega t)} & \text{(symmetric modes)} \\ A\sin(n\pi x_2/d)\,e^{i(kx_1 - \omega t)} & \text{(antisymmetric modes)} \end{cases}$$

Stress

$$\tau_{23}(x_1, x_2, t) = \begin{cases} -B\mu q\,\sin(n\pi x_2/d)\,e^{i(kx_1 - \omega t)} & \text{(symmetric modes)} \\ A\mu q\,\cos(n\pi x_2/d)\,e^{i(kx_1 - \omega t)} & \text{(antisymmetric modes)} \end{cases}$$

$$\tau_{13}(x_1, x_2, t) = \begin{cases} ikB\cos(n\pi x_2/d)\,e^{i(kx_1 - \omega t)} & \text{(symmetric modes)} \\ ikA\sin(n\pi x_2/d)\,e^{i(kx_1 - \omega t)} & \text{(antisymmetric modes)} \end{cases}$$

15.6   Excitation of Shear Horizontal Modes in Flat Layers

In this section we evaluate SH guided wave modes (excited by the wedge technique) in flat layers. Much to our surprise, the solution to this problem does not appear in any of the standard works on linear elastodynamics theory. However, the more difficult problem of Lamb wave generation (by the wedge technique) in isotropic layers is presented by Viktorov (1967) and for generally anisotropic layers by Ditri (1992). Here we solve the SH problem by using spatial Fourier transforms and the excitability functions of the SH modes extracted from the result.

15.6.1   General Considerations

Before beginning with the analysis, a few words must be said concerning the excitation of SH modes using the wedge technique. Since the SH modes are a synthesis of pure out-of-plane shear waves, the tractions applied to the surface of the layer must also have an out-of-plane shear component. This can be achieved only if a viscous couplant (e.g., honey or silicone) is used between the wedge and the layer. In addition, a shear wave transducer must be used as the sending and receiving elements, since a longitudinal wave transducer will produce only in-plane shear (and in-plane normal) tractions and hence will not generate SH modes, which require out-of-plane shear tractions.

15.6.2   Energy Partition into Lamb and SH Modes

The orientation of the shear wave probe on the surface of the wedge is very important in determining how much of the incident energy goes into generating Lamb waves versus how much goes into generating SH waves. In order to specify the orientation of the sending and receiving transducers on the wedges, we attach a local x_1, x_2, x_3 coordinate system to the transducer and a fixed X_1, X_2, X_3 coordinate system to the layer (see Figure 15-4). The transducer coordinate system is simply a rotation of the layer coordinate system by an amount equal to the incident angle about the X_3 axis. That is: $x_3 = X_3$; and x_1, x_2 are in the same plane as X_1, X_2 but rotated by an angle θ_i, where θ_i denotes the incident angle.

If shear wave contact transducers are used, then the particle vibration in the wedge will lie entirely in the (x_1, x_3)-plane. The relative amount of x_1 and x_3 displacements will be a function of angle between the respective axes and the polarization direction of the shear

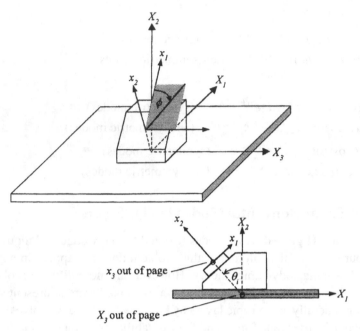

Figure 15-4. Definition of transducer (x_1, x_2, x_3) and layer (X_1, X_2, X_3) coordinate systems; note that ϕ is defined in the (x_1, x_3)-plane.

wave probe. This orientation is specified by the angle ϕ in Figure 15-4. If $\phi = 0°$ then the polarization direction of the shear wave probe coincides with the x_1-axis; for $\phi = 90°$, the polarization direction of the shear wave probe coincides with the x_3-axis.

Assuming that the transducer produces a displacement field of amplitude A, the x_1 and x_3 components of displacement are related to the angle ϕ by

$$u = A\{\cos(\phi)\hat{e}_1 + \sin(\phi)\hat{e}_3\}, \tag{15.20}$$

where $\hat{e}_i$ denotes a unit vector in the x_i direction. The out-of-plane displacement component (u_3) will be responsible for generating the SH modes in the layer, and the in-plane component (u_1) is responsible for generating Lamb waves. Since the energy associated with the incident fields is proportional to the square of the incident displacement amplitudes, the energy partition (E) into SH and Lamb waves will be

$$E_{\text{Lamb}} \propto A^2 \cos^2(\phi) \quad \text{and} \tag{15.21a}$$

$$E_{\text{SH}} \propto A^2 \sin^2(\phi), \tag{15.21b}$$

respectively. Of course, the total energy $E_{\text{tot}} = (E_{\text{Lamb}} + E_{\text{SH}}) \propto A^2$.

Figure 15-5 plots the relative amount of energy that goes into the SH and Lamb wave modes as a function of the orientation of the shear wave transducer on the surface of the wedge. If $\phi = 0°$ then all of the incident energy goes into the Lamb wave modes, since the particle vibration is entirely in the "sagittal" (x_1, x_3)-plane. As the angle ϕ is increased, the amount of energy that goes into the SH modes increases according to $\sin^2(\phi)$. The energy partition between Lamb and SH modes is equal for $\phi = 45°$; for $\phi = 72°$, over 90% of the incident energy goes into the SH modes. To maximize the amount of energy that is

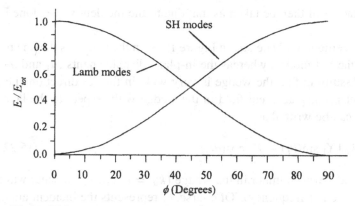

Figure 15-5. Energy partition among the SH and Lamb wave modes as a function of the orientation of the sending and receiving transducers.

put into the SH modes, the angle ϕ should be as close as possible to $90°$. Of course, for $\phi \neq 90°$, both SH and Lamb waves can be generated simultaneously, if that is the goal. It should be noted that we are discussing the amount of energy *available* for the generation of the Lamb and SH modes; the amount of energy that actually goes into the individual modes will be a function of their excitabilities.

15.6.3   Excitability of SH Modes

In this section we address the excitation of SH modes by the wedge technique. This is done by replacing the wedge and incident wave by a known traction distribution on the surface of the layer over the region of contact between the wedge and the layer. The approximate amplitude of each SH wave mode generated by this traction distribution is then found by using the full elasticity equations. A more exact approach would be to solve the two-body (i.e. wedge–layer) problem, enforcing continuity of displacements and tractions across the interface joining the two media. However, this problem is extremely difficult owing to the bounded nature of the wedge and the multiple reflections and complex surfaces that are present.

As mentioned by Viktorov (1967), replacing the wedge by a traction distribution is valid if the acoustic impedance ($w_w \equiv \rho_w c_w$, where c_w represents the shear wave velocity) of the wedge material is much lower than that of the layer material, w_l. This is usually the case because the acoustic impedance of Plexiglas is about 1.3 whereas the acoustic impedance of aluminum is about 8.4. Actually, the statement made by Viktorov (and many of his co-workers on similar problems) is somewhat misleading. It is true that we can always replace the wedge with a system of tractions and find exactly the same solution as if we had solved the wedge–layer problem. In order to do so, however, the tractions used must be those that arise from an exact solution of the wedge–layer problem. That is, the exact wedge–layer solution must be known so that we can specify the true tractions at the interface for the simpler traction-layer problem. Fortunately, some of our recent work shows that the wedge–layer problem can be approximated by a traction–layer problem – where the traction is taken as that arising from the incident wave alone – if the acoustic impedance of the wedge is small in comparison to that of the layer. The specified traction at the wedge–layer interface will then be taken as that due to the incident wave alone in the following analysis.

The out-of-plane displacement field (i.e., u_3 in Figure 15-4) in the wedge is the one responsible for generating the SH modes, whereas the in-plane displacements (u_1 and u_2) generate Lamb waves. Assuming that the wedge is very wide in the X_3 direction (i.e., a strip load), the out-of-plane displacement field in the wedge with respect to the layer coordinates (X_1, X_2, X_3) can be written as

$$u_3(X_1, X_2) = A \exp\{ik_w[X_1 \sin(\theta_i) - X_2 \cos(\theta_i)]\}, \tag{15.22}$$

where k_w represents the shear wavenumber in the wedge ($k_w = \omega/c_w$; $c_w =$ shear wave speed in the wedge, $\omega =$ circular frequency). Of course, θ_i represents the incident angle of the wedge with respect to the normal to the surface of the layer. For a wave of the form (15.22), by using the equation $\tau_{23} = \mu(\partial u_3/\partial X_2)$ it is easy to show that

$$\tau_{23} = -iA\mu_w k_w \cos(\theta_i) \exp\{ik_w[X_1 \sin(\theta_i) - X_2 \cos(\theta_i)]\}. \tag{15.23}$$

Equation (15.23) shows that the magnitude of the traction (produced on the surface of the layer) due to the incident wave is proportional to the cosine of the incident angle θ_i multiplied by some constant. As such, it decreases as the incident angle is increased from 0 to 90 degrees. Now, define the dimensionless factor $R(\theta_i)$ to be the ratio of the magnitude of $\tau_{23}(\theta_i)$ to the magnitude of the incident traction when $\theta_i = 0$, that is, $\tau_{23}(0)$. Direct computation then shows that

$$R(\theta_i) = \cos(\theta_i). \tag{15.24}$$

We are now in a position to formulate the problem at hand. We seek a solution to Navier's displacement equation of motion:

$$\mu\nabla^2 u + (\lambda + \mu)\nabla\nabla \bullet u = \rho\frac{\partial^2 u}{\partial t^2}. \tag{15.25}$$

For all points inside the layer such that the traction vector associated with u satisfies the boundary conditions, we have

$$\tau_{23}(X_1, X_2 = h) = \begin{cases} R(\theta_i)f(X_1)\exp[ik_w \sin(\theta_i)X_1] & \text{if } |X_1| < L, \\ 0 & \text{if } |X_1| > L; \end{cases} \tag{15.26a}$$

$$\tau_{23}(X_1, X_2 = h) \equiv 0. \tag{15.26b}$$

In (15.26a), the function $f(X_1)$ describes the *envelope* of the applied traction distribution (see Figure 15-6) and is related to the pressure distribution across the face of the transducer. The analysis is therefore applicable to any type of traction distribution.

If we assume that the displacement field in the layer has only an X_3 component, and that this component (u_3) is a function only of X_1 and X_2, then Navier's equations reduce to

$$\frac{\partial^2 u_3(X_1, X_2)}{\partial X_1^2} + \frac{\partial^2 u_3(X_1, X_2)}{\partial X_2^2} = \frac{\omega^2}{c_T^2}u_3(X_1, X_2), \tag{15.27}$$

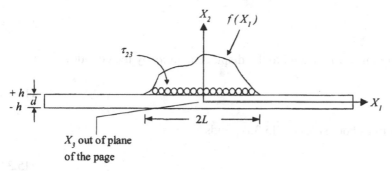

Figure 15-6. Layer geometry showing the applied traction loading (hollow circles denote a direction coming out of the plane of the page).

where $c_T^2 = \mu/\rho$ represents the velocity of bulk shear waves in the layer. (A time variation of the form $e^{-i\omega t}$ has been assumed and thereafter dropped from both sides.)

Equation (15.27) is a partial differential equation for the unknown function $u_3(X_1, X_2)$. In order to turn it into an ordinary differential equation, we will use integral transform techniques. Because the boundary condition specifies a known function of X_1 and is over an infinite domain, we apply the Fourier transform over X_1. The Fourier transform of a function $\phi(X_1, X_2)$, denoted by an over-hat ($\hat{\phi}$), can be taken as

$$\hat{\phi}(\kappa, X_2) \equiv F[\phi] = \frac{1}{2\pi} \int_{-\infty}^{\infty} \phi(X_1, X_2) e^{-i\kappa X_1} \, dX_1. \tag{15.28a}$$

The inverse transform is then given by

$$\phi(X_1, X_2) \equiv \int_{-\infty}^{\infty} \hat{\phi}(\kappa, X_2) e^{i\kappa X_1} \, d\kappa. \tag{15.28b}$$

It is now easy to show, via integration by parts, that the transform of a second-order derivative, $\partial^2\phi/\partial X_1^2$, becomes

$$F\left[\frac{\partial^2\phi}{\partial X_1^2}\right] = -\kappa^2\hat{\phi}, \tag{15.29}$$

assuming that $\phi(X_1, X_2)$ and its first derivative with respect to X_1 vanish sufficiently rapidly as $X_1 \to \pm\infty$. This is actually not the case for our function, $u_3(X_1, X_2)$, which will not vanish at all as $X_1 \to \pm\infty$. We will see that this particular circumstance leads to singularities on the path of the inverse integral. Neglecting this fact for now, we continue by taking the Fourier transform of both sides of (15.27). The result is

$$\frac{d^2\hat{u}_3(\kappa, X_2)}{dX_2^2} + \left(\frac{\omega^2}{c_T^2} - \kappa^2\right)\hat{u}_3 = 0. \tag{15.30}$$

Equation (15.30) has the general solution

$$\hat{u}_3(\kappa, X_2) = A(\kappa)\cos(qX_2) + B(\kappa)\sin(qX_2), \tag{15.31}$$

where $A(\kappa)$ and $B(\kappa)$ are (at this point) arbitrary functions of the transform parameter κ, and where

$$q^2 \equiv \frac{\omega^2}{c_T^2} - \kappa^2. \tag{15.32}$$

The traction component τ_{23} is related to the displacement u_3 by the equation

$$\tau_{23} = \mu \frac{\partial u_3}{\partial X_2}. \tag{15.33}$$

Taking the transform of both sides of (15.33) yields

$$\hat{\tau}_{23} = \mu \frac{\partial \hat{u}_3}{\partial X_2}, \tag{15.34}$$

since the partial derivative with respect to X_2 can be moved out of the integral sign.

Substituting the general solution for $\hat{u}_3$ (i.e., (15.31)) into (15.34), we obtain the trans-formed traction as follows:

$$\hat{\tau}_{23} = \mu q [-A(\kappa) \sin(qX_2) + B(\kappa) \cos(qX_2)]. \tag{15.35}$$

The next step is to transform the boundary conditions (15.26a) and (15.26b). In doing this, we make use of the following well-known property of Fourier transforms:

$$F[\phi(X_1) e^{i\alpha X_1}] = \hat{\phi}(\kappa - a). \tag{15.36}$$

The transformed boundary conditions (15.26a) and (15.26b) can therefore be written as

$$-\mu q A(\kappa) \sin(qh) + \mu q B(\kappa) \cos(q\kappa) = R\hat{f}(\kappa - k_w^0),$$
$$\mu q A(\kappa) \sin(qh) + \mu q B(\kappa) \cos(q\kappa) = 0, \tag{15.37}$$

where $k_w^0 \equiv k_w \sin(\theta_i)$. Equations (15.37) represent two linear, nonhomogeneous equa-tions in the two unknown functions $A(\kappa)$ and $B(\kappa)$, which can be formally solved to yield

$$A(\kappa) = -\frac{R\hat{f}(\kappa - k_w^0)}{2\mu q \sin(qh)} \quad \text{and} \quad B(\kappa) = \frac{R\hat{f}(\kappa - k_w^0)}{2\mu q \cos(qh)}. \tag{15.38}$$

Substituting the functions $A(\kappa)$ and $B(\kappa)$ from (15.38) into the general solution for the transformed displacement field (15.31), we obtain

$$\hat{u}_3(\kappa, X_2) = -\frac{R\hat{f}(\kappa - k_w^0)}{2\mu q \sin(qh)} \cos(qX_2) + \frac{R\hat{f}(\kappa - k_w^0)}{2\mu q \cos(qh)} \sin(qX_2). \tag{15.39}$$

Applying the inverse Fourier transform, according to (15.28b), leads to the formal (trans-formed) solution to the problem:

$$u_3(X_1, X_2)$$
$$= -\frac{R}{2\mu} \int_{-\infty}^{\infty} \left\{ \hat{f}(\kappa - k_w^0) \frac{\cos(qX_2)}{q \sin(qh)} - \hat{f}(\kappa - k_w^0) \frac{\sin(qX_2)}{q \cos(qh)} \right\} e^{i\kappa X_1} \, d\kappa. \tag{15.40}$$

As always when using integral transforms, the hard part of the analysis is to invert the in-tegral to obtain an explicit solution for u_3. However, for the case of (15.40), we will see that this is indeed possible without too much difficulty.

In order to evaluate equation (15.40), residue calculus will be used. The basic idea is to replace the integral that is along the entire (real) κ-axis ($-\infty < \kappa < \infty$) with a contour in the complex κ-plane, which includes the real axis as well as a large semicircle in either the upper or lower half-plane. The direction of the contour (i.e., upper or lower half-plane) is determined by enforcing – on this large semicircle – that the integrand vanishes sufficiently rapidly so as to give zero contribution in the limit as the radius of the semicircle increases beyond bounds. In replacing one contour with another, we must include the contributions of any singularities (poles or other) that we cross in deforming the original contour. The first order of business is to determine the locations and types of singularities of the integrand functions. Before doing this, we can split the solution (15.40) into two parts (representing symmetric and antisymmetric modes) and deal with each one separately. The contribution of the symmetric modes can be written as

$$u_3^s = -\frac{R}{2\mu} \int_{-\infty}^{\infty} \hat{f}(\kappa - k_w^0) \frac{\cos(qX_2)}{q \sin(qh)} e^{i\kappa X_1} \, d\kappa; \tag{15.41}$$

the antisymmetric mode contribution is

$$u_3^a = -\frac{R}{2\mu} \int_{-\infty}^{\infty} \hat{f}(\kappa - k_w^0) \frac{\sin(qX_2)}{q \cos(qh)} e^{i\kappa X_1} \, d\kappa. \tag{15.42}$$

These integrals can be identified as symmetric and antisymmetric modes by noticing that in (15.41) the integrand is even (symmetric) in X_2, whereas in (15.42) the integrand is odd (antisymmetric) in X_2.

15.6.4 Mathematical Considerations

Singularities of the integrands must also be considered. In this section we describe the mathematics behind the evaluation of the integral in (15.41). The key factors in the integration can be identified as follows:

- $\hat{f}(\kappa - k_w^0)$ is the Fourier transform of $f(X_1)e^{i\kappa k_w^0}$;
- κ is the (complex) variable of integration, $\kappa = \kappa_{\mathrm{Re}} + i\kappa_{\mathrm{Im}}$;
- $q^2 \equiv (\omega/c_T)^2 - \kappa^2$ by definition;
- X_1, X_2, X_3, and h are defined via Figure 15-7; and
- u_3 is the displacement in the X_3 direction (the superscript s denotes "symmetric mode").

During complex integration, our concern is drawn to the possibility of singularities and/or multivalued functions in the integrand (see Appendix C for definitions). If a function is *multivalued* then it consists of several functions, each called a *branch*. Branches are separated by a continuum of singularities called a *branch cut*. A singularity common to all branches is called a *branch point*.

Since the Fourier transform and exponential components of the integrand are entire (i.e., they are analytic or have continuous derivatives everywhere on the complex κ-plane), they do not introduce any singularities. The only possible singularities will come from the $q \sin(qh)$ in the denominator of the integrand. Since these singularities will be isolated (poles), no branch cuts or branch points need to be considered. To illustrate this, we examine the term

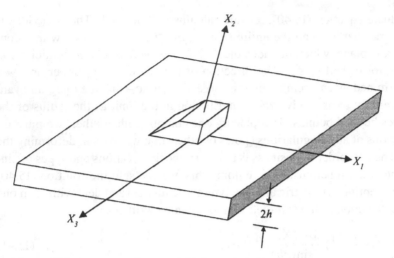

Figure 15-7. Layer coordinate and thickness definitions.

$$\frac{\cos(qX_2)}{q\sin(qh)}.$$ \hfill (15.43)

It will have the obvious singularity at $q = 0$; introducing subscripts to identify particular singularites (since they are isolated), we may write

$$q_0 = 0.$$ \hfill (15.44a)

Because $\sin(qh) = 0$ whenever $qh = \pm n\pi$ for integer $n = 1, 2, 3, \ldots$, there are additional singularities whenever $q = \pm n\pi/h$. Introducing the subscript n for these singularities and casting q in terms of layer thickness $2h$, we have

$$q_n = \pm \frac{n\pi}{2h} \quad \text{for } n \in \{2, 4, 6, \ldots\},$$ \hfill (15.44b)

where the set of allowable values for n has been adjusted for the introduction of the 2 in the denominator.

These are simple poles (first-order singularities). In order to evaluate the type of singularity at $q_0 = 0$ and represent it in terms of the integration variable κ, a Laurent series representation can be used. To simplify notation, X_2 can be taken as $X_2 = h$. First, note that

$$\frac{\cos(qh)}{q\sin(qh)} = \frac{\cot(qh)}{q}.$$

The series for $\cot(z)$ is well known:

$$\cot(z) = \frac{1}{z} - \frac{z}{3} - \frac{z^3}{45} - \frac{2z^5}{945} - \cdots - \frac{2^{2n}B_n z^{2n-1}}{(2n)!} - \cdots \quad (0 < |z| < \pi).$$

The Bernoulli number B_n is given by

$$B_n = \frac{(2n)!}{2^{2n-1}\pi^{2n}}\left(1 + \frac{1}{2^{2n}} + \frac{1}{3^{2n}} + \cdots\right).$$

Letting $z = qh$, we then have

$$\frac{\cot(qh)}{q} = \frac{1}{q}\left(\frac{1}{qh} - \frac{qh}{3} - \frac{q^3h^3}{45} - \cdots\right)$$

or

$$\frac{\cot(qh)}{q} = \frac{1}{q^2h} - \frac{h}{3} - \frac{q^2h^3}{45} - \cdots \qquad (0 < |qh| < \pi). \tag{15.45}$$

Substituting for q^2, from (15.32) we have

$$q^2 = \left(\frac{\omega}{c_T}\right)^2 - \kappa^2 \quad \text{or} \quad q^2 = \left(\frac{\omega}{c_T} + \kappa\right)\left(\frac{\omega}{c_T} - \kappa\right).$$

This shows that $\cos(qh)/q \sin(qh)$ has simple poles at

$$\kappa_0^s = \pm\frac{\omega}{c_T}, \tag{15.46}$$

where (as before) the superscript s means symmetric and the subscript 0 means the poles associated with $q_0 = 0$.

To find the remaining poles – the singularities introduced by $\sin(qh)$ – the definition of q can be used:

$$q_n^2 = \left(\frac{n\pi}{2h}\right)^2 = \left(\frac{\omega}{c_T}\right)^2 - (\kappa_n^s)^2$$

or

$$\kappa_n^s = \pm\sqrt{\left(\frac{\omega}{c_T}\right)^2 - \left(\frac{n\pi}{2h}\right)^2} \quad \text{for } n \in \{2, 4, 6, \ldots\}. \tag{15.47}$$

From a physical point of view, it is important to evaluate the expression under the radical sign. If

$$\left(\frac{\omega}{c_T}\right)^2 > \left(\frac{n\pi}{2h}\right)^2$$

then the poles are real. Using $\omega = 2\pi f$ and (layer thickness) $d = 2h$,

$$\frac{2\pi f}{c_T} > \frac{n\pi}{d}$$

implies that, if

$$n < \frac{2fd}{c_T} \quad \text{for } n \in \{2, 4, 6, \ldots\}$$

or (letting $n' = n/2$) if

$$n' < \frac{fd}{c_T} \quad \text{for } n' \in \{1, 2, 3, \ldots\},$$

then the poles are real. Notice that there can be only a finite number of real poles, since n or n' is limited by the frequency–thickness product fd and the bulk shear wave velocity c_T.

Consider now the exponential term

$$\exp[i\kappa X_1] = \exp[i\kappa_{Re} X_1],$$

which represents a wave propagating in the X_1 direction. Thus, only a finite number (depending on fd and c_T) of symmetric SH mode waves can propagate. Negative values of κ_{Re} would imply propagation in the negative X_1 direction. Since this is not possible because of the angular incidence, negative values of κ_{Re} are not allowed on physical grounds. On the other hand, if

$$\left(\frac{\omega}{c_T}\right)^2 < \left(\frac{n\pi}{2h}\right)^2$$

then the poles are pure imaginary. This implies that

$$\exp[i\kappa X_1] = \exp[i(i\kappa_{Im} X_1)] = \exp[-\kappa_{Im} X_1].$$

Thus, imaginary poles lead to waves that fade away exponentially with X_1 and at a rate determined by κ_{Im}. Note that, if κ_{Im} is negative, then a wave that grows in amplitude with increasing distance is indicated. Because this case has not been physically observed, only positive roots of the expression for the poles of κ should be contemplated (i.e., the poles should be restricted to the real and imaginary axes of the complex κ-plane). Since n can vary from 2 to infinity, and since only a finite number of real poles exist, there are an infinite number of nonpropagating modes.

The integral can now be evaluated using the residue theorem of complex variables. This theorem states that

$$\int_C f(z) = 2\pi i \sum \text{Res},$$

where C encloses the poles of $f(z)$ and where Res denotes the residues at the appropriate poles. For the integral we wish to evaluate, the contour shown in Figure 15-8 can be used. The poles are restricted to the positive axes, as required for a meaningful physical solution. Imagine the contour as having a radius large enough to enclose all the relevant poles. Then, the residue theorem may be applied.

A residue at a simple pole (say, z_0) is defined as

$$\text{Res } z_0 = (z - z_0) f(z)|_{z=z_0}.$$

The residues we are seeking correspond to the poles of:

$$\hat{f}(\kappa - k_w^0) \frac{\cos(qX_2)}{q \sin(qh)} e^{i\kappa X_1}.$$

Note that all functions involved in this expression are analytic functions and hence can be expanded in the Taylor series about the point z_0. Let $N(z)$ denote the numerator and

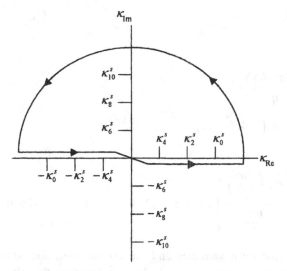

Figure 15-8. Contour for integration of equation (15.41).

$D(z)$ the denominator; then $f(z) = N(z)/D(z)$. Now expand $N(z)$ and $D(z)$ about z_0 in a Taylor series:

$$f(z) = \frac{N(z_0) + N'(z_0)(z - z_0) + N''(z_0)(z - z_0)^2/2! + \cdots}{D(z_0) + D'(z_0)(z - z_0) + D''(z_0)(z - z_0)^2/2! + \cdots}.$$

Since $D(z_0) = 0$, we have

$$f(z) = \frac{N(z_0) + N'(z_0)(z - z_0) + N''(z_0)(z - z_0)^2/2! + \cdots}{D'(z_0)(z - z_0) + D''(z_0)(z - z_0)^2/2! + \cdots}.$$

When we multiply by $(z - z_0)$, we find that

$$(z - z_0)f(z) = \frac{N(z_0) + N'(z_0)(z - z_0) + N''(z_0)(z - z_0)^2/2! + \cdots}{D'(z_0) + D''(z_0)(z - z_0)/2! + \cdots};$$

with $z = z_0$, the $(z - z_0)$ terms vanish, leaving

$$(z - z_0)f(z)|_{z=z_0} = \frac{N(z_0)}{D'(z_0)}.$$

According to this expression, the derivative of $q \sin(qh)$ must be found with respect to the variable of integration, κ. We thereby obtain

$$\frac{\partial}{\partial \kappa} q \sin(qh) = \frac{\partial q}{\partial \kappa} \sin(qh) + q \cos(qh)h \frac{\partial q}{\partial \kappa}$$

$$= \frac{\partial q}{\partial \kappa}(q \cos(qh)h + \sin(qh)).$$

Using the expression for q as a function of κ, we then find

$$q = \sqrt{\left(\frac{\omega}{c_T}\right)^2 - \kappa^2} \quad \text{and}$$

$$\frac{\partial q}{\partial \kappa} = \frac{1}{2}\frac{-2\kappa}{\sqrt{(\omega/c_T)^2 - \kappa^2}} = -\frac{\kappa}{q}.$$

As a result,

$$\frac{\partial}{\partial \kappa} q \sin(qh) = -\left(h\cos(qh) + \frac{\sin(qh)}{q}\right)\kappa,$$

$$\kappa = \kappa_n^s \quad \text{and} \quad q_n h = \frac{n\pi}{2} \quad \text{for } n \in \{2, 4, 6, \ldots\}.$$

We thus arrive at

$$\frac{\partial}{\partial \kappa} q \sin(qh) = -h\left(\cos\left(\frac{n\pi}{2}\right) + \frac{\sin\left(\frac{n\pi}{2}\right)}{\frac{n\pi}{2}}\right)\kappa_n^s, \quad n \in \{2, 4, 6, \ldots\}. \tag{15.48}$$

For the allowed values of n, the sine term vanishes and the cosine term alternates between $+1$ and -1 as n increases. Since n is always even, $(-1)^{n/2-1}$ describes this alternating sign character (with consideration for the already present -1). With this notation,

$$\frac{\partial}{\partial \kappa} q \sin(qh) = (-1)^{n/2-1}h\kappa_n^s \quad \text{for } n \in \{2, 4, 6, \ldots\}. \tag{15.49}$$

If $n = 0$ then the cosine term approaches 1 (as does the sine term, which behaves essentially like $(\sin x)/x$). For the $n = 0$ case,

$$\frac{\partial}{\partial \kappa} q \sin(qh) = -2h\kappa_0^s. \tag{15.50}$$

Recalling that the residue(s) is (are) given by

$$\text{Res} = \frac{\hat{f}(\kappa - k_w^0)\cos(qX_2)e^{i\kappa X_1}}{\dfrac{\partial}{\partial \kappa}q\sin(qh)},$$

we have

$$\sum \text{Res} = \frac{\hat{f}(\kappa - k_w^0)e^{i\kappa_0 X_1}}{-2h\kappa_0^s} + \sum_{n=2,4,6,\ldots}^{N}\frac{\hat{f}(\kappa - k_w^0)\cos(qX_2)e^{i\kappa_0 X_1}}{(-1)^{n/2-1}h\kappa_n^s},$$

where N is the largest even integer less than $2fd/c_T$.

Applying the residue theorem to

$$u_3^s = -\frac{R}{2\mu}\int_{-\infty}^{\infty}\hat{f}(\kappa - k_w^0)\frac{\cos(qX_2)}{q\sin(qh)}e^{i\kappa X_1}\,d\kappa,$$

we have

$$u_3^s = -\frac{R}{2\mu}2\pi i\sum\text{Res} \quad \text{or}$$

$$u_3^s = \frac{R\pi i}{\mu}\left\{\frac{\hat{f}(\kappa_0^s - k_w^0)e^{i\kappa_0^s X_1}}{2h\kappa_0^s}\right.$$

$$\left. + \sum_{n=2,4,6,\ldots}^{N}(-1)^{n/2}\frac{\hat{f}(\kappa_n^s - k_w^0)\cos(qX_2)e^{i\kappa_n^s X_1}}{h\kappa_n^s}\right\}. \tag{15.51}$$

Assuming that the displacement u is separable in spatial and time parts (i.e., assuming $u = u_x(X_1, X_2)u_t(t)$), we then have

$$u_3^s(X_1, X_2, t) = u_3^s(X_1, X_2)e^{-i\omega t}.$$

The analysis for the second integral (in (15.42), corresponding to the antisymmetric modes) is very similar to that just presented, so only results will be given. The new function whose singularities must be located is

$$\frac{\sin(qX_2)}{q\cos(qh)}. \tag{15.52}$$

This function is regular at $q = 0$ because the $\sin(qX_2)$ in the numerator and the q in the denominator both approach zero as $q \to 0$. Physically, this is a result of the fact that there is no (nontrivial) $n = 0$ solution to the dispersion equation for antisymmetric SH modes. The only singularities are therefore simple poles located at the roots of the equation,

$$\cos(qh) = 0. \tag{15.53}$$

This is the dispersion equation for SH modes in free layers. The solutions to (15.53) can be written as

$$q_n = \frac{n\pi}{2h} \quad \text{for } n \in \{1, 3, 5, \ldots\} \tag{15.54}$$

or, equivalently, as

$$\kappa_n^a = \pm\sqrt{\left(\frac{\omega}{c_T}\right)^2 - \left(\frac{n\pi}{2h}\right)^2} \quad \text{for } n \in \{1, 3, 5, \ldots\}. \tag{15.55}$$

It can also be easily shown that

$$\lim_{\kappa \to \kappa_n^a} \frac{\partial[\cos(qh)]}{\partial\kappa} = \frac{(-1)^{(n-1)/2}\kappa_n^a h}{q_n} \quad \text{for } n \in \{1, 3, 5, \ldots\}. \tag{15.56}$$

The contribution of the antisymmetric modes (for $X_1 > L$) can therefore be written as

$$u_3^a(X_1, X_2, t)$$
$$= \frac{R\pi i}{\mu}\left\{\sum_{n=1,3,5,\ldots}^{\infty} (-1)^{(n+1)/2}\frac{\hat{f}(\kappa_n^a - k_w^0)\sin(n\pi X_2/2h)e^{i\kappa_n^a X_1}}{h\kappa_n^a}\right\}e^{-i\omega t}. \tag{15.57}$$

15.6.5   Excitability Functions

The addition of equations (15.51) and (15.57) represents the explicit solution to the posed problem. It can be seen that the field in the loaded layer has been found in the form of a pair of infinite series. One series is over the roots of the symmetric SH mode dispersion equation, and the other is over the roots of the antisymmetric mode SH dispersion equation. Physically, the solution can be thought of as a sum of the SH modes in the free layer multiplied by certain amplitudes that depend on both frequency (i.e., the point on the dispersion curve) and the Fourier transform of the load function $f(X_1)$.

We now put the final solution in a slightly different form by simply changing the notation a bit. First recall that the X_3 displacement field component of the nth symmetric SH mode in a free layer could be written as

$$u^s_{3n} = \cos(q^s_n X_2) \exp[i(\kappa^s_n X_1 - \omega t)],$$
(15.58)

whereas the X_3 displacement field component of the nth antisymmetric SH mode in a free layer was of the form

$$u^a_{3n} = \cos(q^a_n X_2) \exp[i(\kappa^a_n X_1 - \omega t)].$$
(15.59)

We then write the solution of the forced loading problem as an explicit summation over all symmetric and antisymmetric modes:

$$u_3 = \sum_{n=0,2,4,\ldots}^{\infty} A_n u^s_{3n} + \sum_{n=1,3,5,\ldots}^{\infty} B_n u^a_{3n}.$$
(15.60)

Comparing (15.60) with equations (15.51) and (15.57), we see that

$$A_0 = \frac{\pi i}{2\mu h \kappa^s_0} R \hat{f}(\kappa^s_0 - k^0_w),$$

$$A_n = \frac{(-1)^{n/2}\pi i}{\mu h \kappa^s_n} R \hat{f}(\kappa^s_n - k^0_w) \quad \text{for } n \in \{2, 4, 6, \ldots\},$$
(15.61)

$$B_n = \frac{(-1)^{(n+1)/2}\pi i}{\mu h \kappa^a_n} R \hat{f}(\kappa^a_n - k^0_w) \quad \text{for } n \in \{1, 3, 5, \ldots\}.$$

The expansion amplitudes in (15.61) are the product of three terms. The $\hat{f}$ term represents the Fourier transform of the load function and thus has nothing to do with which mode is being excited. The R term (recall that $R \equiv \cos(\theta_i)$) is dependent only upon the incident angle and not the mode chosen. These two factors are therefore functions of the source characteristics.

The remaining term concerns the mode being excited (and at which point on its dispersion curve it is excited) and has nothing to do with the source. This term is therefore called the *excitability function* of mode n and is denoted by $\Phi_n(\kappa)$. We have the following explicit expressions for the excitability functions:

$$\Phi^s_0(\kappa^s_0) = \frac{\pi i}{2\mu h \kappa^s_0},$$

$$\Phi^s_n(\kappa^s_n) = \frac{(-1)^{n/2}\pi i}{\mu h \kappa^s_n} \quad \text{for } n \in \{2, 4, 6, \ldots\},$$
(15.62)

$$\Phi^a_n(\kappa^a_n) = \frac{(-1)^{(n+1)/2}\pi i}{\mu h \kappa^a_n} \quad \text{for } n \in \{1, 3, 5, \ldots\}.$$

It can be seen that the excitability functions are indeed functions of the position on the mode's dispersion curve, since they are functions of the wavenumber of the mode κ. For

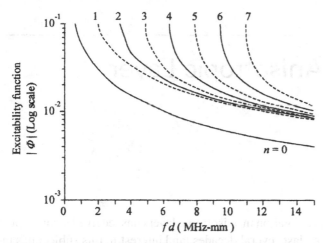

Figure 15-9. Magnitude of excitation functions for first four symmetric (*n* even) and antisymmetric (*n* odd) modes versus fd.

given material properties, we can numerically evaluate the excitability functions as functions of the position on the respective dispersion curves (i.e., as functions of fd). Using the definition of $\kappa_n^{s,a}$, the magnitude of the excitability functions can be written as

$$|\Phi_n| = \frac{1}{\varepsilon_n \mu} \left| \sqrt{\left(\frac{fd}{c_T}\right)^2 - \left(\frac{n}{2}\right)^2} \right|, \tag{15.63}$$

where $\varepsilon_0 = 2$ and $\varepsilon_n = 1$ for $n > 0$.

The magnitude of the excitability functions of the first four symmetric (even n) and antisymmetric (odd n) propagating SH modes are plotted in Figure 15-9. As can be seen, the excitability function of any mode decreases as fd increases. This is related to the fact that some modes carry more energy and hence require less amplitude for excitation.

15.7 Exercises

1. Calculate cutoff frequencies for horizontal shear waves in an aluminum plate of thickness 1 mm.

2. Plot horizontal shear phase and group velocity dispersion curves for an aluminum plate of thickness 1 mm.

3. Discuss a method of generating both vertical and horizontal shear waves in a structure.

4. Define "excitability" with respect to the capability of producing horizontal shear waves in an aluminum plate.

5. Are horizontal shear waves all nondispersive in character? Explain.

6. Describe the displacement field for a nondispersive ($n = 0$) SH plate mode.

7. Why can't the phase velocities of the SH modes be smaller than the bulk shear velocity c_T?

16

Waves in an Anisotropic Layer

16.1 Introduction

The problem of elastic wave propagation in anisotropic layers has received a fair amount of attention in the literature of the last several decades, and interest in this subject has increased even more recently. This is undoubtedly due, at least in part, to the increased use of composite materials in many facets of the design of structures. Composite materials, which are mechanically anisotropic, offer many benefits over more conventional materials – for example, a higher stiffness-to-weight ratio. This advantage of composites is in turn due to the fact that their mechanical properties, such as elastic modulus, can be tailored to be high in the directions that are expected to see high loads while remaining considerably lower in other directions. This directional dependence of the mechanical properties of composites classifies them as anisotropic media.

The benefits of using composites, however, comes at the cost of a more complicated mechanical response to applied loads, static or dynamic. The anisotropic nature of the solid introduces many interesting wave phenomena not observed in isotropic bodies: a directional dependence of wave speed, a difference between phase and group velocity of the waves, wave skewing, three different wave speeds instead of two, and many somewhat more subtle differences. An understanding of the nature of waves in plates made of anisotropic materials is certainly required if one wants to use these materials effectively in structure design or if one wants to inspect them using ultrasonic methods.

In this chapter we will investigate the key points in the propagation of elastic waves in free anisotropic plates (i.e., plates with traction-free surfaces). At the outset, we refer the reader to Solie and Auld (1973), which contains an excellent exposition of the general methods of analyzing such wave motions. We will see that even though the anisotropic case is somewhat more involved than its isotropic plate counterpart, many of the final results will be qualitatively the same. For example, we will still find that the waves are governed by a dispersion equation whose roots define the possible modes in the structure. In the case of generally anisotropic media, however, the dispersion equation is much more complicated than for the case of an isotropic layer, and it is usually left in the form of a determinant being set equal to zero.

An excellent review and state-of-the-art report that exceeds the basic introductory material presented here can be found in the textbook by Nayfeh (1995). Other related reading can be found in Auld (1990), Balasubramaniam (1989), Cheng and Zhang (1966), Datta et al. (1988b,c), Fedorov (1968), Green (1982), Helbig (1984), Lekhnitskii (1981), Li and Thompson (1990), Mal (1988a,b), Mal and Rajapakse (1990), Mal and Ting (1988), Merkulov (1963), Nemat-Nasser and Yamada (1981), Rokhlin, Bolland, and Adler (1986a),

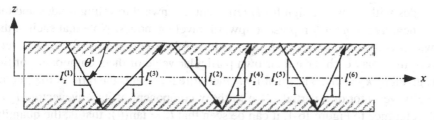

Figure 16-1. Upward and downward traveling partial waves in an anisotropic layer, where each partial wave has an associated propagation angle θ and a z component of the wavevector, $k_z = k l_z$. (Note that $l_z^{(i)} = \tan(\theta^i)$.)

Rokhlin and Wang (1991a,b), Rose et al. (1990), Scott and Miklowitz (1967), Shah and Datta (1982), Solie and Auld (1973), and Thompson, Lee, and Smith (1987).

16.2 Analysis

The propagation of elastic waves in anisotropic media, as discussed in previous chapters, is governed by the equations of motion. For zero body forces, these can be written as

$$C_{ijkl} = \frac{\partial^2 u_k}{\partial x_j x_l} = \rho \frac{\partial^2 u_i}{\partial t^2}. \tag{16.1}$$

Recall that, for isotropic media, we could now introduce the scalar and vector Helmholtz potentials to reduce the foregoing set of three coupled partial differential equations into two separate equations: one governing the propagation of longitudinal modes and the other governing the propagation of shear waves. In general, this is not possible for anisotropic media, not because the vector displacement field $\bar{u}(x, t)$ cannot be decomposed via the Helmholtz decomposition (any suitable field can be) but rather because making such a substitution will no longer split (16.1) into two simpler equations. So rather than using potentials, the most general method of treating the propagation of elastic waves in anisotropic layers uses the *partial wave* technique (Solie and Auld 1973). The essence of the technique is to attempt to satisfy the governing equation (16.1) – and the appropriate boundary conditions – by taking a superposition of three upward traveling plane wave modes (i.e., one quasi-longitudinal and two quasi-transverse) and three downward traveling plane wave modes; see Figure 16-1.

Each of the six waves is termed a partial wave because they all combine to give a single guided wave mode of the layer. Of course, each of the individual modes is assumed to satisfy the governing equation (16.1); as a result, the sum of the six modes is assumed to do so (since the equation is homogeneous).

The amplitudes of the six partial waves must be chosen to ensure that the appropriate boundary conditions are satisfied at the upper and lower surfaces of the layer. For the present case, the boundary conditions will have tractions that vanish. The mathematical form of each of the partial waves shown in Figure 16-1 is

$$u_i = \alpha_i \exp[ik(x + l_z z)] \exp[-i\omega t], \tag{16.2}$$

where the modes with a positive sign for l_z represent downward traveling modes and the modes with a negative sign for l_z represent upward traveling modes. Note that each of the six partial waves is assumed to have the same value of the x component of the wavevector k, whereas they may each have their own particular value of the z component of the wavevector, given by $k_z = kl_z$. The quantity l_z therefore represents the ratio of the z component of the wavevector of a partial wave to the x component of its wavevector: $l_z = k_z/k_x$. With reference to Figure 16-1, it can be seen that $l_z = \tan(\theta)$; that is, the quantity l_z corresponds to the angle of propagation of the partial wave relative to the layer surfaces.

Requiring the x component of the wavevector of all the modes to be equal at this point is actually not necessary, and the more general form (of unequal z and x components of the wavevector) could be assumed. When we try to satisfy the boundary conditions, however, all of the x components will have to be equal. This is the anisotropic medium's equivalent to Snell's law for isotropic media. We shall therefore assume from the outset that the x component of the wavevector of each of the six partial waves is equal.

As mentioned, because the governing equation (16.1) is homogeneous, each of the partial waves must satisfy it. If the assumed form of displacement, (16.2), is substituted into (16.1), the result is

$$[C_{ijkl}k_j k_l - \rho\omega^2 \delta_{ik}]\alpha_k = 0, \tag{16.3}$$

where – for the particular geometry shown in Figure 16-1 – the components of the wavevector are $k_1 \equiv k_x \equiv k$, $k_z \equiv k_3 = l_z k$, and $k_y \equiv k_2 = 0$. We assume that the wave has no wavevector component in the y (or "2") direction; see Figure 16-2. This does not mean, however, that the individual partial waves will not have y components of displacement. In general, they will have all three displacement components. The particular form chosen for the individual partial waves is independent of the y-coordinate. This means that all fields (displacement, stress, etc.) will be independent of the y-coordinate and, as a result, that the problem is one of plane strain in the (x, z)-plane (the $(1, 3)$-plane).

It is important to note that, in anisotropic media such as composites, the properties of Lamb waves (e.g., dispersion curves) are very dependent upon the direction of propagation of the waves in the plane of the layer. For instance, the Lamb wave dispersion curves will, in general, be functions of the propagation direction ϕ relative to the x-coordinate axis (part (a) of Figure 16-2). When we assume that the partial waves have only x and z components of the wavevector, we are immediately assuming that the partial waves are propagating in the direction $\phi = 0$ in Figure 16-2(a) – that is, along the x-coordinate direction. For $\phi \neq 0$, the partial waves would have three nonzero wavevector components. However, there is no loss in generality in assuming propagation along the x-axis. This is due to the fact that we are free to choose any coordinate system we wish. If we wanted to investigate possible Lamb waves for propagation directions $\phi \neq 0$, we could simply imagine our original (x, y, z)-coordinate system rotated about the z-axis by an amount equal to ϕ (part (b) of Figure 16-2). The propagation direction would then be once again aligned with our x-axis, as we have been assuming. However, this rotation of the coordinate system will necessitate a corresponding transformation of the elastic moduli C_{ijkl} from the old coordinate system to the new (rotated) coordinate system.

Expression (16.3) is a set of three homogeneous algebraic equations for the quantities $\alpha_1, \alpha_2, \alpha_3$, which must be simultaneously satisfied if each of the partial waves is to satisfy

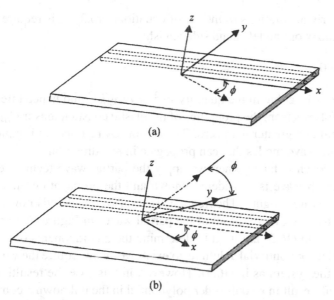

(a)

(b)

Figure 16-2. Lamb wave propagation considerations in an anisotropic plate. (a) Lamb wave propagation in the direction making
an angle ϕ with respect to the original x-axis, where the original
(x, y, z)-coordinate system is assumed to be aligned with the crystallographic axes of the material of the layer; in this case, all three
components of the wavevector of the partial waves are nonzero.
(b) Rotated coordinate system with the new x-axis along the propagation direction; in this case, all partial waves will have only x
and z components of the wavevector.

(16.1). All partial wave components have equal x components of the wavevector k, which
also represents the wavenumber of the Lamb wave mode. It is important to note that the
Lamb wave mode resulting from the superposition of the partial waves has an x component only. That is, the displacement field of any Lamb wave mode in the layer can be
represented in the form

$$u_i(x, z, t) = \xi_i(z)e^{ik(x-ct)}, \tag{16.4}$$

where k is the same for all components; specifically, k is equal to the x component of all
of the partial waves that form the Lamb wave. The vector $\xi_i(z)$, which will be different for
different modes, determines how the displacements caused by the Lamb wave vary with
depth in the layer. The *modal displacement field*, $\xi_i(z)$, will also vary with frequency for
any given Lamb wave mode.

Because k is the wavenumber of the Lamb wave modes, it follows that $c = \omega/k$ is the
phase velocity of the Lamb wave modes resulting from the superposition of partial waves
with common x wavevector components k. A Lamb wave mode has only an x component of the wavevector ($k_x \equiv k$), so the magnitude of its wavevector – its *wavenumber* –
is also equal to k.

The typical method of solving (16.3) is to evaluate the known propagation direction
(and hence the wavevector components k_i) and thence to seek permissible values of c that

satisfy the equation. For nontrivial solutions to the set of equations (16.3), it is required that the determinant of the matrix on the left-hand side vanish:

$$\det[C_{ijkl}k_jk_l - \rho\omega^2\delta_{ik}] = 0. \tag{16.5}$$

Once expanded, this equation is bi-cubic in the quantity $\rho c^2 = \rho(\omega/k)^2$ and hence three values will result for ρc^2. It follows from the properties of the elastic constant tensor C_{ijkl} that all of the roots will be real and greater than zero. These roots correspond, in infinite media, to the three permissible wave modes that can propagate in any direction.

A somewhat different approach is taken when we apply the partial wave technique. The wavenumber k of the Lamb wave is considered known and the propagation directions of each of the partial waves are sought. (The frequency ω is also considered known.) Hence, the x components of the wavevector of all of the partial waves in Figure 16-1 are considered known, and (16.3) and (16.5) are used to determine the z components of the wavevector. Of course, in order for nontrivial solutions to exist, we again require the vanishing of the determinant of the system as in (16.5). However, in this case the resulting expanded form will, in general, result in a sixth-order polynomial in the unknown z component of the wavevector of the partial waves.

The expanded form can be written as

$$Al_z^6 + Bl_z^5 + Cl_z^4 + Dl_z^3 + El_z^2 + Fl_z + G = 0, \tag{16.6}$$

where the six coefficients (A through G) are functions of the density and elastic constants of the layer as well as the chosen phase velocity c of the Lamb wave mode. Expressions for the constants, which are rather lengthy, can be found in Balasubramaniam (1989). In general, the six roots of this equation will be different, meaning that the six partial waves will propagate with different angles to the z-axis shown in Figure 16-1. The roots may be complex, but since the coefficients of the equation are real, it follows that any complex roots must occur in complex conjugate pairs.

For certain material symmetries, the sixth-order equation governing the z components of the wavevector of the partial waves is considerably simplified. If, for instance, the layer material exhibits a plane of symmetry about its mid-thickness (i.e., the material is symmetric about $z = 0$), then it is termed a *monoclinic* material and only 13 of the 21 elastic constants are nonzero. In this case, it can be shown by examination of the constants A through G that only the terms containing even powers of l_z will be nonzero. That is, for monoclinic or higher-symmetry materials, the constants $B = D = F = 0$. For monoclinic materials, the equation for l_z, the z components of the wavevector, therefore reduces to

$$Al_z^6 + Cl_z^4 + El_z^2 + G = 0. \tag{16.7}$$

It should be noted that many materials exhibit monoclinic or higher material symmetry. For example, unidirectionally reinforced composites are transversely isotropic, containing only five independent elastic constants. Transverse isotropy includes monoclinic symmetry and even additional symmetry, so (16.7) applies. More generally, symmetrically laminated composites are often considered as orthotropic media, which also are more symmetric than monoclinic and so (16.7) applies to them as well. Orthotropic materials contain nine independent elastic moduli.

Although there are sixth-order powers of l_z in (16.7), the formula actually represents a cubic equation in the quantity l_z^2. This simplification, induced by monoclinic symmetry, has direct physical significance. Because (16.7) is a cubic equation in l_z^2, it will have three roots for l_z^2 and hence the six roots for l_z will occur in three pairs related to each other by a change of sign. If we denote the first three roots by $l_z^{(1)}$, $l_z^{(2)}$, and $l_z^{(3)}$, then the last three will be $l_z^{(4)} = -l_z^{(1)}$, $l_z^{(5)} = -l_z^{(2)}$, and $l_z^{(6)} = -l_z^{(3)}$. Each pair of roots corresponds to a pair of partial waves; one of each pair represents a downward propagating wave and the other an upward propagating wave at the same angle to the z-axis.

Once we have found the six values of l_z that satisfy equation (16.7), the polarization vector corresponding to each root can be found by resubstituting each root into (16.7). The six polarization vectors obtained thereby are denoted $\alpha^{(1)}, \alpha^{(2)}, \dots, \alpha^{(6)}$.

A linear combination of the six partial waves is then taken with the goal of satisfying traction-free boundary conditions on the upper and lower surfaces of the layer. The displacement field is thus expressed as

$$u_j = \sum_{n=1}^{6} C_n \alpha_j^{(n)} \exp[ik(x + l_z^{(n)} z)]. \tag{16.8}$$

Requiring the traction to vanish on the upper and lower surfaces of the layer is equivalent to the conditions

$$T_{xz} = T_{yz} = T_{zz} = 0. \tag{16.9}$$

Hence the remaining part of the analysis is to compute, from the displacement field in (16.8), the required traction components in (16.9) and then to set those components equal to zero for both $z = +h/2$ and $z = -h/2$. We must calculate the traction components by applying the generalized Hooke's law,

$$T_{ij} = C_{ijkl} \varepsilon_{kl}, \tag{16.10}$$

where the ε_{kl} are components of the strain tensor. Using the strain–displacement relations, we have

$$\varepsilon_{kl} = \frac{1}{2} \left\{ \frac{\partial u_k}{\partial x_l} + \frac{\partial u_l}{\partial x_k} \right\}. \tag{16.11}$$

The symmetry of the elastic constant tensor, $C_{ijkl} = C_{ijlk}$, then gives the required traction components as follows:

$$T_{xz} \equiv T_{13} = C_{13kl} \frac{\partial u_k}{\partial x_l},$$

$$T_{yz} \equiv T_{23} = C_{23kl} \frac{\partial u_k}{\partial x_l}, \tag{16.12}$$

$$T_{zz} \equiv T_{33} = C_{33kl} \frac{\partial u_k}{\partial x_l}.$$

Of course, once the traction components are calculated according to (16.12), they will be functions of the six unknown partial wave amplitudes $C_1, C_2, \ldots, C_6$. Enforcing the boundary conditions will therefore lead to a set of six homogeneous equations in the six unknown amplitudes, which can be written as

$$B_{ij}(\rho, C_{ijkl}, hk)C_j = 0. \tag{16.13}$$

The matrix $[B]$ will contain information concerning the density and thickness of the layer, as well as the preassigned x component of wavenumber k. For nontrivial solutions to exist, the determinant of $[B]$ must be set equal to zero. This defines the dispersion equation of the anisotropic layer,

$$\det[B(\rho, C_{ijkl}, hk, \omega)] = 0. \tag{16.14}$$

Equation (16.14) represents an implicit relation between frequency ω and wavenumber k. The complete dispersion curves of the structure can be obtained by solving (16.14) for a set of values for ω.

We now present detailed expressions for the dispersion equations. Consider first an orthotropic plate (see Pelts and Rose 1996). In this case, the Christoffel equation has the cubic form. For convenience, we introduce $\lambda \equiv l_z$. The possible values of λ are roots of the equation

$$a_1\lambda^6 + a_2\lambda^4 + a_3\lambda^2 + a_4 = 0, \tag{16.15}$$

where

$$a_1 = p_4 p_5,$$

$$a_2 = p_4 p_5 A_{54} + p_5 A_{62} + p_4 A_{16} + k^2[p_4(p_5 + p_7)^2 \cos^2\phi + p_5(p_4 + p_8)^2 \sin^2\phi],$$

$$a_3 = p_5 A_{62} A_{54} + p_4 A_{16} A_{54} + A_{16} A_{62}$$
$$\qquad + k^2(p_5 + p_7)^2 A_{62} \cos^2\phi + k^2(p_4 + p_8)^2 A_{16} \sin^2\phi$$
$$\qquad + k^4[2(p_5 + p_7)(p_4 + p_8) - p_9]p_9 \cos^2\phi \sin^2\phi,$$

$$a_4 = A_{54}(A_{16} A_{62} - k^4 p_9^2 \cos^2\phi \sin^2\phi);$$

$$p_1 = C_{11}/C_{33}, \quad p_2 = C_{22}/C_{33}, \quad p_3 = C_{12}/C_{33},$$

$$p_4 = C_{44}/C_{33}, \quad p_5 = C_{55}/C_{33}, \quad p_6 = C_{66}/C_{33},$$

$$p_7 = C_{13}/C_{33}, \quad p_8 = C_{23}/C_{33}, \quad p_9 = p_3 + p_6;$$

$$\chi^2 = \rho\omega^2/C_{33};$$

$$A_{16} = \chi^2 - k^2(p_1 \cos^2\phi + p_6 \sin^2\phi),$$

$$A_{62} = \chi^2 - k^2(p_6 \cos^2\phi + p_2 \sin^2\phi),$$

$$A_{54} = \chi^2 - k^2(p_5 \cos^2\phi + p_4 \sin^2\phi).$$

In these expressions, ϕ is the angle shown in Figure 16-2 and $0 \leq \phi < 2\pi$; ρ is plate density.

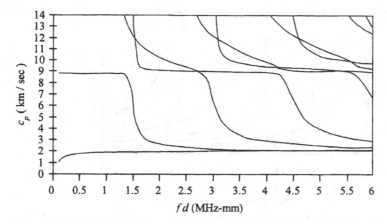

Figure 16-3. Dispersion curves for unidirectional composite in fiber direction (x-axis direction, $0°$).

The dispersion equation of the orthotropic layer for the symmetric case is

$$F_s(k, \phi) = 0, \tag{16.16}$$

where

$$F_s(k, \phi) = \begin{vmatrix} F_{11}^s & F_{12}^s & F_{13}^s \\ F_{21}^s & F_{22}^s & F_{23}^s \\ F_{31}^s & F_{32}^s & F_{33}^s \end{vmatrix},$$

$$F_{1n}^s = (\lambda_n^2 T_n^{(1)} - T_n^{(3)}) \sinh(\lambda_n d/2),$$

$$F_{2n}^s = (\lambda_n^2 T_n^{(2)} - T_n^{(3)}) \sinh(\lambda_n d/2),$$

$$F_{3n}^s = \lambda_n(T_n^{(3)} + p_7 T_n^{(1)} k^2 \cos^2 \phi + p_8 T_n^{(2)} k^2 \sin^2 \phi) \cosh(\lambda_n d/2);$$

and where, for $n = 1, 2, 3$,

$$T_n^{(1)} = p_9(p_4 + p_8)k^2 \sin^2 \phi + (p_5 + p_7)(p_4 \lambda_n^2 + A_{62}),$$

$$T_n^{(2)} = p_9(p_5 + p_7)k^2 \cos^2 \phi + (p_4 + p_8)(p_5 \lambda_n^2 + A_{16}),$$

$$T_n^{(3)} = (p_5 \lambda_n^2 + A_{16})(p_4 \lambda_n^2 + A_{62}) - p_4^2 k^4 \cos^2 \phi \sin^2 \phi.$$

For the antisymmetric case, the dispersion equation can be written as

$$F_a = (k, \phi) = 0. \tag{16.17}$$

All components of equation (16.17) can be determined from the aforelisted formulas, wherein the index s should be replaced by a and in which "sinh" and "cosh" should be interchanged.

16.3 Sample Results

Some sample theoretical results are presented in Figures 16-3 to 16-6. Note the unusual plateaus in certain directions. Dispersion curves are presented for the unidirectional

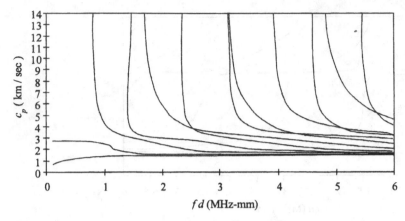

Figure 16-4. Dispersion curves for unidirectional composite in y-axis direction (90°), perpendicular to the fiber direction.

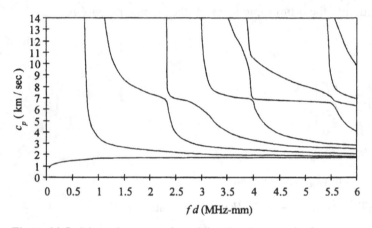

Figure 16-5. Dispersion curves for unidirectional composite for antisymmetric modes at 45°.

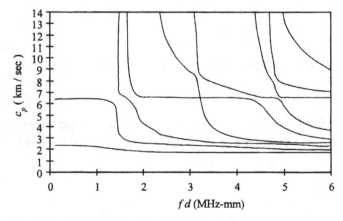

Figure 16-6. Dispersion curves for unidirectional composite for symmetric modes at 45°.

composite with elastic constants as follows: $C_{11} = 128.2$ GPa, $C_{22} = C_{33} = 14.95$ GPa, $C_{44} = 3.81$ GPa, $C_{55} = C_{66} = 6.73$ GPa, $C_{12} = C_{13} = 6.9$ GPa, $C_{23} = C_{22} - 2C_{44} = 7.33$ GPa, and $\rho = 1.58$ g/cm^3.

16.4 Exercises

1. Outline an experimental procedure to produce dispersion curves for an anisotropic structure. What precautions and assumptions are necessary? Be specific with regard to transducer design and instrumentation parameters.

2. How would you proceed to solve a problem of skew angle versus launch angle in the wave vector direction for a composite plate?

3. How would you proceed to solve a problem of guided waves in a multilayer composite plate?

4. Develop an equation for group velocity computation for the phase velocity result in Figure 16-3.

5. Derive the Rayleigh–Lamb dispersive equation for a unidirectional composite for planes of symmetry.

6. Determine the Rayleigh velocity for an anisotropic half-space for different directions of propagation, based on Figures 16.3–16.6.

7. Describe wave field phenomena for cutoff frequencies in a plate.

8. What are the limitations of the problem parameters in considering a composite as a homogeneous anisotropic plate?

17

Elastic Constant Determination

17.1 Introduction

Elastic constant values or "mechanical properties" are often difficult to obtain, particularly for anisotropic media. Nevertheless, these values are critically necessary in engineering applications and are used, for example, in finite element and boundary element analyses of materials and structures. For years, establishing these values for isotropic media relied on destructive testing. However, tensile tests with appropriate strain gage recordings can provide us with reliable isotropic elastic constants for the material being tested. Elastic wave measurements can also be used whereby solving a fairly simple inverse problem (via longitudinal and shear wave velocity measurements) can provide us with an accurate determination of the appropriate elastic constants.

For anisotropic media, determining the elastic constants is more complex. Destructive testing can still be used to determine some of the constants, and special cube-cutting procedures followed by destructive testing can be used to obtain additional constants. See for example Ditri and Rose (1993), Papadakis et al. (1991), Rose et al. (1991), Sachse (1974), Tauchert and Guzelsu (1972), Van Buskirk, Cowin, and Carter (1986), and Zimmer and Cost (1970). Additional information on elastic constant measurements can be found in Chu, Degtyar, and Rokhlin (1994), Hosten (1991), Kawashima et al. (1997), Kinra et al. (1994), Kline (1992), Lee, Kim, and Achenbach (1995), Littles, Jacobs, and Zureick (1997), and Rokhlin and Wang (1992). Quite often in engineering practice, testing is not performed; rather, "effective modulus" theories are used to calculate the elastic constants. See for example Aboudi (1981), Achenbach (1976), Chou, Carleone, and Hsu (1972), and Hashin (1962).

This chapter discusses our preferred method of obtaining the elastic constant values – by making elastic wave propagation measurements. There is much ongoing work in this area. For one-sided and nondestructive approaches, see Castagnede et al. (1990), Every and Sachse (1990), Every et al. (1991), Hosten, Deschamps, and Tittmann (1987), Karim, Mal, and Bar-Cohen (1990), Pilarski and Rose (1989), Rokhlin and Chimenti (1990), and Rose et al. (1990, 1991). This topic is not a closed one in the sense that it is difficult and often controversial as to how these measurements and constants can be determined. Bulk waves and even guided waves can be used to assist this measurement process; one approach is outlined in the following pages. See Rose et al. (1991) for more details.

Some of this material is included in Chapter 3 and Appendix B. A slightly different approach is used here to facilitate a better appreciation of composite materials and of the measurement techniques that can be used in elastic constant determination.

17.2 Background

The term "mechanical properties" can be given a more concrete definition. It has been concluded on the basis of numerous physical observations (dating back to Robert Hooke in 1660) that the components of the stress tensor at a point in a continuum can, to a first (linear) approximation, be written as a linear combination of the strain tensor components at the same point. (See Love 1944a,b for an interesting account of the history of elasticity theory.) For generally anisotropic media, this law takes the form

$$\sigma_{ij}(x) = C_{ijkl}(x)\varepsilon_{kl}(x) \tag{17.1}$$

and is known as Hooke's (or generalized Hooke's) law. In (17.1), σ_{ij} denotes the ij component ($i, j \in \{1, 2, 3\}$) of the stress tensor with reference to a Cartesian coordinate system, and ε_{kl} represents the components of strain at the same point. The strain components can be expressed, to a first (linear) approximation, in terms of the particle displacement gradients $\partial u_i(x)/\partial x$ as

$$\varepsilon_{ij}(x) \equiv \frac{1}{2}\left(\frac{\partial u_i}{\partial x_j} + \frac{\partial u_j}{\partial x_i}\right). \tag{17.2}$$

The coefficients C_{ijkl} appearing in Hooke's law are known as the elastic *moduli* (or *constants* or *stiffnesses*) of the material and are, in general, functions of position within any given body. More specifically, for wave propagation problems the C_{ijkl} are called the *adiabatic* elastic constants (as opposed to the isothermal constants; see Bhatia 1985). It can be shown that, for any material that has a positive definite strain energy function, there are at most 21 independent elements out of the 81 C_{ijkl} coefficients (see Green and Zerna 1992). Because of the symmetry of the stress tensor in nonpolar media ($\sigma_{ij} = \sigma_{ji}$) and since the strain tensor is defined as being symmetric ($\varepsilon_{kl} = \varepsilon_{lk}$), the following symmetries can always be assumed for the elements C_{ijkl}:

$$C_{ijkl} = C_{jikl} = C_{ijlk}. \tag{17.3}$$

If, in addition, the material admits a positive definite strain energy function, then we also have the following symmetry:

$$C_{ijkl} = C_{klij}. \tag{17.4}$$

It is important to remember that equations (17.1) and (17.2) are pointwise relations – that is, they are valid at any point in a continuum. The elastic stiffnesses may (and, in almost all real cases, do) vary from point to point in a body. This is certainly true for composite materials, which by definition are made of a mixture of two or more materials having different properties.

Some materials (e.g., quartz, salt, sapphire, graphite, and most pure metals such as aluminum, copper, steel, etc.) have internal symmetry in their atomic structure. These symmetries on the molecular scale manifest themselves macroscopically in the elastic constants by reducing the number of independent constants from 21 to some lesser number,

Table 17-1. *Symmetry classes and number of independent constants*

System	Number of independent constants
Triclinic	21
Monoclinic	13
Orthorhombic (orthotropic)	9
Tetragonal	6 or 7
Trigonal	6 or 7
Hexagonal (transversely isotropic)	5
Cubic	3
Isotropic	2

depending on the amount of symmetry. Crystallographers have categorized these symmetries into the systems displayed in Table 17-1 (see also Appendix B and Auld 1990).

In Table 17-1, the parenthetical names are more common in elasticity literature than the crystallographic name given to the system, where any material that is not isotropic is termed (mechanically) anisotropic. These crystal symmetries hold strictly only for single crystals whose elastic moduli do not vary from point to point within the material or at specific points in nonhomogeneous media. Many composite materials, however, are modeled using one of the listed symmetries. The reason (and justification) for doing so will be discussed in what follows.

It should be noted that the elastic moduli of any medium are not entirely arbitrary but rather are restricted by the constraint of positive definiteness. Physically, this condition ensures that the strain energy density of the medium is an absolute minimum (arbitrarily chosen as zero) when the strain tensor vanishes identically. It can be shown that, for this to occur (more generally, for any matrix to be positive definite), all of the determinants formed from the matrix by striking out the row and column associated with any element (or combination of elements) along the main diagonal must be greater than zero; these are the *principal subdeterminants* of the matrix (see Perlis 1991). One immediate consequence of this theorem is that all of the elastic moduli along the principal diagonal (i.e., C_{ii} for $i \in \{1, 2, \ldots, 6\}$, no sum) must be positive. We will show that these constants actually represent the squared wave speed of a plane wave in some direction and hence should physically be positive as well.

17.3   Elastic Wave Propagation in Anisotropic Media

The propagation characteristics of elastic waves in anisotropic media are determined by the elastic stiffnesses of the material. This can be seen by developing the governing equation for elastic wave propagation from the basic equations of elasticity. As demonstrated previously, the equations of motion for any continuum with no body forces can be written in the form

$$\frac{\partial \sigma_{ij}}{\partial x_j} = \rho \frac{\partial^2 u_i}{\partial t^2}. \tag{17.5}$$

Using Hooke's law (equation (17.1)), the strain–displacement relation (17.2), and the symmetry of the elastic moduli for arbitrary media (equation (17.3)), the divergence of the stress tensor, $\partial \sigma_{ij}/\partial x_j$, can be expressed as

$$\frac{\partial \sigma_{ij}}{\partial x_j} = \frac{\partial}{\partial x_j}\left(C_{ijkl} \frac{\partial u_k}{\partial x_l} \right)$$

$$= \frac{\partial C_{ijkl}}{\partial x_j} \frac{\partial u_k}{\partial x_l} + C_{ijkl} \frac{\partial^2 u_k}{\partial x_j \, \partial x_l}. \tag{17.6}$$

Substituting this into (17.5) results in the governing equation for wave propagation in inhomogeneous elastic media:

$$\frac{\partial C_{ijkl}}{\partial x_j} \frac{\partial u_k}{\partial x_l} + C_{ijkl} \frac{\partial^2 u_k}{\partial x_j \, \partial x_l} = \rho \frac{\partial^2 u_i}{\partial t^2}. \tag{17.7}$$

If the medium is homogeneous, then the elastic constants are not functions of position within the body and hence their derivatives (with respect to position) vanish. In this case, we get the governing equation for waves in homogeneous elastic media,

$$C_{ijkl} \frac{\partial^2 u_k}{\partial x_j \, \partial x_l} = \rho \frac{\partial^2 u_i}{\partial t^2}, \tag{17.8}$$

where each of the C_{ijkl} are constants.

From a mathematical point of view, equations (17.7) and (17.8) each represent three partial differential equations for the three unknown displacement field components. However, the solutions to (17.8) have a property that is not true of solutions to (17.7). In particular, all solutions to (17.8) are *nondispersive* because the velocity of propagation of a wave governed by that equation will not be a function of its frequency; in general, this is not true for the solutions to (17.7).

The reason why solutions to (17.8) are nondispersive lies in the structure of the differential equations themselves. Notice that only second-order derivatives appear in (17.8), whereas first- and second-order derivatives appear in (17.7). In addition, recall that the C_{ijkl} in (17.8) are constants. Therefore, if we substitute any twice-differentiable function of the form

$$u_k = \phi_k \left[\frac{2\pi}{\lambda} (\hat{n} \cdot \bar{x} - ct) \right] \tag{17.9}$$

into equation (17.8) (where $\hat{n}$ represents a unit vector in the propagation direction) with the hope of finding an expression for the phase velocity c such that our assumed form (17.9) satisfies (17.8), we are led to

$$C_{ijkl} \hat{n}_j \hat{n}_l \phi_k'' = \rho c^2 \phi_i'', \tag{17.10}$$

where the prime denotes differentiation with respect to $\hat{n} \cdot \bar{x} - \omega t$ and where we have used that

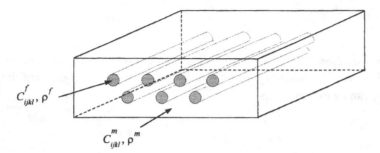

Figure 17-1. Schematic drawing of a unidirectional fiber-reinforced composite; fiber properties are denoted by C^f_{ijkl} and matrix properties by C^m_{ijkl}.

$$\frac{\partial \phi_k}{\partial x_m} = \frac{2\pi}{\lambda} n_m \phi'_k \quad \text{and} \quad \frac{\partial \phi_k}{\partial t} = -\frac{2\pi}{\lambda} c \phi'_k.$$

Given $\phi''_i = \delta_{ik} \phi''_k$ (where δ_{ik} denotes the Kronecker delta), we can write (17.10) in the form

$$(C_{ijkl} \hat{n}_j \hat{n}_l - \rho c^2 \delta_{ik}) \phi''_k = 0. \tag{17.11}$$

For a solution other than $\phi''_k \equiv 0$, we must require that

$$\det(C_{ijkl} \hat{n}_j \hat{n}_l - \rho c^2 \delta_{ik}) = 0. \tag{17.12}$$

Equation (17.12) is the sought-after (implicit) expression for the phase velocity c. As can be seen, λ (the wavelength when ϕ_k is periodic) does not appear in the equation and hence the phase velocity is independent of wavelength, that is, the solutions are nondispersive. The key to the success of this procedure is that the ϕ''_k term appeared on both sides of equation (17.10) and thus effectively cancelled out. If we had a first-derivative term in the equation (such as with (17.7)), then this would not happen (since there would be a ϕ'_k term generated that would not combine with the ϕ''_k terms) and the dispersion equation analogous to (17.12) would then contain λ explicitly.

Expression (17.12) may be recognized as Christoffel's equations governing the propagation of plane harmonic waves in anisotropic media. The derivation here shows that Christoffel's equations actually govern any plane wave solutions, not only the time-harmonic ones. See Section 3.3.

17.4 Composite Materials and Effective Moduli

Composite materials are, by definition, inhomogeneous. That is, they are a mixture of two or more materials with different material properties. As such, they have no elastic constants as strictly defined; instead, each constituent has its own elastic constants. This is shown in Figure 17-1, a schematic cross-sectional view of a fiber-reinforced composite. The fibers and matrix each have their own elastic constants, denoted (resp.) by C^f_{ijkl} and C^m_{ijkl}.

In addition to their elastic constants, each phase has its own mass density. Although the elastic constants vary from point to point in a composite, the expressions (17.1) and

(17.2) continue to hold at each point in the composite if the appropriate elastic constants are used (i.e., C_{ijkl}^m if the point is in the matrix and C_{ijkl}^f if the point is in a fiber). We can define the (position-dependent) elastic constants of the composite as being equal to C_{ijkl}^f when x falls within a fiber or to C_{ijkl}^m when x falls within the matrix.

With this interpretation, equation (17.7) would govern the waves propagating in the composite. However, the equation must be understood in a generalized sense because the derivatives of the C_{ijkl} are undefined at the interfaces between fiber and matrix, where the C_{ijkl} are discontinuous. Another approach is to apply (17.8) to the fiber and matrix domains individually and then apply continuity conditions at the interfaces. The final results – dispersion and attenuation – will remain the same. As mentioned before, using an equation of the form (17.7) will in general lead to dispersive solutions. The presence of discontinuities in the elastic constants will also lead to scattering centers within the composite and hence to attenuation.

There are, however, instances – especially when using low-frequency waves – of the observed dispersion and attenuation effects being weak or nonexistent. In such cases, the composite behaves as a homogeneous but anisotropic material. This generally happens when the frequency is well below any of the internal resonance frequencies associated with the microstructure of the composite. Equivalently, the wavelength of the waves should be large when compared with such characteristic dimensions of the composite as lamina thickness, fiber diameter, and interdiameter spacing. In this case it is not the separate fiber and matrix properties that govern the propagation of waves but rather some weighted average of both fiber and matrix properties. Such average properties are usually called the *effective elastic constants* of the composite.

Likewise, a static (or low-frequency) measurement will supply information not about individual stiffnesses but instead about the effective stiffness. Because the composite is composed of different constituent materials in possibly different amounts, the operative elastic constant in a simple expression like $\sigma = E\varepsilon$ will be frequency-dependent. However, when the frequency of excitation is well below any of the internal resonances of the structure, the operative (effective) elastic constant is a function of the elastic constants of the constituent materials and also of their relative amounts in the composite. The one-dimensionality of the foregoing example can easily be generalized to anisotropic media (with 21 elastic constants), where each of the elastic constants reduces to an effective elastic constant when the frequency is low enough. These effective elastic constants are those that would be measured in static tests and thus are useful in finite element modeling or other numerical techniques for predicting the behavior of the composite under load. It should be noted that expressions are available for the (anisotropic) effective elastic constants of composites of various symmetry classes based on various assumptions concerning the deformation that the composite undergoes.

17.5 Measurement of Effective Moduli

Having given a more concrete meaning to the term "effective moduli," we now discuss how elastic waves can be used to determine the constants for anisotropic materials. It should be mentioned that statistical methods also exist to determine the full elasticity tensor (see Hayes 1969). The composite is therefore modeled as a homogeneous anisotropic elastic continuum with effective elastic constants given by C_{ijkl}, and equation (17.8) governs the propagation of elastic waves through the medium.

Consider the propagation of plane time-harmonic waves (i.e., waves of the form described by equation (3.25)):

$$u_i = A_i \exp\{i(k_j x_j - \omega t)\}. \tag{17.13}$$

This represents a wave of constant amplitude A, traveling in the direction of $\bar{k}$ with frequency ω (ω is assumed small enough so that the composite can be considered homogeneous). Equivalently, the wavelength is assumed larger than any internal structure of the composite, whose corresponding wavelength $\lambda = 2\pi/k$ (or call it $\xi_k \hat{e}_k$) is polarized along the vector $\bar{u}_m$.

The form of (17.13) simply corresponds to a particular choice of ϕ in (17.9) in the form of an exponential function. Therefore, the equations that followed (17.9) – in particular, (17.11) and (17.12) – are immediately applicable to this case as well. For our particular choice of ϕ, these equations become

$$(C_{ijkl}\hat{n}_j\hat{n}_l - \rho c^2 \delta_{ik})\xi_k = 0 \quad \text{and} \tag{17.14}$$

$$\det(C_{ijkl}\hat{n}_j\hat{n}_l - \rho c^2 \delta_{ik}) = 0. \tag{17.15}$$

Equations (17.14) and (17.15) govern the phase velocity of waves of the form (17.13) when propagating in the direction $\hat{n}$.

Following the development in Chapter 3, the acoustical (or Christoffel) tensor Γ_{im} can be obtained with components given by the relation

$$\lambda_{im} = \Gamma_{im} = C_{iklm}n_k n_l. \tag{17.16}$$

Because of the summation over repeated subscripts, the RHS of (17.16) actually contains nine terms for any choice of the free indices. Because of the symmetry characteristics of the effective elastic moduli (see (17.3) and (17.4)), the matrix of acoustical tensor components is also symmetric; that is, $\Gamma_{ij} = \Gamma_{ji}$. We therefore have only six independent components of Γ, which can be written as

$$\Gamma_{11} = C_{11}n_1^2 + C_{66}n_2^2 + C_{55}n_3^2 + 2C_{16}n_1 n_2 + 2C_{56}n_2 n_3 + 2C_{15}n_1 n_3,$$

$$\Gamma_{22} = C_{66}n_1^2 + C_{22}n_2^2 + C_{44}n_3^2 + 2C_{26}n_1 n_2 + 2C_{24}n_2 n_3 + 2C_{46}n_1 n_3,$$

$$\Gamma_{33} = C_{55}n_1^2 + C_{44}n_2^2 + C_{33}n_3^2 + 2C_{45}n_1 n_2 + 2C_{34}n_2 n_3 + 2C_{35}n_1 n_3,$$

$$\Gamma_{12} = C_{16}n_1^2 + C_{26}n_2^2 + C_{45}n_3^2$$
$$\qquad + (C_{12} + C_{66})n_1 n_2 + (C_{25} + C_{46})n_2 n_3 + (C_{14} + C_{56})n_1 n_3, \tag{17.17}$$

$$\Gamma_{13} = C_{15}n_1^2 + C_{46}n_2^2 + C_{35}n_3^2$$
$$\qquad + (C_{14} + C_{56})n_1 n_2 + (C_{36} + C_{45})n_2 n_3 + (C_{13} + C_{55})n_1 n_3,$$

$$\Gamma_{23} = C_{56}n_1^2 + C_{24}n_2^2 + C_{34}n_3^2$$
$$\qquad + (C_{25} + C_{46})n_1 n_2 + (C_{23} + C_{44})n_2 n_3 + (C_{36} + C_{45})n_1 n_3.$$

In (17.17), reduced notation has been used for the elastic moduli: the combination of indices 11, 22, 33, 23, 13, and 12 are replaced by the single indices 1, 2, 3, 4, 5, and 6, respectively (e.g., $C_{1233} = C_{63}$). See Chapter 3 and Appendix B.

We may conclude that the phase velocity (actually ρc^2) of a plane time-harmonic wave of the form (17.13), when propagating in the direction $\hat{n}$, is an eigenvalue of the acoustical tensor for that direction. Also, the polarization vector of the mode (u_m) is equal to the eigenvector of $\Gamma(\hat{n})$ corresponding to the appropriate eigenvalue. There are, of course, three such eigenvalues and corresponding eigenvectors for any given propagation direction $\hat{n}$.

Because the acoustical tensor is real and symmetric, its eigenvalues are real (as they should be, since they represent the square of a wave speed) and its eigenvectors corresponding to different eigenvalues are orthogonal (see Perlis 1991). The eigenvectors corresponding to repeated eigenvalues can always be orthogonalized. This property of Γ ensures that the polarization vectors of the three modes permissible in any direction are orthogonal to each other, even in generally anisotropic media. Of the three modes, one is termed quasi-longitudinal and the other two are termed quasi-transverse. For certain propagation directions, the three waves will actually become pure longitudinal and transverse, in the sense that their polarization vectors will be (resp.) along and perpendicular to the unit vector $\hat{n}$. It can be shown that there are at least three directions in any anisotropic solid in which the waves become pure in this sense. See Kolodner (1966) and Chapter 3.

17.6   Sample Problem for an Orthotropic Medium

We now consider an example of orthotropic media in an attempt to measure the nine elastic constants. For given elastic constants and any chosen direction, we can solve Christoffel equations for the velocity of the three modes propagating in that direction. For example, recall that an orthotropic solid has nine independent elastic moduli (see Appendix B or Table 17-1). With reference to the material's natural or crystallographic coordinate system, the elastic constant matrix can be written as

$$[C] = \begin{bmatrix} C_{11} & C_{12} & C_{13} & 0 & 0 & 0 \\ C_{12} & C_{22} & C_{23} & 0 & 0 & 0 \\ C_{13} & C_{23} & C_{33} & 0 & 0 & 0 \\ 0 & 0 & 0 & C_{44} & 0 & 0 \\ 0 & 0 & 0 & 0 & C_{55} & 0 \\ 0 & 0 & 0 & 0 & 0 & C_{66} \end{bmatrix}. \tag{17.18}$$

Imagine the propagation of a plane wave of the form (17.13) in the (x_1, x_2)-plane. In this case $n_3 = 0$, and the formula for calculating the phase velocities (using Christoffel equations) becomes

$$\begin{vmatrix} C_{11}n_1^2 + C_{66}n_2^2 - \rho c^2 & (C_{12} + C_{66})n_1 n_2 & 0 \\ (C_{12} + C_{66})n_1 n_2 & C_{66}n_1^2 + C_{22}n_2^2 - \rho c^2 & 0 \\ 0 & 0 & C_{55}n_1^2 + C_{44}n_2^2 - \rho c^2 \end{vmatrix} = 0. \tag{17.19}$$

The determinant can then be expanded to yield

$$\{C_{55}n_1^2 + C_{44}n_2^2 - \rho c^2\}\{(C_{11}n_1^2 + C_{66}n_2^2 - \rho c^2)(C_{66}n_1^2 + C_{22}n_2^2 - \rho c^2)$$
$$- (C_{12} + C_{66})^2 n_1^2 n_2^2\} = 0. \qquad (17.20)$$

Setting the first factor equal to zero (and letting $n_1 = \cos(\theta)$ and $n_2 = \sin(\theta)$), we have

$$\rho c_T^2 = C_{55}\cos^2(\theta) + C_{44}\sin^2(\theta). \qquad (17.21)$$

After appropriate computation, we see that this represents a pure transverse wave whose polarization vector is entirely along x_3. Setting the second factor in (17.20) equal to zero yields the phase velocity of the other two modes. We define A and C as follows:

$$A \equiv C_{66} + C_{11}\cos^2(\theta) + C_{22}\sin^2(\theta),$$
$$C \equiv [C_{11}\cos^2(\theta) + C_{66}\sin^2(\theta)][C_{66}\cos^2(\theta) + C_{22}\sin^2(\theta)] \qquad (17.22)$$
$$- (C_{12} + C_{66})^2 \cos^2(\theta)\sin^2(\theta).$$

Then, the velocities can be written as

$$2\rho c_{QL}^2 = A + \sqrt{A^2 - 4C}, \qquad (17.23)$$

$$2\rho c_{QT}^2 = A - \sqrt{A^2 - 4C}, \qquad (17.24)$$

where c_{QL} denotes the phase velocity of the longitudinal or quasi-longitudinal mode and c_{QT} the phase velocity of the transverse or quasi-transverse mode. The polarization vectors of both of these modes remain in the (x_1, x_2)-plane for any angle θ.

Note that only six independent constants appear in the expressions (17.21)–(17.24) for the phase velocities. Therefore, no matter how many measurements are made in the (x_1, x_2)-plane, only six of the elastic constants can be determined. This same phenomenon occurs in generally anisotropic media, where fifteen such constants can be obtained from propagation in a single plane (see Ditri 1994b).

We now consider propagation along the x_1-axis (i.e. $\theta = 0$) and use c_{ij} to denote the phase velocity of a mode propagating in the x_i direction and polarized in the x_j direction. We thus obtain

$$\rho c_{13}^2 = C_{55} \quad \text{(by (17.21)),}$$
$$\rho c_{11}^2 = C_{11} \quad \text{(by (17.23) with (17.22)),} \qquad (17.25)$$
$$\rho c_{12}^2 = C_{66} \quad \text{(by (17.24) with (17.22)).}$$

Measuring the phase velocities of these three modes therefore gives the effective elastic constants C_{11}, C_{55}, and C_{66}. For propagation in the x_2 direction ($\theta = \pi/2$), it can likewise be shown that we can determine C_{44} and C_{22} as follows:

$$\rho c_{23}^2 = C_{44} \quad \text{(by (17.21)),}$$
$$\rho c_{22}^2 = C_{22} \quad \text{(by (17.23) with (17.22)),} \qquad (17.26)$$
$$\rho c_{21}^2 = C_{66} \quad \text{(by (17.24) with (17.22)).}$$

Note that C_{66} has, in effect, been measured twice: once from a shear wave propagating in the x_1 direction and polarized in the x_2 direction, and once from a shear wave propagating in the x_2 direction and polarized in the x_1 direction. The equality of these phase velocities is implied by the original assumption of orthotropy; hence, if the measured values differ significantly, then the material cannot be considered effectively orthotropic. In this case, monoclinic symmetry may better model the material behavior since the equality of shear wave speeds is not implied thereby.

At this point, we have obtained five of the six elastic constants associated with the (x_1, x_2)-plane, each of which lies on the main diagonal of the C matrix. The only constant that remains is C_{12}. In order to determine this constant, we must measure the phase velocity of a quasi-longitudinal or a quasi-transverse mode that does not propagate along a principal direction – that is, a mode that has both an n_1 and an n_2 component. A convenient choice is $\theta = 45°$. This measurement, along with the previously measured constants, will yield the value of C_{12} using equations (17.23) or (17.24) with (17.22).

We must now find three additional constants. When we evaluate propagation in the (x_1, x_3)-plane (i.e. $n_2 = 0$), the Christoffel equations become

$$\begin{vmatrix} C_{11}n_1^2 + C_{55}n_3^2 - \rho c^2 & 0 & (C_{13} + C_{55})n_1 n_3 \\ 0 & C_{66}n_1^2 + C_{44}n_3^2 - \rho c^2 & 0 \\ (C_{13} + C_{55})n_1 n_2 & 0 & C_{55}n_1^2 + C_{33}n_3^2 - \rho c^2 \end{vmatrix} = 0. \quad (17.27)$$

The determinant can be expanded to yield

$$\{C_{66}n_1^2 + C_{44}n_3^2 - \rho c^2\}\{(C_{11}n_1^2 + C_{55}n_3^2 - \rho c^2)(C_{55}n_1^2 + C_{33}n_3^2 - \rho c^2) \\ - (C_{13} + C_{55})^2 n_1^2 n_3^2\} = 0. \quad (17.28)$$

Setting the first factor equal to zero (and letting $n_1 = \cos(\theta)$ and $n_3 = \sin(\theta)$), we have

$$\rho c_T^2 = C_{66}\cos^2(\theta) + C_{44}\sin^2(\theta), \quad (17.29)$$

which can be shown to represent a pure transverse wave with polarization vector entirely along x_2. Setting the second factor in (17.28) equal to zero yields the phase velocity of the other two modes. This time, define E and F as

$$E \equiv C_{55} + C_{11}\cos^2(\theta) + C_{33}\sin^2(\theta),$$

$$F \equiv [C_{11}\cos^2(\theta) + C_{55}\sin^2(\theta)][C_{55}\cos^2(\theta) + C_{33}\sin^2(\theta)] \quad (17.30)$$
$$- (C_{13} + C_{55})^2 \cos^2(\theta)\sin^2(\theta).$$

Then, the velocities can be written as

$$2\rho c_{QL}^2 = E + \sqrt{E^2 - 4F}, \quad (17.31)$$

$$2\rho c_{QT}^2 = E - \sqrt{E^2 - 4F}; \quad (17.32)$$

as before, c_{QL} denotes the phase velocity of the quasi-longitudinal mode and c_{QT} that of the quasi-transverse mode. The polarization vectors of both of these modes remain in the (x_1, x_3)-plane for any angle θ.

It should be noted that only an additional two elastic constants appear in these expressions; therefore, no matter how many measurements are made in this plane, only these two additional constants will be obtained. Together with the data from the (x_1, x_2)-plane, this leaves one constant (C_{23}) undetermined. In the case of general anisotropy, it can be

Table 17-2. *Bulk wave measurement protocol for orthotropic media*

Measurement number	Velocity	$\hat{n}$	ξ	Constant
1	c_{11}	x_1	x_1	C_{11}
2	c_{22}	x_2	x_2	C_{22}
3	c_{33}	x_3	x_3	C_{33}
4	c_{23}	x_2	x_3	C_{44}
5	c_{13}	x_1	x_3	C_{55}
6	c_{12}	x_1	x_2	C_{66}
7	c_{QL} or c_{QT}	45° to x_1 and x_2	(x_1, x_2)-plane	C_{12}
8	c_{QL} or c_{QT}	45° to x_1 and x_3	(x_1, x_3)-plane	C_{13}
9	c_{QL} or c_{QT}	45° to x_2 and x_3	(x_2, x_3)-plane	C_{23}

shown that inclusion of data from a second plane yields only an additional five elastic constants, which – together with the fifteen derived from the first plane – leaves one constant undetermined (Ditri 1994b). Which constant is left undetermined can be established using a formula developed by Norris (1988).

Again, we choose to propagate along the principal directions x_1 and x_3. For propagation along x_1, we arrive at equations (17.26) and therefore add no new information. For propagation along x_3 we find

$$\rho c_{32}^2 = C_{44} \quad \text{(by (17.29)),}$$

$$\rho c_{33}^2 = C_{33} \quad \text{(by (17.31) with (17.30)),} \tag{17.33}$$

$$\rho c_{31}^2 = C_{55} \quad \text{(by (17.32) with (17.30)).}$$

Observe that we have now obtained all of the elastic constants along the main diagonal of C. As was the case for the off-diagonal constant C_{12}, in order to determine the off-diagonal element C_{13} we must measure the phase velocity of either a quasi-longitudinal or a quasi-transverse wave propagating in a non–principal axis direction – that is, one for which both n_1 and n_3 are nonzero. Choosing $n_1 = n_3$ (and hence $\theta = 45°$) is appropriate for this purpose.

For a complete specification of the matrix C, the only elastic constant that remains to be determined is C_{23}. As might be expected, this constant can be obtained from data taken in the (x_2, x_3)-plane. In fact, this constant too can be determined only by measuring the phase velocity of a quasi-longitudinal or quasi-transverse wave propagating in a non–principal axis direction – that is, one for which both n_2 and n_3 are nonzero. Choosing $n_2 = n_3$ (and hence $\theta = 45°$) will enable the determination of C_{23} and thus complete the specification of C. This time, $n_1 = 0$. Again, consider propagation of a plane wave of the form (17.13) in the (x_2, x_3)-plane and expand the Christoffel equation to obtain an expression with C_{23}; then solve for C_{23}.

All the constants are now known. Table 17-2 summarizes the minimum number of measurements that could be made and the resulting elastic moduli.

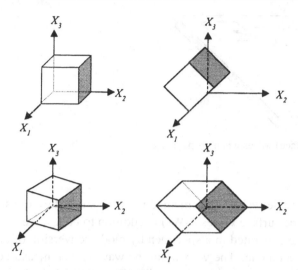

Figure 17-2. Orientation of the four cubes necessary to implement the cube-cutting protocol.

17.7   Common Measurement Protocols

17.7.1   Cube Cutting

The measurement protocol just described (and summarized in Table 17-2) is actually a common method of determining C for orthotropic media. This method is commonly referred to as the cube-cutting technique. In order to make the required measurements, cubes are cut from the original specimen at various orientations to its principal axes. Four such cubes are necessary to enable all the measurements in Table 17-2, and their orientations are as shown in Figure 17-2. One cube is cut along the principal axes so that the normal to each face is in a principal direction. The three other cubes are cut so that the normal to one face is in a principal direction while normals to the other two faces make an angle of 45° with the other two principal axes.

The cube-cutting method can be very accurate, and its results for low-frequency waves are very near those measured statically or predicted by good static effective modulus theories (see Ditri and Rose 1993). However, one major disadvantage of the cube-cutting technique is that it is destructive in nature (since cubes must be cut from the sample). Another disadvantage is the cost of correctly preparing the cubes from a bulk specimen.

17.7.2   One-Sided Approach

A measurement protocol requiring access to only one side of the specimen has been developed in an attempt to measure all nine elastic moduli nondestructively (see Rose et al. 1990). Limiting access to a single plane reduces the number of measurements that can be made using the bulk waves that were previously described. Yet if this limited information is supplemented with guided wave velocity information, then (in theory) it can be

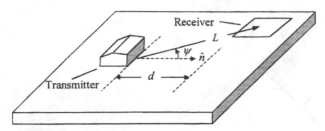

Figure 17-3. Skewing of SSL (or surface) waves when propagating in non–principal axis directions.

shown that all nine elastic moduli are derivable. One method is to use subsurface quasi-longitudinal (SSL) and Rayleigh-type surface waves (SWs) in addition to bulk waves.

Both the SSL waves and SWs are generated in a specimen by mode conversion of an incident longitudinal wave at a critical angle. The velocity of the wave is then measured by receiving it at two surface positions spaced a significant distance apart. The velocity of either wave type depends, of course, on the elastic moduli of the specimen.

The inclusion of SSL data adds little computational difficulty if it is assumed that the phase velocity of SSL waves in a given direction is the same as that of a bulk quasi-longitudinal wave propagating in the same direction. In this case, equations similar to (17.25) and (17.32) can be used to relate the measured phase velocities to the elastic moduli. Care must be taken, however, to measure the phase velocity of the waves and not their group velocity. For complete determination of the nine elastic moduli of orthotropic media, the one-sided technique requires SSL waves to be propagated in non–principal axis directions on the surface. This will lead to skewing of the wave packet as it propagates and thus, in order to maximize the amplitude of the received signal, the receiving transducer should be offset from the sender by some amount (see Figure 17-3).

If Δt denotes the time taken by the wave to travel from the sender to the receiver, then the group velocity of the wave is given by

$$c_g = \frac{L}{\Delta t},\tag{17.34}$$

where L represents the travel path of the wave. The phase velocity can be found by using an analogous formula – but with the separation distance d used instead of travel distance L:

$$c_p = \frac{d}{\Delta t}.\tag{17.35}$$

This follows from the relation $c_p = c_g \cos(\psi)$, where ψ is the skew angle.

The inclusion of surface waves in the measurement protocol increases the complexity of computation, since the velocity of surface waves in anisotropic media is a complicated function of the elastic constants. (Recall that, even for propagation on isotropic media, a cubic equation must be solved for the surface wave velocity.) Nonetheless, a measurement protocol and associated inverse formulas have been worked out by Rose et al. (1990) for particular choices of propagation directions. The equations theoretically allow the determination of all nine elastic moduli from single side access; in practice, though, small errors in velocity measurements (and/or propagation directions) can lead to erroneous results for some of the (off-diagonal) elastic moduli (see Rose et al. 1991). It should also be mentioned that the use of SSL or SW data to determine the effective moduli is valid only if the moduli are constant with depth, because the energy of the SSL and SW modes is confined to the surface of the specimen and so these modes carry information about this surface layer only.

17.7.3   Other Approaches

We should note that other nondestructive ultrasonic methods exist for determining the effective elastic moduli of anisotropic media. One of the most popular methods is the *leaky Lamb wave* (LLW) technique. In this technique, the dispersion curves of a sample are experimentally generated by immersing the sample in a liquid tank and insonifying it with an acoustic beam at various incident angles and frequencies. Theoretical dispersion curves are then "fitted" to the experimental data by adjusting the elastic constants of the layer in the theoretical model until good agreement is obtained between the theoretically and experimentally generated curves. In most cases, this procedure involves iteratively minimizing the squares of the differences between the theoretically and experimentally generated dispersion curves. The accuracy of the technique is usually reported to be quite good, but care must be taken to include a large enough portion of the dispersion curves (since it is known that different regions of the curves are more sensitive to certain elastic moduli than others). Dispersion curves from different sagittal planes must also be included because – as mentioned in the section on bulk waves – information from only a single plane is used to derive a limited number of constants.

The LLW technique is not very efficient, since most of the collected data is redundant (i.e., neighboring points on a given mode's dispersion curve). The method also requires relatively complex numerical manipulation (requiring a computer) to enable theoretical determination of the dispersion curves of an anisotropic layer (or multilayer), given its elastic moduli.

17.8   Exercises

1. For a simple isotropic disk, outline a procedure for using wave propagation measurements to calculate the modulus of elasticity, modulus of rigidity, and Poisson's ratio.

2. Outline a wave velocity measurement protocol to determine the five elastic constants of a transversely isotropic unidirectional graphite epoxy plate.

3. As a simple example of the concept of effective elastic moduli, consider the two-mass–two-spring system shown in Figure 17-4. If we make the analogies of force for stress ($F \leftrightarrow \sigma$), displacement for strain ($x \leftrightarrow \varepsilon$), and spring constant for elastic constant ($K \leftrightarrow E$), then we have the formally similar equations of:

$$F(t) = Kx(t) \quad \leftrightarrow \quad \sigma(t) = E\varepsilon(t).$$

Our operative elastic constant $K(t)$ is then given in terms of the forcing function $F(t)$ and the displacement of mass M_1 as

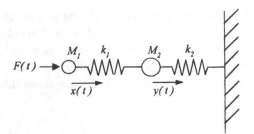

Figure 17-4. Exercise 3.

$$K(t) = \frac{F(t)}{x_f(t)}, \tag{17E.1}$$

where $x_f(t)$ represents the displacement $x(t)$ due to the applied load $F(t)$ (i.e., $x(t)$ minus any homogeneous solution that can exist even if $F(t) = 0$).

(a) Applying Newton's second law of motion to masses M_1 and M_2 individually, show that the governing equations for the system can be written as

$$M_1\frac{\partial^2 x}{\partial t^2} + k_1(x - y) = F(t),$$
$$M_2\frac{\partial^2 y}{\partial t^2} + (k_1 + k_2)y - k_1 x = 0. \tag{17E.2}$$

(b) Show (or disprove) that, for a time-varying periodic forcing function of the form $F(t) = \cos(\omega t)$, the solution for $x(t)$ can be written as

$$x(t) = x_H(t) - \left\{\frac{d\cos(\omega t)}{a\omega^4 - b\omega^2 + c}\right\}, \tag{17E.3}$$

where $x_H(t)$ is the general solution of

$$a\frac{d^4 x}{dt^4} + b\frac{d^2 x}{dt^2} + cx = 0 \tag{17E.4}$$

and where

$$a = M_1 M_2, \qquad b = k_1(M_1 + M_2) + k_2 M_1,$$
$$c = k_1 k_2, \qquad d = M_2\omega^2 - (k_1 + k_2). \tag{17E.2}$$

(*Hint:* The first equation can be solved for $y(t)$ and substituted into the second.) Identifying the second term in (17E.3) with the $x_f(t)$ in (17E.1) results in

$$K(\omega) = -\frac{a\omega^4 - b\omega^2 + c}{M_2\omega^2 - (k_1 + k_2)}. \tag{17E.5}$$

Evaluate the case when ω is small. In this context, "small" means that the terms containing ω in (17E.5) are negligible in comparison to those that do not (though such a value of ω may in fact be quite large with respect to 1).

(c) Show that, as $\omega \to 0$, the operative elastic constant of the system (17E.5) approaches the value that would be seen by a static load:

$$\lim_{\omega \to 0} K(\omega) = \frac{k_1 k_2}{k_1 + k_2}. \tag{17E.6}$$

This example shows that, for frequencies well below any of the internal resonances of the structure (the actual meaning of small ω in part (b)), the material responds as though it had an effective elastic constant – stiffness, in this example – equal to that which would be measured statically. The same conclusion holds true for elastic waves in composites, where the relevant internal resonance frequencies are determined by fiber diameter, lamina thickness, fiber spacing, or any other internal length scale associated with the composite.

4. Show that $\Gamma(\hat{n})$, the acoustic tensor for the propagation direction $\hat{n}$, is symmetric. (*Hint:* Use the symmetry characteristics of C_{ijkl}.) It follows from this simple fact that the phase velocity of a plane wave propagating in the direction $\hat{n}$ is real and that the polarization vectors of the three wave modes permissible in any direction $\hat{n}$ are orthogonal.

5. For orthotropic media, show that – in order to determine an elastic moduli of the form C_{ij} with $i \neq j$ (i.e., one that does not lie on the main diagonal) by the cube-cutting technique – we must choose a propagation direction $\hat{n}$ such that both $\hat{n}_i$ and $\hat{n}_j$ are nonzero. (*Hint:* Examine the elements of the acoustic tensor to see their dependence on these elastic moduli.)

6. Discuss a method of guided wave measurements and analysis that might be capable of accurately determining elastic constants.

7. For the orthotropic sample problem discussed in Section 17.6, find the expression for C_{23}.

18

Waves in Viscoelastic Media

18.1 Introduction

This chapter outlines basic concepts and analysis of viscoelasticity and its impact on wave propagation. Even though the attenuation due to viscoelastic effects has plagued investigators in ultrasonic NDE for years, little progress has been made in the study of viscoelasticity, especially in guided wave analysis. Attenuation is often neglected in wave propagation analysis. Yet reality calls for an understanding of attenuation principles as a function of material properties, distance, and frequency. In this chapter we present one of many possible approaches and discuss a few sample problems.

General elasticity theory assumes that, during deformation, a material stores energy with no dissipation. But many modern artificial materials (in particular, polymers and composites) dissipate a great deal of energy during deformation. The behavior of these materials combines the energy-storing features of elastic media and the dissipating features of viscous liquids; such materials are called *viscoelastic*. Stresses for viscoelastic materials are functions of strains and derivatives of strains over time. If the stresses and strains and their derivation over time are related linearly, then the material has properties of linear viscoelasticity. Viscoelastic materials are very sensitive to temperature changes. We will now describe two well-known viscoelastic models for uniaxial stress: Maxwell and Kelvin–Voight.

18.2 The Maxwell Model

Hooke's law describes a linear elastic material as

$$\sigma = G\varepsilon. \tag{18.1}$$

Newton's law for a viscous liquid is

$$\sigma = \eta \frac{d\varepsilon}{dt}, \tag{18.2}$$

where η is the coefficient of viscosity. Viscoelastic materials combine the features of liquid and elastic solids, which is reflected in the model of Figure 18-1.

As shown in the figure, a spring and dashpot are in *series*. The total strain ε is thus

$$\varepsilon = \varepsilon_1 + \varepsilon_2. \tag{18.3}$$

Taking the derivative of (18.3) with respect to time, we obtain

$$\dot{\varepsilon} = \dot{\varepsilon}_1 + \dot{\varepsilon}_2. \tag{18.4}$$

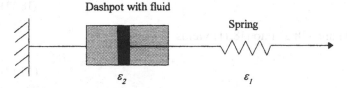

Figure 18-1. Maxwell model.

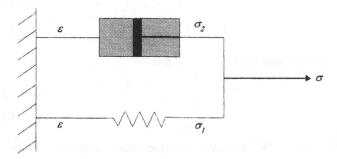

Figure 18-2. Kelvin–Voight solid.

Using substitutions from (18.1) and (18.2), the terms $\dot{\varepsilon}_1$, $\dot{\varepsilon}_2$ in (18.4) become

$$\dot{\varepsilon} = \frac{\dot{\sigma}}{G} + \frac{\sigma}{\eta}. \tag{18.5}$$

In wave propagation problems, all field variables (displacements, stresses, strains) are expressed as a harmonic function of time: thus, we have

$$\varepsilon = \hat{\varepsilon}e^{i\omega t} \quad \text{and} \quad \sigma = \hat{\sigma}e^{i\omega t}. \tag{18.6}$$

We desire an expression of the form

$$\hat{\sigma} = \hat{G}(\omega)\hat{\varepsilon}, \tag{18.7}$$

where

$$\hat{G}(\omega) = \hat{G}_1 + i\hat{G}_2. \tag{18.8}$$

Substituting (18.6) into (18.5), we therefore discover

$$\hat{G}_1 = \frac{G\omega^2}{G\eta^2 + \omega^2}, \tag{18.9}$$

$$\hat{G}_2 = \omega\frac{G^2\eta}{G\eta^2 + \omega^2}. \tag{18.10}$$

18.3 The Kelvin–Voight Model

We now examine the Kelvin–Voight solid depicted in Figure 18-2. A spring and dashpot are in *parallel* as shown. The total stress σ is split into σ_1 for a spring and σ_2 for a dashpot in such a way that ε is the same for the spring and the dashpot. Hence,

$$\sigma = \sigma_1 + \sigma_2. \tag{18.11}$$

Substituting equations (18.1) and (18.2) into (18.11) yields

$$\sigma = G\varepsilon + \eta \frac{d\varepsilon}{dt}. \tag{18.12}$$

Again, we desire an expression of the form

$$\hat{\sigma} = G^*(\omega)\hat{\varepsilon}, \tag{18.13}$$

$$G^*(\omega) = G_1^* + iG_2^*. \tag{18.14}$$

For the harmonic solution, from (18.6) and (18.12) we find that

$$G_1^* = G, \tag{18.15}$$

$$G_2^* = \omega\eta. \tag{18.16}$$

From equations (18.1), (18.7), and (18.13), we can see that the stress–strain relations using harmonic conditions for a viscoelastic case indeed have the same form as in elasticity theory, except that the elastic modulus is complex and depends on frequency. This relation of the viscoelastic solution with the solution of the corresponding elastic problem is known as the *correspondence principle* (see Auld 1990). The real part of the elastic modulus represents the capacity to store energy, while the imaginary part represents the loss of energy. For general anisotropic materials,

$$\sigma_{jk} = C_{jklm}\varepsilon_{lm}, \tag{18.17}$$

where the C_{jklm} are complex and depend on frequency.

18.4 Horizontal Shear Waves in a Viscoelastic Orthotropic Plate

18.4.1 Background

We will now analyze a specific sample problem of excitation and propagation of shear horizontal (SH) waves in a viscoelastic orthotropic plate. We will evaluate the differences in the phase velocity curves and wave structure diagrams between elastic and viscoelastic cases. Changes in wave propagation behavior depend on the mode attenuation associated with the complex elastic moduli. We will consider various models of viscoelastic material in exploring the possibility of using SH wave propagation in an anisotropic viscoelastic layer.

Modern composites are viscoelastic and have a generally anisotropic character. Manufacturing these materials with the aid of ultrasonic sensors includes cure monitoring and *in situ* inspection of material changes and discontinuities. Applications of theoretical and experimental results on the use of ultrasonic waves for monitoring the curing process in viscoelastic media are presented by Eder and Rose (1995) and also by Rokhlin et al. (1986b), who consider shear horizontal wave source excitation of an orthotropic viscoelastic plate. The problem of Lamb wave propagation in an isotropic viscoelastic layer with stress-free surfaces has also been evaluated by several investigators. In their study of this problem, Weiss (1959) and Tamm and Weiss (1961) assume that the elastic constants are complex and independent of frequency. Coquin (1964) develops an approximation method for materials with small losses and frequency-dependent elastic moduli. Chervinko and Savchenkov (1986) use numerical techniques for investigating low-compressibility materials with real Poisson's ratio and frequency-dependent complex shear moduli.

18.4.2 Problem Formulation and Solution

We shall now discuss a viscoelastic orthotropic plate (occupying the region $|z| \leq d/2$, $|x| < \infty$, $|y| < \infty$) with a time-harmonic shear stress loading $p(x)e^{i\omega t}$ on the upper surface in the finite area $|x| \leq b$. The lower surface of the plate is stress-free, and $\omega = 2\pi f$ is the angular frequency of vibration. The equation of motion for the viscoelastic case is found from the elastic case when we replace the real elastic moduli with a complex frequency dependence $C_{kl}(\omega)$. For SH waves, we have the following governing equation (the time factor $e^{i\omega t}$ has been suppressed for ease of discourse):

$$C_{66}(\omega)\frac{\partial^2 u_y}{\partial x^2} + C_{44}(\omega)\frac{\partial^2 u_y}{\partial z^2} + \rho\omega^2 u_y = 0, \tag{18.18}$$

$$C_{kl}(\omega) = C_{kl}^r(\omega) + iC_{kl}^i(\omega), \tag{18.19}$$

where $u_y = u_y(x, z)$ (see Auld 1990).

The appropriate boundary conditions for the plate are

$$\sigma_{yz} = p(x)e^{i\omega t} \quad \text{at } z = d/2, \ |x| \leq b, \tag{18.20a}$$

$$\sigma_{yz} = 0 \quad \text{at } z = d/2, \ |x| > b, \tag{18.20b}$$

$$\sigma_{yz} = 0 \quad \text{at } z = -d/2, \ |x| \geq 0. \tag{18.20c}$$

The complex moduli represent the capability of storing energy under elastic deformation, where $C_{kl}^r(\omega)$ is the real part and energy damping $C_{kl}^i(\omega)$ is the imaginary part. We now introduce the following dimensionless variables:

$$\zeta = z/d, \quad \xi = x/d, \quad a = b/d. \tag{18.21}$$

Applying a Fourier transform to equation (18.18) with respect to the x-axis and satisfying boundary conditions (18.20) (see Achenbach 1984), one discovers an expression for the displacement field in the form

$$u_y(\xi, \zeta, \omega) = \frac{d}{2\pi C_{44}(\omega)} \int_{-\infty}^{\infty} \left[\frac{\sinh(\lambda\zeta)}{\lambda\cosh(\lambda/2)} + \frac{\cosh(\lambda\zeta)}{\lambda\sinh(\lambda/2)}\right] T(\alpha)e^{-i\alpha\xi}\, d\alpha, \tag{18.22a}$$

where

$$\lambda^2 = (C_{66}(\omega)\alpha^2 - s^2)C_{44}^{-1}(\omega) \quad (s^2 = \rho(\omega d)^2), \tag{18.22b}$$

$$T(\alpha) = \frac{1}{2}\int_{-a}^{a} p(\gamma)e^{i\alpha\gamma}\, d\gamma. \tag{18.22c}$$

The formal solution (18.22a) is the same as in the elastic case. When evaluating the integrals, one must take into account the poles of the integrands. Poles for the symmetric part of the solution (first term in (18.22a)) are

$$\alpha_n = \sqrt{\{s^2 - C_{44}(\omega)(2\pi n)^2\}C_{66}^{-1}(\omega)} \quad (n = 0, 1, 2, \ldots); \tag{18.23a}$$

for the antisymmetric part (second term in (18.22a)), the poles are

$$\beta_n = \sqrt{\{s^2 - C_{44}(\omega)[\pi(2n+1)^2]\}C_{66}^{-1}(\omega)} \quad (n = 0, 1, 2, \ldots). \tag{18.23b}$$

In an elastic solid, for each particular value ω, one calculates from equations (18.23) a finite number of real poles and an infinite number of imaginary poles that are symmetric to the coordinate directions. For the viscoelastic case, α_n and β_n are always complex wavenumbers.

Waves in viscoelastic material decay in the direction of propagation. Accordingly, we attach signs to the imaginary parts of the wavenumbers α_n and β_n, where the imaginary part of a wavenumber is the attenuation coefficient. Applying the residue theory to (18.22a) yields

$$u_y(\xi, \zeta, \omega) = -\frac{id}{C_{66}(\omega)}$$

$$\times \left\{ \frac{T(\alpha_0)}{2\alpha_0}e^{-i\alpha_0\xi} + \sum_{n=1}^{\infty}\left[\frac{\cos(2\pi n\zeta)}{\cos(\pi n)}\frac{T(\alpha_n)}{2\alpha_n}e^{-i\alpha_n\xi} \right.\right.$$

$$\left.\left. + \frac{\sin(\pi(2n-1)\zeta)}{\cos\pi(n-1)}\frac{T(\beta_n)}{2\beta_n}e^{-i\beta_n\xi} \right]\right\}, \tag{18.24}$$

where $|\xi| > a$. The summation in (18.24) is over poles with negative imaginary parts for positive ξ and over poles with positive imaginary parts for negative ξ.

All calculations are carried out for $p(x) = 1$ and for two cases of viscoelastic solids. The first case is the Voight solid, which has an elastic moduli given by Coquin (1964) and Auld (1990):

$$C_{kl}(\omega) = C_{kl}^r + i\omega C_{kl}^i. \tag{18.25}$$

For the second case, the real and imaginary parts of the moduli are independent of the frequency. This assumption approximates the response of some materials over a fairly wide frequency range.

18.4.3   Numerical Results

The units of C_{kl} are given in gigapascals (GPa); data are taken from experimental results in Deschamps and Hosten (1992). For the Voight solid:

$$C_{44}^r = 6.15, \quad C_{44}^i = 0.02, \quad C_{66}^r = 3.32, \quad C_{66}^i = 0.009; \quad \rho = 1.5 \text{ g/cm}^3, \quad a = 1.5.$$

For the elastic solid, $C_{44}^i = C_{66}^i = 0$.

The phase–velocity dispersion curves for the Voight material are plotted in Figure 18-3; Figure 18-4 graphs phase velocity when the elastic moduli are independent of frequency. In part (b) of Figure 18-4, dispersion phase velocity curves overlap the dispersion curves for the elastic plate.

The character of the viscosity effect is illustrated in Figure 18-5, whose graphed curves show the change in phase velocity from the Voight material (solid curves) to the elastic

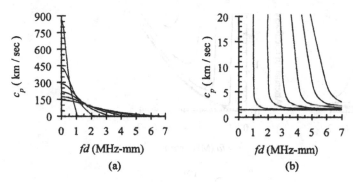

Figure 18-3. Phase velocity curves for a Voight viscoelastic solid, where (a) and (b) show the same curves using different scales.

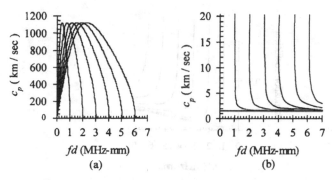

Figure 18-4. Phase velocity curves for a viscoelastic solid when the complex elastic moduli are independent of frequency, where (a) and (b) show the same curves using different scales.

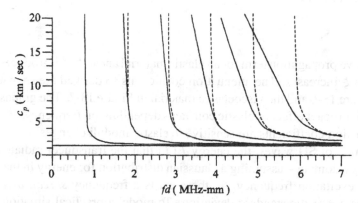

Figure 18-5. Phase velocity curves for the Voight model (solid lines) and the elastic solid (dashed lines), where the first two modes overlap each other for viscoelastic and elastic cases.

solid (dashed curves). The first two modes overlap each other for the viscoelastic and elastic cases.

We now consider the wave field representation, using equation (18.24). We can decompose the complex wavenumbers α_n and β_n as $k = k_{\mathrm{Re}} + ik_{\mathrm{Im}}$. Every term of (18.24) can thus be written as

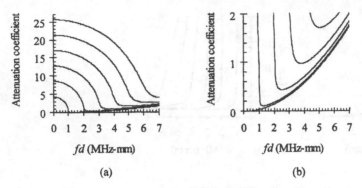

Figure 18-6. Attenuation coefficient for Voight solid, where (a) and (b) show the same curves using different scales.

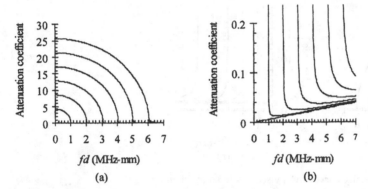

Figure 18-7. Attenuation coefficient for viscoelastic solid when the complex elastic moduli are independent of frequency, where (a) and (b) show the same curves using different scales.

$$u_y \sim A e^{ik\xi} = A e^{ik_{\mathrm{Re}}\xi} e^{-k_{\mathrm{Im}}\xi},$$

where $e^{ik_{\mathrm{Re}}\xi}$ is a typical wave propagation term for an elastic material and $e^{-k_{\mathrm{Im}}\xi}$ introduces attenuation while distance ξ increases. The attenuation coefficients so derived are shown for the Voight solid in Figure 18-6 and the viscoelastic material in Figure 18-7. The graphs in Figure 18-7 show the significant effect of elastic constants depending on frequency: the phase velocity and attenuation coefficients are sensitive to elastic moduli changes.

In the practical excitation of SH waves, the energy $E(f)$ that the transducer radiates will spread in the frequency domain – assuming a Gaussian distribution for energy in that domain. Hence, for any excitation frequency f_0, there exists a frequency *spread area* $(f_0 - 3\sigma, \ f_0 + 3\sigma)$, where σ is the standard deviation. To model a practical situation, one must express the resultant amplitude for any f_0 as follows:

$$\bar{u}_y(\xi, \zeta, f_0) = \int_{f_0-3\sigma}^{f_0+3\sigma} u_y(\xi, \zeta, 2\pi f) E(f)\, df. \tag{18.26}$$

Figure 18-8 graphs the resulting dimensionless magnitude $M = |\bar{u}_y(\xi, d/2, f)|$ in the frequency domain (where $d = 1$, $x = 7$, $a = 1.5$, $\sigma = 0.03$). The peaks in this figure are associated with cutoff frequencies for phase velocity curves from Figure 18-3 (elastic solid). Viscosity makes the peaks much smoother for the viscoelastic solid (with

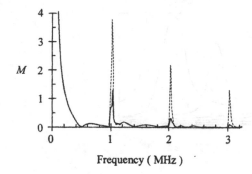

Figure 18-8. Magnitude–frequency relationship for independent viscoelastic solid, showing frequency-elastic moduli (solid line) and elastic material (dotted line).

frequency-independent moduli); for the Voight material, there is no such relation because $C_{kl}(\omega)$ elevates with increasing frequency.

18.4.4   Summary

In this section we have examined the viscoelastic effects of guided SH waves; our results can be summarized as follows. The phase velocity curves of viscoelastic material approach the corresponding curves for the elastic case for a small imaginary part of the complex elastic moduli. For an elastic plate, attenuation coefficients are smallest in the neighborhood of the cutoff frequency. The amplitude peaks reflect clear cutoff frequencies for the elastic case; the peaks for viscoelastic material are smaller. We have demonstrated also the source influence on wave excitation.

The practical significance of this work is demonstrated, for example, in powder metal injection molding, where transducers can be used to excite guided waves in the part being manufactured. Selecting points of minimum attenuation can be achieved experimentally by frequency and angle tuning.

18.5   Lamb Waves in a Viscoelastic Layer

In this section, we evaluate waves in a two-dimensional viscoelastic isotropic layer with traction-free boundary conditions.

We assume that x is in the direction of wave propagation in the layer of thickness d. Just as for the viscoelastic case, we will use complex wavenumbers and wave velocities. The real part of the wavenumber represents the propagation of the wave, and the imaginary part describes the wave attenuation. According to equation (18.16) for the Kelvin–Voight model, we assume that the attenuation of bulk waves varies linearly with frequency and hence defines the two damping parameters (per unit frequency) η_L and η_T for longitudinal and shear (transverse) bulk waves, respectively. Complex wavenumbers and bulk velocities can then be introduced as follows (see Chimenti and Nayfeh 1989):

$$k_L^* = \omega/c_L + i\eta_L\omega/(2\pi), \qquad k_T^* = \omega/c_T + i\eta_T\omega/(2\pi)$$

$$(c_L^* = \omega/k_L^*, \ c_T^* = \omega/k_T^*).$$

(18.27)

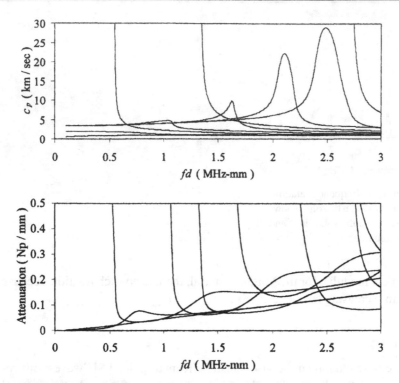

Figure 18-9. Phase velocity and attenuation dispersion curves for a viscoelastic plate with bulk damping parameters $\eta_L = 0.02$ Np/MHz-mm and $\eta_T = 0.07$ Np/MHz-mm; bulk velocity for the appropriate elastic media are $c_L = 2.7$ Km/s and $c_T = 1.1$ Km/s.

As before, c_L and c_T are longitudinal and shear (transverse) wave velocities in the elastic media.

The analytical solution of this wave propagation problem for a viscoelastic material can be found in the same fashion as that used for the elastic case: simply replace real wavenumbers and bulk velocities with complex wavenumbers and velocities, according to equation (18.27). We may begin by stating the stress-free boundary conditions as

$$\sigma_x = \sigma_{xy} = 0 \quad \text{at } y = \pm d/2. \tag{18.28}$$

This leads to the complex Rayleigh–Lamb dispersion equation found in (8.31) and (8.32):

$$\frac{\tan(qd/2)}{\tan(pd/2)} + \left\{ \frac{4pqk^2}{(q^2 - k^2)^2} \right\}^{\pm 1} = 0$$

$$(p^2 = (\omega/c_L^*)^2 - k^2, \ q^2 = (\omega/c_T^*)^2 - k^2). \tag{18.29}$$

Now consider the root $k = \alpha + i\beta$ of equation (18.29). For each root, the components of the displacement field can be written as

$$u_k(x, y) = U_k(y, k)e^{i(kx-\omega t)} = U_k(y, k)e^{i(\alpha x - \omega t)}e^{-\beta x}, \tag{18.30}$$

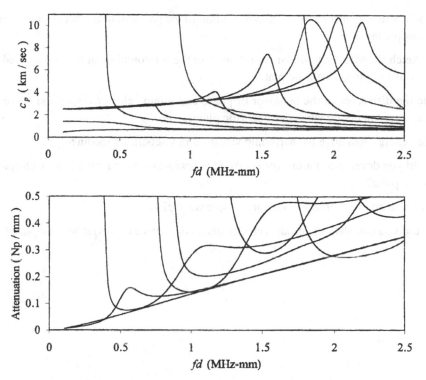

Figure 18-10. Phase velocity and attenuation dispersion curves for viscoelastic plate with bulk damping parameters $\eta_L = 0.1 \, \text{Np/MHz-mm}$ and $\eta_T = 0.2 \, \text{Np/MHz-mm}$; $c_L = 1.8 \, \text{Km/s}$ and $c_T = 0.8 \, \text{Km/s}$.

where $u_1 = u_x$ and $u_2 = u_y$. As shown in (18.30), the amplitude of the guided wave will decay as $e^{-\beta x}$ after propagating a distance x. Then, phase velocity values may be obtained as $c_p = \omega/\alpha$ for an appropriate value k.

Because equation (18.29) is itself complex, it does not have complex conjugate roots. This leads to a violation of the symmetry of the frequency spectrum with respect to the system of coordinates. The elastic spectrum is evaluated as asymptotic for a viscoelastic material when both η_L and η_T approach zero. The concept of a cutoff frequency does not have meaning for viscoelastic material, because $k = 0$ is not a root of the equation (18.29) for $\omega > 0$.

As we can see from Figure 18-9, the phase velocity and attenuation dispersion curves for the viscoelastic case have maximum curvature near the cutoff frequencies of the elastic spectrum. One can therefore generate a number of low-attenuation guided wave modes for long-distance propagation by using minimum attenuation values. Two cases of viscoelastic materials are graphed in Figure 18-9 and Figure 18-10, which show that only the A0 mode attenuation varies linearly with the frequency change.

18.6   Exercises

1. For normal incidence of 10-MHz longitudinal waves onto a plane interface between water and epoxy, find the difference between reflection factors determined for lossless versus attenuative media.

2. How would you formulate and develop a solution to the problem of a viscoelastic layer on an isotropic layer?

3. Make a sketch showing the boundary conditions for the horizontal shear transducer loading of an orthotropic viscoelastic plate.

4. Describe the difference in the behavior of phase and group velocity dispersion curves for Lamb waves in elastic versus viscoelastic plates.

5. How does energy loss arise for harmonic vibration in viscoelastic media?

6. How could you develop equations for horizontal shear wave propagation in an isotropic viscoelastic plate?

7. Why are there no cutoff frequencies for a viscoelastic plate?

8. How would you calculate the group velocity dispersion curve for a viscoelastic plate?

19

Stress Influence

19.1 Introduction

Almost all manufactured items contain residual stresses, which may result either from the manufacturing process itself or from any machining the part undergoes during its final processing. The presence of residual stress can be beneficial to the structure, particularly if it is of a compressive nature on the surfaces: such stresses can increase the item's "fatigue life," since the stresses tend to prevent the growth of any small surface scratches – in essence, closing them. In contrast, surface tension would tend to cause small surface cracks to grow, decreasing the fatigue life of the component.

Whether beneficial or detrimental, the presence of residual stress is a phenomenon that must be properly understood and compensated for when designing any structure. Unfortunately, predicting the amount and nature of residual stresses that might arise from manufacturing or machining is usually quite difficult because (at least local) plastic deformation of the material is usually involved. Although there have been many recent theoretical advances, it must be admitted that fundamental issues regarding the plastic flow of materials remain poorly understood. The problem of experimentally measuring the residual stress in a component is also difficult, though several methods have been developed and are currently being used for this purpose.

One of the most promising techniques for measuring residual as well as applied stresses is with the use of ultrasonic waves. Linear elastodynamic theory (covered in previous chapters) does not allow for residual stresses to affect any of the properties of the ultrasonic waves, but incorporating nonlinearity in the equations does. The goal of this chapter is to outline briefly the influence of residual or applied stresses on the propagation of small-amplitude ultrasonic waves.

The theory of wave propagation in stressed media has a long history, dating back to the time of Cauchy (1828), who made some advances in formulating the basic equations. Since then, there have been numerous theoretical and experimental publications on the topic. Although the basics of acoustoelasticity (i.e., the effect of stress on wave propagation) will be only briefly outlined in this chapter, more comprehensive coverage can be found in the references. See, for example, Ditri (1997), Ditri and Hongerholt (1996), Hirao, Fukuoka, and Hori (1981), Hirao, Fukuoka, and Murakami (1992), Hughes and Kelley (1951), Husson (1985), Husson and Kino (1982), Ogden and Sotiropoulos (1995), Sotiropoulos and Sifniotopoulos (1995), Tokuoka and Iwashimizu (1968), Toupin and Bernstein (1961), and Tverdokhlebov (1983). Also, see Kline and Jiang (1996) and Lavrentyev, Degtyar, and Rokhlin (1996).

19.2 Linear Theory

Recall that, when we formulated the equations governing the propagation of small-amplitude elastic waves in initially undeformed media, we made use of the following:

(1) dynamic equilibrium (Newton's second law),

$$\sigma_{ij,j} = \rho \ddot{u}_i; \tag{19.1}$$

(2) the infinitesimal (linear) strain–displacement relation,

$$\varepsilon_{ij} = \tfrac{1}{2}(u_{i,j} + u_{j,i}); \tag{19.2}$$

(3) a linear stress–strain (or constitutive) relation – namely, the generalized Hooke's law,

$$\sigma_{ij} = C_{ijkl}\varepsilon_{kl}. \tag{19.3}$$

When the stresses and strains are eliminated from these expressions, one is led to the previously derived equations governing the displacements of a linearly elastic body:

$$C_{ijkl}\frac{\partial^2 u_i}{\partial x_j\,\partial x_l} = \rho\frac{\partial^2 u_k}{\partial t^2}. \tag{19.4}$$

One could proceed to investigate particular types of solutions to (19.4), such as a plane wave of the form

$$u_k = A_k \exp[i(\bar{k} \cdot \bar{r} - \omega t)]. \tag{19.5}$$

Since we have already investigated this type of wave, we know that the characteristic equation that results from substituting (19.5) into (19.4) leads to three permissible values of phase velocity, $c_p = \omega/k$, for any choice of the propagation direction $r/|\bar{r}|$. Note that, since the body was initially assumed to be stress-free, the phase velocities obtained by solving the characteristic equation are independent of any stress applied to the body.

Thus, for infinite isotropic materials, it is well known that the longitudinal wave velocity is given by

$$c_L = \sqrt{\frac{\lambda + 2\mu}{\rho}} \tag{19.6}$$

and that the transverse wave velocity is given by

$$c_T = \sqrt{\frac{\mu}{\rho}}, \tag{19.7}$$

where λ and μ are the two Lamé constants of the isotropic body and ρ denotes its constant mass density.

When there is an initial deformation (and hence stress) present in the body, two possibilities arise. The first possibility is that the initial deformation and deformation gradients are so small that the linear strain–displacement relation (19.2) holds (the second possibility will be covered in the next section). In this case, if a linear version of Hooke's law also holds, then all of the basic governing equations remain linear.

As long as we use linear equations, we can use the principle of superposition to study the propagation of waves in the initially (infinitesimally) deformed medium. If the medium is originally subjected to a displacement field then that field, which we denote by $u^0(x, t)$, must satisfy the governing equation (19.4) as follows:

$$C_{ijkl} \frac{\partial^2 u_i^0}{\partial x_j \, \partial x_l} = \rho \frac{\partial^2 u_k^0}{\partial t^2}. \tag{19.8}$$

Owing to the initial (infinitesimal) deformation, the body will have acquired internal stresses $\sigma_{ij}^0(\bar{x}, t)$ and strains $\varepsilon_{ij}^0(\bar{x}, t)$ that are related to each other by (linear) Hooke's law. If we now superimpose upon the original deformation an additional infinitesimal displacement field of the form (19.5) or any other form, it is easy to see that this field also must satisfy the governing equation (19.4). That is, if

$$u_k^{\text{total}} = u_k^0 + u_k \tag{19.9}$$

is substituted into (19.4) then – because of the governing equation's linearity – the result would be

$$C_{ijkl} \frac{\partial^2 u_i^0}{\partial x_j \, \partial x_l} + C_{ijkl} \frac{\partial^2 u_i}{\partial x_j \, \partial x_l} = \rho \frac{\partial^2 u_k^0}{\partial t^2} + \rho \frac{\partial^2 u_k}{\partial t^2}. \tag{19.10}$$

Since the original deformation $u^0(x, t)$ satisfies (19.8), it follows that

$$C_{ijkl} \frac{\partial^2 u_i}{\partial x_j \, \partial x_l} = \rho \frac{\partial^2 u_k}{\partial t^2}. \tag{19.11}$$

We have thus arrived at the conclusion that – if the body is initially deformed infinitesimally and if (linear) Hooke's law relates the stresses and strains – then the propagation of waves is governed by the same equation as if the medium were originally unstressed. Hence, for the case of original infinitesimal deformation, the wave speeds and all other characteristics of the waves that can propagate in the body remain unchanged. The net displacement, stress, and strain fields in the body are simply the algebraic sum of the initial fields plus any additional fields that are caused by the passing of the waves.

We have been speaking of an "initial" and a "superimposed" deformation, but neither the application order nor the number of such deformation fields present has any effect on the foregoing conclusions. In other words, given linear strain–displacement and stress–strain relations, the principle of superposition holds. One consequence of this linearity is that any number of small-amplitude waves can exist simultaneously in a body without interacting – that is, the properties of any individual wave are as if it alone were present in the body.

19.3 Nonlinear Theory

The second possibility alluded to earlier with reference to the pre-deformation is that the deformation is large enough to invalidate the linear strain–displacement relation (19.2).

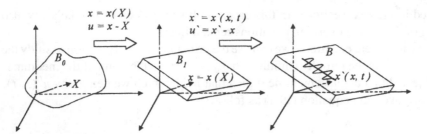

Figure 19-1. Deformation of the original configuration B_0 into the configuration B_1 and finally into B. In going from B_0 to B_1, possibly finite displacements occur (as, e.g., in the shaping operations typical of manufacturing). Configuration B is obtained from B_1 by superimposing on the latter a small-amplitude wave.

Recall that the strain–displacement relation was derived by neglecting second-order terms that, for small displacement gradients, were negligible when compared with the linear terms (see Malvern 1969). The full strain–displacement relation can be written as

$$E_{ij} = \frac{1}{2}\left[\frac{\partial u_i}{\partial X_j} + \frac{\partial u_j}{\partial X_i} + \frac{\partial u_k}{\partial X_i}\frac{\partial u_k}{\partial X_j}\right], \tag{19.12}$$

where summation is implied over the repeated subscript k.

The uppercase letter E is used to differentiate the full or finite strain tensor from the infinitesimal strains ε. The strain tensor components defined by (19.12) are the components of the Lagrangian strain tensor, since they are considered functions of the original coordinates X_i (as opposed to the current coordinates x_i) of a point in the body. For infinitesimal displacements, there was no need to distinguish between the two sets of variables (i.e., between x_i and X_i) when differentiating since they were essentially the same. It should also be noted that the strain–displacement relation given in (19.12) is complete and exact – and not simply a second-order approximation of the true relation.

The theory of finite elasticity is based on the use of this (or similar) finite strain tensors. These theories therefore include the capability of modeling problems in which large displacements take place before rupture occurs – as, for example, in the deformation of rubber. Finite elasticity, like any theory incorporating nonlinear behavior, is an intricate and often very difficult subject. However, one need not look past the strain–displacement relation to realize that the superposition principle, which was found to hold for infinitesimal deformations, will not hold for finite deformation.

For finite elasticity problems, one must clearly differentiate between different states or *configurations* of a body. The configuration of the body before any deformation takes place is typically called the "initial" or "undeformed" state, and subsequent configurations or states of the body are obtained from the undeformed state by specifying a displacement field or a mapping between points of the old and new configurations. Physically, the different states of a body can correspond to the same piece of material at different stages during its production.

Shown in Figure 19-1, for example, is a schematic representation of the deformation of an originally undeformed body B_0 into the deformed configuration B_1. In the original configuration, B_0, the position vector to any material particle in the body is specified by its position vector X relative to the origin of some suitably chosen Cartesian coordinate system. The state B_0 could represent a piece of metal prior to working it into the physical shape represented by B_1. The initial piece of material is then shaped into that represented by B_1 via such mechanical processing techniques as rolling, forging, and so forth.

Mathematically, we consider the body B_1 to be obtained from body B_0 by a continuous movement of the material particles comprising B_0 into new locations until finally the material points are arranged in space as depicted by B_1. This deformation from B_0 to B_1 can be represented by a set of expressions giving the new location of any particle x in B_1 as a function of its original location X in B_0. These relations can be written symbolically as

$$\bar{x} = \bar{x}(\bar{X}). \tag{19.13}$$

Note that the mapping in (19.13) need not be infinitesimal: the displacement of any point, given by $u(X) = x(X) - X$, or any of its derivatives need not be small compared with unity. In most manufacturing processes, the initial deformations are actually quite large.

If we wish to investigate the effect of this pre-deformation on the propagation of small-amplitude elastic waves, then yet another configuration is required. The presence of the wave and its associated displacements again moves the particles of the body (now, B_1) into new time-dependent positions denoted, say, by the new position vector $x'(x, t)$. The current configuration, denoted by B, is the result of superimposing the displacement field of the elastic wave onto B_1. Mathematically, this means that another set of functions are given that specify the new locations of the material particles when the wave propagates on the body B_1. The mapping of the locations of material particles in B_1 to their new locations in B (due to the presence of the wave) can be written as

$$\bar{x}' = \bar{x}'(\bar{x}, t), \tag{19.14}$$

where it is usually assumed that the displacements taking the body from B_1 to B are infinitesimal. That is, the displacement vector $u'(x', t) = x'(x, t) - x$ is assumed to be small enough that any products of the form $u_i' u_j'$ (or derivatives of the form $u_{i,j}' u_{l,m}'$) can be neglected in comparison with first-order terms or unity. This is what we mean by small-amplitude waves. In practice, the amplitude of any waves generated by ultrasonic equipment for the purpose of NDE causes extremely small displacements of the body on which they propagate, typically of the order $1-10 \times 10^{-9}$ m.

It can be shown that the *infinitesimal* strain tensor, which results from the successive deformations depicted in Figure 19-1, is the sum of the two individual infinitesimal strain tensors. That is, if deformations from B_0 to B_1 and from B_1 to B can both be represented by small displacements (and gradients), then the net strain tensor – going from the initial configuration B_0 to the final configuration B – is simply the sum of the two individual strain tensors (in going from B_0 to B_1 and from B_1 to B). In other words, the principle of superposition holds. This can be seen by directly calculating the infinitesimal strain tensors for the two successive deformations and using that, for infinitesimal motion, we have

$$\frac{\partial}{\partial \bar{X}} \approx \frac{\partial}{\partial \bar{x}} \approx \frac{\partial}{\partial \bar{x}'}; \tag{19.15}$$

that is, the derivatives with respect to initial, intermediate, and current coordinates can be freely interchanged.

However, the *finite* strain tensor that results from the superposition of the two displacement fields is not simply the sum of the finite strain tensors for the individual fields: it will contain cross-product terms due to the squared terms in its definition. This coupling of the "initial" and superimposed displacement fields is one mechanism that leads to dependence of the small-amplitude wave velocity (i.e., the superimposed displacements in going from B_1 to B) on the initial displacements to which the body was subjected (i.e., those that arise from manufacturing or machining processes in going from B_0 to B_1).

One could go on to investigate the details of the superposition of an infinitesimal time-dependent displacement field upon a static, finite displacement field. That is, one could derive the equations governing the superimposed displacements. The essential requirement is developing equations that govern the motion of a particle (in the current configuration B) induced by the internal stresses – and possibly body forces – acting on it. The stresses acting on particles in B include, of course, the original stresses induced in the body in going from B_0 to B_1 as well as the new stresses induced by the wave and taking the body from state B_1 to its current state B. In fact, the requisite analysis has been performed by many authors in several publications and textbooks (see especially Green and Zerna 1992). The resulting theory is rather involved, and many formulations are found in the literature that differ from one another in the types of motion allowed (i.e., finite or infinitesimal) as well as in the constitutive relations used.

Before listing some of the main acoustoelastic results obtained over the years, we briefly mention that there is another mechanism that leads to a stress dependence of the velocity of small-amplitude waves – namely, a nonlinear stress–strain relation. In fact, this is the most commonly assumed cause of the dependence of velocity on stress. Rather than using (linear) Hooke's law, the stress is assumed to depend not only linearly on the strain at a point but also on the squares of the strain components. Such a relationship is commonly written in the form

$$\sigma_{ij} = C_{ijkl} E_{kl} + K_{ijklmn} E_{kl} E_{mn}, \tag{19.16}$$

where the K_{ijklmn} are the components of a sixth-order tensor and are often referred to as the "third-order elastic constants." Using a nonlinear constitutive relation of the form (19.16) reveals the stress dependence of elastic wave velocities even for small-amplitude waves superimposed on an initial small-amplitude static deformation. Most approaches to acoustoelasticity incorporate both types of nonlinearity discussed here. That is, the finite strain tensor is used and carried along in all developments, and a nonlinear stress–strain relation essentially equivalent to (19.16) is also used.

It should be noted that, in theoretical work on acoustoelasticity, the constitutive relation is rarely written explicitly as in (19.16). Instead, most theoretical work deals with a strain energy density function, $W(E)$, which is a function of the finite strain tensor E. The components of the stress tensor are then obtained from the energy density function by differentiation with respect to the corresponding strain component, yielding the stress–strain relation

$$\tilde{T}_{ij} = \frac{\partial W(E)}{\partial E_{ij}}, \tag{19.17}$$

where $\tilde{T}_{ij}$ represents the second Piola–Kirchhoff stress tensor (see Malvern 1969).

To allow for a nonlinear stress–strain relation as in (19.16), the elastic energy function should contain at least third-order terms in the finite strain. See, for example, Toupin and Bernstein (1961), who use a potential function of the form

$$W(E) = \tfrac{1}{2} C_{ijkl} E_{ij} E_{kl} + \tfrac{1}{6} C_{ijklmn} E_{ij} E_{kl} E_{mn} + \cdots , \tag{19.18}$$

which leads to a constitutive relation of the form (19.16).

19.4 Summary of Existing Results

Many researchers have developed formulas for the change in wave speed of various types of small-amplitude elastic waves that is caused by the presence of stress. In this section we summarize the formulas concerning such wave-speed changes due to the presence in originally isotropic media of either a uniaxial applied stress or hydrostatic pressure. The results are essentially taken from Hughes and Kelly (1951), although our symbol and sign convention for the applied stress are different: we denote the uniaxial stress by σ and consider it positive if tensile and negative if compressive. (In their formula, Hughes and Kelly use the symbol T and consider compressive stresses as positive and tensile stresses as negative.)

Before listing the results, some notation should be clarified. The velocity of a wave propagating in a uniaxially stressed medium is a function of two "directions" or relative orientations:

(i) the direction of propagation of the wave relative to the direction of the uniaxially applied stress; and
(ii) the direction of the polarization vector of the wave relative to the applied stress.

In the formulas to follow, the velocities will be given three subscripts (as in V_{ijk}). The first index denotes the direction of propagation of the wave, the second index denotes the direction of polarization of the wave, and the third index denotes the direction in which the uniaxial stress acts. For example, V_{111} denotes a wave that propagates in the x_1 direction, is polarized in the x_1 direction, and whose stress is aligned along the x_1 direction – in other words, a longitudinal wave that propagates in the direction of the applied stress. The velocity V_{112}, on the other hand, represents a longitudinal wave that propagates perpendicular to the applied stress direction.

In these formulas, the second-order elastic constants of the medium are denoted by λ and μ, where $K_0 = \lambda + \tfrac{2}{3}\mu$. The third-order elastic constants, which are those originally used by Murnaghan (1951), are denoted by l, m, and n. The term ρ_0 denotes the density of the medium before any deformation takes place, and σ represents the applied uniaxial stress (positive values signify tension; negative, compression).

(1) Longitudinal waves propagating in the direction of the applied stress:

$$\rho_0 V_{111}^2 = \lambda + 2\mu + \frac{\sigma}{3K_0}\left[\frac{\lambda + \mu}{\mu}\{4\lambda + 10\mu + 4m\} + \lambda + 2l\right]. \tag{19.19}$$

(2) Longitudinal waves propagating perpendicular to the direction of the applied stress:

$$\rho_0 V_{113}^2 = \lambda + 2\mu + \frac{\sigma}{3K_0}\left[2l - \frac{2\lambda}{\mu}\{\lambda + 2\mu + m\}\right].$$ (19.20)

(3) Shear waves propagating in the direction of the applied stress, polarized perpendicular to the applied stress:

$$\rho_0 V_{131}^2 = \mu + \frac{\sigma}{3K_0}\left[\frac{\lambda n}{4\mu} + 4\lambda + 4\mu + m\right].$$ (19.21)

(4) Shear waves propagating perpendicular to the applied stress, polarized perpendicular to the applied stress:

$$\rho_0 V_{132}^2 = \mu + \frac{\sigma}{3K_0}\left[m - \frac{\lambda + \mu}{2\mu}n - 2\lambda\right].$$ (19.22)

(5) Shear waves propagating perpendicular to the applied stress, polarized parallel to the applied stress:

$$\rho_0 V_{133}^2 = \mu + \frac{\sigma}{3K_0}\left[m + \frac{\lambda n}{4\mu} + \lambda + 2\mu\right].$$ (19.23)

(6) Longitudinal waves propagating in a medium subjected to a pressure P (compressive P is considered positive in the formula, and propagation can be in any direction):

$$\rho_0 V_{11P}^2 = \lambda + 2\mu - \frac{P}{3K_0}[6l + 4m + 7\lambda + 10\mu].$$ (19.24)

(7) Shear waves propagating in a medium subjected to a pressure P (compressive P is considered positive in the formula, and propagation can be in any direction):

$$\rho_0 V_{13P}^2 = \mu - \frac{P}{3K_0}\left[3m - \frac{n}{2} + 3\lambda + 6\mu\right].$$ (19.25)

In addition to the stress dependence of the bulk wave speeds given here, work has also been done on determining the change in speed of other types of elastic waves, such as surface and interface waves. The results cannot be written as simply as for the bulk waves, so we refer the reader to the relevant references (e.g., Husson 1985; Sotiropoulos and Sifniotopoulos 1995).

19.5 Exercises

1. What is the difference in the coordinate system considered for small versus finite deformations of an elastic body?

2. What assumptions are required to develop a linear strain–displacement relation?

3. Describe a procedure to derive the higher- (third-) order elastic constant tensor.

4. Describe an approach for measuring third-order constants experimentally.

5. What conditions are required of a material by ultrasonic methods of estimating stress level in a structure?

6. What wave propagation techniques might be studied to estimate stress gradients in a structure?

20

Introduction to Boundary Element Modeling

20.1 Introduction

Solutions to many problems in wave propagation can be difficult to obtain, so numerical computational procedures are often required. Finite difference techniques and the finite element method (FEM) have been used for decades in examining stress wave propagation characteristics in solids. See Al-Nassar, Datta, and Shah (1991), Koshiba, Karakida, and Suzuki (1984), Talbot and Przemieniecki (1975), and Verdict, Gien, and Burger (1992). In this chapter we discuss a technique called the boundary element method (BEM).

The boundary element method uses classical mathematical methods similar to those for the FEM, converting volume integrals to surface integrals with the aid of special Green's functions – the so-called fundamental solutions. The BEM is particularly useful for solving both bulk and guided wave scattering problems, and it is computationally more efficient than the FEM. For more information on boundary element techniques, see Achenbach (1992), Banerjee, Ahmad, and Manolis (1987), Bond (1990), Brebbia, Umetani, and Trevelyan (1985), Cruse and Rizzo (1968), Dominguez (1985), and Kobayashi (1988).

In this chapter we review briefly the boundary element method with emphasis on the kinds of wave propagation problem in solids that can be treated with this technique. For example, "meshing" in FEM calls for the placement of small-volume elements throughout the body. In BEM, elements are placed over the boundary of the solid being studied and over defects or inhomogeneities within the structure. Even though elements are placed only on the boundary, a unique interior solution is obtained when satisfying all boundary conditions and the governing wave equations. The pre-processing step of meshing – especially considering three-dimensional analysis – is thereby much easier to carry out.

Figure 20-1 compares a physical model of ultrasonic wave propagation in a plate with the conceptual (mathematical) model used for the BEM analysis. In part (a) of the figure, the excitation transducer on the left introduces tractions on the plate surface that produce wave (displacement) propagation within the plate. These waves travel through the interior of the plate and also cause surface displacements. The receiving transducer on the right detects surface displacements; this is the response to the excitation from the left-hand ultrasonic input. The propagation interior to the plate can be modeled and studied, but of more interest and importance – for ultrasonic NDE – are the excitation and its associated response. However, this is not to discount the effects of interior propagation on that response. Parts (b) and (c) of Figure 20-1 illustrate the boundary considerations for an interior and an exterior BEM model, showing waves inside a plate and around a void, respectively.

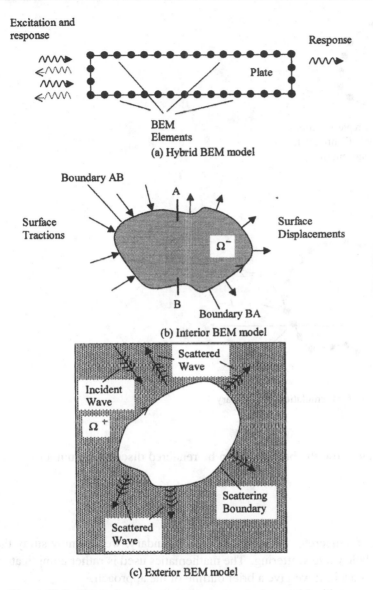

Figure 20-1. Physical wave propagation and scattering model with boundary element method concepts.

In Figure 20-1(b), the shaded area (labeled Ω^-) represents the interior of the plate. The BEM assumes that the behavior of the plate conforms to the laws of elastodynamics, both within its interior and along its boundary. Further, the method postulates that if the boundary is divided into discrete elements then the parts of the partitioned boundary are mathematically related: effects on one part can be evaluated as effects on another part. For example, consider the boundary going counterclockwise from B to A; this portion of the boundary can be considered as acted upon by the tractions produced by the left-hand transducer. Going from A to B, the right-hand portion of the boundary can be considered as an example of responding to the boundary tractions of the left-hand side.

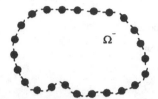

Figure 20-2. Discrete representation
of the boundary shown in Figure 20-1,
with elements of constant length.

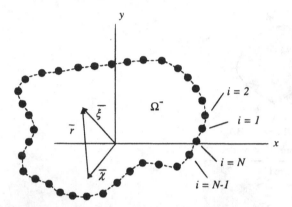

Figure 20-3. Definitions for formulation of boundary
integral equations.

Figure 20-2 illustrates how the boundary can be rendered discrete for numerical implementation.

20.2 Analysis

Given the concept of a discrete representation of a boundary, we can now study the BEM formulation for bulk wave scattering. The mathematics used is rather complicated, both in concept and in notation; we give a brief outline of the approach.

The first important relationship is the elastodynamic equation of motion, where we assume that overall displacements are the sums of incident and scattered displacements. Thus we have

$$\sigma_{ij,j} + \rho\omega^2 u_i = 0,$$

$$u_i^{\text{total}} = u_i^{\text{incident}} + u_i^{\text{scattered}},$$

where σ_{ij} = stress, ρ = density, ω = angular frequency, and u_i = displacement in the i direction.

The coordinate system and two vectors, $\bar{\chi}$ and $\bar{\xi}$, are defined as shown in Figure 20-3. Using these definitions, fundamental solutions are further defined as the Green's functions at the point $\bar{\chi}$ when a unit point load is applied at the point $\bar{\xi}$. Fundamental solutions will be designated with an asterisk, and the subscripts k, i denote that the solution was obtained in the i direction when the unit point load was in the k direction. The fundamental solution is needed for converting a volume integral to a surface integral; the solution here is considered for isotropic media (fundamental solutions for anisotropic media are considerably more complex).

In order to formulate the basic boundary integral equation, we use the reciprocity theorem (Achenbach, Gautesen, and McMaken 1982; Kobayashi 1987), the method of weighted residuals (Brebbia, Tells, and Wrobel 1984; Kobayashi 1987), and such algebraic operations as integration by parts. This basic boundary equation, where Γ represents the smooth boundary, is

$$\frac{1}{2}u_i(\bar{\xi}) + \int_\Gamma t_{ki}^*(\bar{\xi}, \bar{\chi})u_i \, d\Gamma = \int_\Gamma u_{ki}^*(\bar{\xi}, \bar{\chi})t_i \, d\Gamma. \tag{20.1}$$

Using this integral equation, a matrix formulation is possible through the assumption that the displacements u_i are constant at each boundary element i (refer to Figure 20-3).

We now present the fundamental solutions u_{ki}^* and t_{ki}^* from Brebbia et al. (1984) and Kobayashi (1987):

$$u_{ki}^*(\bar{\xi}, \bar{\chi}) = A(\hat{U}_1\delta_{ki} - \hat{U}_2 r_{,k} r_{,i}),$$

$$\begin{aligned}
t_{ki}^*(\bar{\xi}, \bar{\chi}) = GA\Bigg[&\left\{ \left(\delta_{ki}\frac{\partial r}{\partial n} + n_k r_{,i}\right) + \frac{\lambda}{G}n_i r_{,k} \right\}\frac{d\hat{u}_1}{dr} \\
&- \left\{ \left(\delta_{ki}\frac{\partial r}{\partial n} + n_k r_{,i}\right) + 2\left(n_i r_{,k} - 2r_{,k} r_{,i}\frac{\partial r}{\partial n}\right) + \alpha\frac{\lambda}{G}n_i r_{,k} \right\}\frac{\hat{u}_2}{r} \\
&- \left\{ 2r_{,k} r_{,i}\frac{\partial r}{\partial n} + \frac{\lambda}{G}n_i r_{,k} \right\}\frac{d\hat{u}_2}{dr} \Bigg],
\end{aligned}$$

where

$$\sigma_{kij,j}^* + \rho\omega^2 u_{ki}^* = -\delta_{ki}(\bar{\xi}, \bar{\chi}) \quad \text{and} \quad t_{ki}^*(\bar{\xi}, \bar{\chi}) = \sigma_{kij}^* n_j.$$

Note that $\delta_k(\bar{\xi}, \bar{\chi})$ is the representation of the unit point load in terms of the unit delta function, t denotes traction, and n_j is the jth component of the outward normal to the boundary of the body. Also, $\bar{\xi}$ is a position vector of the unit point load, $\bar{\chi}$ is a position vector of the field point, and $\bar{r}$ is the distance vector between $\bar{\xi}$ and $\bar{\chi}$, $|\bar{r}| = |\bar{\xi} - \bar{\chi}|$.

Set

$$\alpha = \begin{bmatrix} 1 & (2-D) \\ 2 & (3-D) \end{bmatrix} \quad \text{and} \quad A = \begin{bmatrix} i/4G & (2-D) \\ 1/4\pi G & (3-D) \end{bmatrix}.$$

The components $\hat{U}_1$ and $\hat{U}_2$ of the fundamental solution may be written as

$$\hat{U}_1 = H_0^{(1)}(k_T r) - \frac{1}{k_T r}H_1^{(1)}(k_T r) + \left(\frac{k_L}{k_T}\right)^2\frac{1}{k_L r}H_1^{(1)}(k_L r),$$

$$\hat{U}_2 = -H_2^{(1)}(k_T r) + \left(\frac{k_L}{k_T}\right)^2 H_2^{(1)}(k_L r),$$

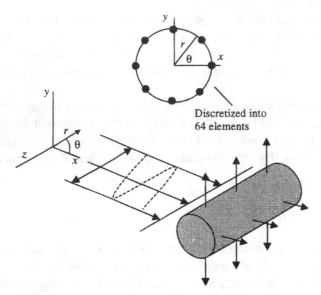

Figure 20-4. Coordinate system and discretization for the
BEM problem of SH waves impinging on a cylindrical hole
embedded in an infinite medium.

where c_L is the longitudinal wave velocity, c_T is the transverse wave velocity, $H_l^{(m)}$ is the
mth-kind Hankel function of order l, $k_L = \omega/c_L$, and $k_T = \omega/c_T$.

The discrete form of equation (20.1) is as follows:

$$
\begin{bmatrix} \frac{1}{2} & 0 \\ 0 & \frac{1}{2} \end{bmatrix} \begin{bmatrix} u_{2p-1} \\ u_{2p} \end{bmatrix} = \sum_{q=1}^{N} \int_{\Gamma_q} \begin{bmatrix} u^*_{2p-1,2q-1} & u^*_{2p-1,2q} \\ u^*_{2p,2q-1} & u^*_{2p,2q} \end{bmatrix} \begin{bmatrix} t_{2q-1} \\ t_{2q} \end{bmatrix} d\Gamma_q
$$

$$
- \sum_{q=1}^{N} \int_{\Gamma_q} \begin{bmatrix} t^*_{2p-1,2q-1} & t^*_{2p-1,2q} \\ t^*_{2p,2q-1} & t^*_{2p,2q} \end{bmatrix} \begin{bmatrix} u_{2q-1} \\ u_{2q} \end{bmatrix} d\Gamma_q. \qquad (20.2)
$$

The index $p = 1, 2, 3, \ldots, N$ denotes that a unit point load is applied to the pth boundary
element.

Displacements can be represented as polynomial or quadratic displacements. A matrix formulation is also possible for the boundary element integral equation with quadratic
displacements. See Cho (1995) for an alternative derivation of the boundary integral equations. It should be noted that there are singularities that could arise within the solutions
to the boundary integral equations, but they do not present an insurmountable difficulty.

20.3 Bulk Wave Sample Problems

To verify and validate the BEM software used to generate two-dimensional boundaries
for scatterers, we have studied two well-documented sample problems with known analytical solutions (Pao and Mow 1973): scattering from a cylindrical hole embedded in
an infinite medium when impinged upon by (1) horizontally and (2) vertically polarized
shear waves. Figure 20-4 illustrates the formulation of the first problem.

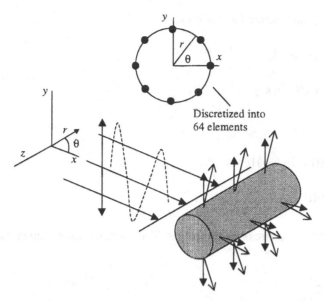

Figure 20-5. Coordinate system and discretization for the BEM problem of SV waves impinging on a cylindrical hole embedded in an infinite medium.

From the analytical solution and from physical reasoning, it is known that no mode conversions will occur at the hole; hence, this is a rather simple test of the BEM. For computational convenience, the radius of the hole is assumed to be 1. The agreement between the BEM results and the analytical results was shown to be excellent.

Figure 20-5 illustrates the formulation of the second problem. The scattering of incident *vertically* polarized shear is somewhat different because mode conversions do occur at the hole. As a result, both longitudinal and vertical shear waves share the same particle displacement plane.

We have shown how the boundary element method for bulk wave scattering is conceptualized and how the formulation can be cast in a matrix setting. Most importantly, we have demonstrated that the BEM computer code can handle mode conversion problems, a critical need when considering guided wave problems. For more detail, see Cho (1995).

20.4 Hybrid Modeling

In this section, we integrate the BEM code with the normal mode expansion technique (NMET) to produce a hybrid formulation that is useful for studying guided wave scattering problems. The NMET is necessary (a) because the incident modes are scattered and (b) because all of the modes possible via the dispersion equation can be produced at a scattering center.

This section, too, is mathematically intense. Only a sketch of the reasoning will be presented here. The guided wave model equations are split into symmetric and antisymmetric modes by grouping sine and cosine terms. Amplitude ratios are defined by dividing equations through by A_1 and A_2, respectively.

The symmetric mode may be characterized as follows:

$$\hat{u}_x = \frac{u_x}{A_1} = [(ik\cos k_l y) + \alpha(k_t \cos k_t y)]e^{i(kx-\omega t)},$$

$$\hat{u}_y = \frac{u_y}{A_1} = [-(k_l \sin k_l y) - \alpha(ik \sin k_t y)]e^{i(kx-\omega t)};$$

(20.3)

for the antisymmetric mode, we have

$$\hat{u}_x = \frac{u_x}{A_2} = [(ik\sin k_l y) - \beta(k_t \sin k_t y)]e^{i(kx-\omega t)},$$

$$\hat{u}_y = \frac{u_y}{A_2} = [(k_l \cos k_l y) - \beta(ik \cos k_t y)]e^{i(kx-\omega t)}.$$

(20.4)

The same procedure is applied to the stress relationships. The normalization constants α and β are given as

$$\alpha = \frac{A_4}{A_1} = \frac{2ikk_l \sin k_l(d/2)}{(k^2 - k_t^2)\sin k_t(d/2)},$$

$$\beta = \frac{A_3}{A_2} = \frac{-2ikk_l \cos k_l(d/2)}{(k^2 - k_t^2)\cos k_t(d/2)}.$$

These expressions for the displacements (and stresses) are then multiplicatively renormalized with power ratios. It is critical to note that, once $\bar{k}$ is found, the displacements (and stresses) are well-defined as functions of x and y.

The normalized equations for the displacements and tractions (via stresses) are available for use in the matrix formulation of the BEM displayed as equation (20.2). Incident mode displacements are then defined as

$$u^I = \alpha^p \left\{ \begin{array}{c} \bar{u}_x^p \\ \bar{u}_y^p \end{array} \right\} e^{ik^p x},$$

(20.5)

where the index p represents the pth mode. Because displacements have been normalized, $\bar{u}_x^p$ and $\bar{u}_y^p$ actually represent the normalized displacement modal functions for x, y directions with α^p the amplitude of the incident pth mode.

The incident mode can be scattered at the geometrical boundaries that define a guided wave scattering problem, so a representation for back-scattered (reflected) and forward-scattered (transmitted) displacements is necessary. The representation must also consider the mode conversions that occur at the scattering center as well as any superpositions that might occur. Consider that J reflected modes and L transmitted modes are possible; we then have

$$u^{BS} = \sum_{j=1}^{J} \beta^j \left\{ \begin{array}{c} \bar{u}_x^j \\ \bar{u}_y^j \end{array} \right\} e^{-ik^j x} \quad \text{(backward scattering)},$$

$$u^{FS} = \sum_{l=1}^{L} \beta^l \left\{ \begin{array}{c} \bar{u}_x^l \\ \bar{u}_y^l \end{array} \right\} e^{ik^l x} \quad \text{(forward scattering)}.$$

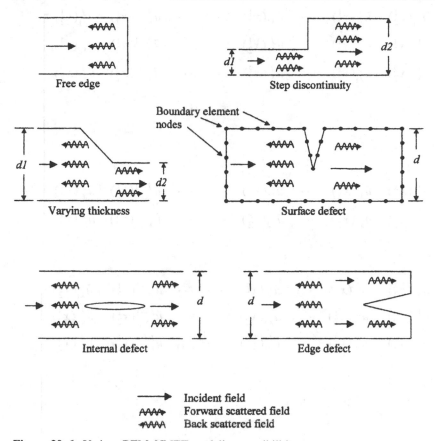

Incident field

AAAA▸ Forward scattered field

◂AAAA Back scattered field

Figure 20-6. Various BEM–NMET modeling possibilities.

A similar approach is used to define tractions. The terms β^j and β^l represent the amplitudes of the jth reflected mode and the lth transmitted mode, respectively. Thus, total displacements are the sum of the incident mode p and, depending on the problem, reflected and/or transmitted modes. Several examples of scattering problems are illustrated in Figure 20-6.

The total displacement field can be defined as the superposition of incident and scattered fields on the left and right cross-sectional boundaries (resp. Γ_1 and Γ_2):

$$u = \begin{cases} u^I + u^{BS} & \text{on } \Gamma_1, \\ u^{FS} & \text{on } \Gamma_2. \end{cases} \tag{20.6}$$

Discretizing Γ_1 and Γ_2, the cross-sectional boundaries of the waveguide with (resp.) k and m nodal points may be expressed as

$$\{u\}_{2k \times 1}^{\Gamma_1} = [\bar{u}]_{2k \times J}^I \{\alpha^p \delta_{pj} e^{ik^j x_1}\}_{J \times 1} + [\bar{u}]_{2k \times J}^{BS} \{\beta^j e^{-ik^j x_1}\}_{J \times 1} \text{ on } \Gamma_1, \tag{20.7}$$

$$\{u\}_{2m \times 1}^{\Gamma_2} = [\bar{u}]_{2m \times L}^{FS} \{\beta^l e^{ik^l x_1}\}_{L \times 1} \text{ on } \Gamma_2, \tag{20.8}$$

where

$$[\bar{u}]^{I,\text{BS}} = \begin{bmatrix} \bar{u}^1_{x_1}(x^1_2) & \bar{u}^2_{x_1}(x^1_2) & \cdots & \bar{u}^j_{x_1}(x^1_2) & \cdots & \cdots & \bar{u}^{J-1}_{x_1}(x^1_2) & \bar{u}^J_{x_1}(x^1_2) \\ \bar{u}^1_{x_2}(x^1_2) & \bar{u}^2_{x_2}(x^1_2) & \cdots & \bar{u}^j_{x_2}(x^1_2) & \cdots & \cdots & \bar{u}^{J-1}_{x_2}(x^1_2) & \bar{u}^J_{x_2}(x^1_2) \\ \bar{u}^1_{x_1}(x^2_2) & \bar{u}^2_{x_1}(x^2_2) & \cdots & \bar{u}^j_{x_1}(x^2_2) & \cdots & \cdots & \bar{u}^{J-1}_{x_1}(x^2_2) & \bar{u}^J_{x_1}(x^2_2) \\ \vdots & \vdots & \vdots & \vdots & \cdots & \cdots & \vdots & \vdots \\ \vdots & \vdots & \vdots & \vdots & \ddots & & \vdots & \vdots \\ \vdots & \vdots & \vdots & \vdots & & \ddots & \vdots & \vdots \\ \bar{u}^1_{x_1}(x^k_2) & \bar{u}^2_{x_1}(x^k_2) & \cdots & \bar{u}^j_{x_1}(x^k_2) & \cdots & \cdots & \bar{u}^{J-1}_{x_1}(x^k_2) & \bar{u}^J_{x_1}(x^k_2) \\ \bar{u}^1_{x_2}(x^k_2) & \bar{u}^2_{x_2}(x^k_2) & \cdots & \bar{u}^j_{x_2}(x^k_2) & \cdots & \cdots & \bar{u}^{J-1}_{x_2}(x^k_2) & \bar{u}^J_{x_2}(x^k_2) \end{bmatrix}_{2k \times J}$$

on Γ_1 and, on Γ_2,

$$[\bar{u}]^{I,\text{FS}} = \begin{bmatrix} \bar{u}^1_{x_1}(x^1_2) & \bar{u}^2_{x_1}(x^1_2) & \cdots & \bar{u}^l_{x_1}(x^1_2) & \cdots & \cdots & \bar{u}^{L-1}_{x_1}(x^1_2) & \bar{u}^L_{x_1}(x^1_2) \\ \bar{u}^1_{x_2}(x^1_2) & \bar{u}^2_{x_2}(x^1_2) & \cdots & \bar{u}^l_{x_2}(x^1_2) & \cdots & \cdots & \bar{u}^{L-1}_{x_2}(x^1_2) & \bar{u}^L_{x_2}(x^1_2) \\ \bar{u}^1_{x_1}(x^2_2) & \bar{u}^2_{x_1}(x^2_2) & \cdots & \bar{u}^l_{x_1}(x^2_2) & \cdots & \cdots & \bar{u}^{L-1}_{x_1}(x^2_2) & \bar{u}^L_{x_1}(x^2_2) \\ \vdots & \vdots & \vdots & \vdots & \cdots & \cdots & \vdots & \vdots \\ \vdots & \vdots & \vdots & \vdots & \ddots & & \vdots & \vdots \\ \vdots & \vdots & \vdots & \vdots & & \ddots & \vdots & \vdots \\ \bar{u}^1_{x_1}(x^m_2) & \bar{u}^2_{x_1}(x^m_2) & \cdots & \bar{u}^l_{x_1}(x^m_2) & \cdots & \cdots & \bar{u}^{L-1}_{x_1}(x^m_2) & \bar{u}^L_{x_1}(x^m_2) \\ \bar{u}^1_{x_2}(x^m_2) & \bar{u}^2_{x_2}(x^m_2) & \cdots & \bar{u}^l_{x_2}(x^m_2) & \cdots & \cdots & \bar{u}^{L-1}_{x_2}(x^m_2) & \bar{u}^L_{x_2}(x^m_2) \end{bmatrix}_{2m \times L}$$

Following Fung (1965), the normalized surface traction can be defined as "the product of the boundary outward normal and the normalized stress modal functions." That is,

$$\bar{t}^n = \left\{ \begin{array}{c} \bar{t}^n_{x_1} \\ \bar{t}^n_{x_2} \end{array} \right\} = \left\{ \begin{array}{c} n_{x_1}\bar{\sigma}^n_{x_1 x_1} \\ n_{x_1}\bar{\sigma}^n_{x_1 x_2} \end{array} \right\}, \tag{20.9}$$

where $n_{x_2} = 0$ on $\Gamma_{1,2}$ and $n_{x_1} = -1$ and 1 on Γ_1 and Γ_2, respectively.

In a similar way, the total boundary tractions can be defined by a linear superposition of the normalized traction modal functions for the incident and scattered fields:

$$\{t\}^{\Gamma_1}_{2k \times 1} = [\bar{t}]^I_{2k \times J} \{\alpha^p \delta_{pj} e^{ik^j x_1}\}_{J \times 1} + [\bar{t}]^{\text{BS}}_{2k \times J} \{\beta^j e^{-ik^j x_1}\}_{J \times 1} \quad \text{on } \Gamma_1, \tag{20.10}$$

$$\{t\}^{\Gamma_2}_{2m \times 1} = [\bar{t}]^{\text{FS}}_{2m \times L} \{\beta^l e^{ik^l x_1}\}_{L \times 1} \quad \text{on } \Gamma_2, \tag{20.11}$$

where

$$[\bar{t}]^{I,\text{BS}} = \begin{bmatrix} \bar{t}^1_{x_1}(x_2^1) & \bar{t}^2_{x_1}(x_2^1) & \cdots & \bar{t}^j_{x_1}(x_2^1) & \cdots & \cdots & \bar{t}^{J-1}_{x_1}(x_2^1) & \bar{t}^J_{x_1}(x_2^1) \\ \bar{t}^1_{x_2}(x_2^1) & \bar{t}^2_{x_2}(x_2^1) & \cdots & \bar{t}^j_{x_2}(x_2^1) & \cdots & \cdots & \bar{t}^{J-1}_{x_2}(x_2^1) & \bar{t}^J_{x_2}(x_2^1) \\ \bar{t}^1_{x_1}(x_2^2) & \bar{t}^2_{x_1}(x_2^2) & \cdots & \bar{t}^j_{x_1}(x_2^2) & \cdots & \cdots & \bar{t}^{J-1}_{x_1}(x_2^2) & \bar{t}^J_{x_1}(x_2^2) \\ \vdots & \vdots & \vdots & \vdots & \cdots & \cdots & \vdots & \vdots \\ \vdots & \vdots & \vdots & \vdots & \ddots & & \vdots & \vdots \\ \vdots & \vdots & \vdots & \vdots & & \ddots & \vdots & \vdots \\ \bar{t}^1_{x_1}(x_2^k) & \bar{t}^2_{x_1}(x_2^k) & \cdots & \bar{t}^j_{x_1}(x_2^k) & \cdots & \cdots & \bar{t}^{J-1}_{x_1}(x_2^k) & \bar{t}^J_{x_1}(x_2^k) \\ \bar{t}^1_{x_2}(x_2^k) & \bar{t}^2_{x_2}(x_2^k) & \cdots & \bar{t}^j_{x_2}(x_2^k) & \cdots & \cdots & \bar{t}^{J-1}_{x_2}(x_2^k) & \bar{t}^J_{x_2}(x_2^k) \end{bmatrix}_{2k \times J}$$

on Γ_1 and, on Γ_2,

$$[\bar{t}]^{I,\text{FS}} = \begin{bmatrix} \bar{t}^1_{x_1}(x_2^1) & \bar{t}^2_{x_1}(x_2^1) & \cdots & \bar{t}^l_{x_1}(x_2^1) & \cdots & \cdots & \bar{t}^{L-1}_{x_1}(x_2^1) & \bar{t}^L_{x_1}(x_2^1) \\ \bar{t}^1_{x_2}(x_2^1) & \bar{t}^2_{x_2}(x_2^1) & \cdots & \bar{t}^l_{x_2}(x_2^1) & \cdots & \cdots & \bar{t}^{L-1}_{x_2}(x_2^1) & \bar{t}^L_{x_2}(x_2^1) \\ \bar{t}^1_{x_1}(x_2^2) & \bar{t}^2_{x_1}(x_2^2) & \cdots & \bar{t}^l_{x_1}(x_2^2) & \cdots & \cdots & \bar{t}^{L-1}_{x_1}(x_2^2) & \bar{t}^L_{x_1}(x_2^2) \\ \vdots & \vdots & \vdots & \vdots & \cdots & \cdots & \vdots & \vdots \\ \vdots & \vdots & \vdots & \vdots & \ddots & & \vdots & \vdots \\ \vdots & \vdots & \vdots & \vdots & & \ddots & \vdots & \vdots \\ \bar{t}^1_{x_1}(x_2^m) & \bar{t}^2_{x_1}(x_2^m) & \cdots & \bar{t}^l_{x_1}(x_2^m) & \cdots & \cdots & \bar{t}^{L-1}_{x_1}(x_2^m) & \bar{t}^L_{x_1}(x_2^m) \\ \bar{t}^1_{x_2}(x_2^m) & \bar{t}^2_{x_2}(x_2^m) & \cdots & \bar{t}^l_{x_2}(x_2^m) & \cdots & \cdots & \bar{t}^{L-1}_{x_2}(x_2^m) & \bar{t}^L_{x_2}(x_2^m) \end{bmatrix}_{2m \times L}$$

Rearranging (20.7) and (20.8) with respect to the scattered amplitudes β^j and β^l yields

$$\{\beta^j e^{-ik^j x_1}\}_{J \times 1} = [\bar{u}^{-1}]^{\text{BS}}_{J \times 2k}\{u\}^{\Gamma_1}_{2k \times 1} - [\bar{u}^{-1}]^{\text{BS}}_{J \times 2k}[\bar{u}]^I_{2k \times J}\{\alpha^P \delta_{pj} e^{ik^j x_1}\}_{J \times 1} \text{ on } \Gamma_1, \quad (20.12)$$

$$\{\beta^l e^{ik^l x_1}\}_{L \times 1} = [\bar{u}^{-1}]^{\text{FS}}_{L \times 2m}\{u\}^{\Gamma_2}_{2m \times 1} \text{ on } \Gamma_2, \quad (20.13)$$

where $[\bar{u}^{-1}]$ denotes the generalized complex inverse matrix of $[\bar{u}]$,

$$[\bar{u}^{-1}] = ([\bar{u}^*]^T [\bar{u}])^{-1} [\bar{u}^*]^T,$$

and $[\bar{u}^*]^T$ denotes the transpose of the complex conjugate of $[\bar{u}]$ (Lancaster and Tismenetsky 1985). Substituting (20.12) and (20.13) into (20.10) and (20.11), we obtain

$$\{t\}^{\Gamma_1}_{2k \times 1} = [\bar{t}]^I_{2k \times J}\{\alpha^P \delta_{pj} e^{ik^j x_1}\}_{J \times 1} + [\bar{t}]^{\text{BS}}_{2k \times J}[\bar{u}^{-1}]^{\text{BS}}_{J \times 2k}\{u\}^{\Gamma_1}_{2k \times 1}$$

$$- [\bar{t}]^{\text{BS}}_{2k \times J}[\bar{u}^{-1}]^{\text{BS}}_{J \times 2k}[\bar{u}]^I_{2k \times J}\{\alpha^P \delta_{pj} e^{ik^j x_1}\}_{J \times 1} \text{ on } \Gamma_1, \quad (20.14)$$

$$\{t\}^{\Gamma_2}_{2m \times 1} = [\bar{t}]^{\text{FS}}_{2m \times L}[\bar{u}^{-1}]^{\text{FS}}_{L \times 2m}\{u\}^{\Gamma_2}_{2m \times 1} \text{ on } \Gamma_2. \quad (20.15)$$

Finally, we obtain $2k$ and $2m$ relations between boundary tractions and displacements on the left and right cross-sectional boundaries of a waveguide, respectively (instead of the $2k$ and $2m$ boundary conditions obtained previously).

The preceding matrix formulation can be written in matrix operator form as

$$[H]U = [G]T,$$

where $[H]$ is the matrix operator on displacements and $[G]$ is the matrix operator on tractions. From these considerations of mode conversions and forward and back scattering, an alternative matrix formulation is possible – namely,

$$[C] = [A][X],$$

where $[C]$ is a constant vector, $[A]$ is a coefficient matrix containing the operators $[H]$ and $[G]$, and $[X]$ is a vector of unknown displacements.

Gaussian elimination is used to solve for $[X]$. After these unknown displacements have been found, the amplitudes β^j and β^l of the forward- and back-scattered fields become available. Using these along with the incident mode amplitude α^p, we can determine the reflection and transmission coefficients as follows:

$$R^{jp} = \beta^j/\alpha^p \quad \text{(reflection coefficient),}$$
$$T^{lp} = \beta^l/\alpha^p \quad \text{(transmission coefficient).}$$

In the next section, these coefficients will be used to evaluate the scattering of guided waves from selected scatterers.

20.5 The Energy Principle

We can evaluate the BEM–NMET code by using a conservation of energy principle: the sum of the energies contained in the forward- and back-scattered fields must equal the energy of the incident field. Applying this principle confirms the BEM–NMET formulation. Because the input energy induced by an incident mode to a waveguide must be equal to the sum of all multimode scattering energy (in terms of reflection on the left cross-sectional boundary Γ_1 and of transmission on the right cross-sectional boundary Γ_2), it follows that

$$E_p^I = \sum_{j=1}^{J} E_j^{BS} + \sum_{l=1}^{L} E_l^{FS}.$$

Here E_p^I is the energy flux transported by the incident pth mode through the left side boundary Γ_1, and E_j^{BS} and E_l^{FS} represent the energy fluxes transported by the jth reflected mode and the lth transmitted mode through the left side boundary Γ_1 and the right side boundary Γ_2, respectively; J and L denote the total number of possible propagating modes on (resp.) Γ_1 and Γ_2.

Once the wave structure of each mode is normalized by the relative power ratio between the incident pth mode and a scattered mode, the normalized transported energy fluxes can be simply expressed as the squares of the normalized displacement amplitude with respect to α^p, the incident pth mode amplitude, as follows:

$$\bar{E}_p^I = [\alpha^p/\alpha^p]^2 = 1,$$

$$\bar{E}_j^{BS} = [\beta^j/\alpha^p]^2 = [R^{jp}]^2,$$

$$\bar{E}_l^{FS} = [\beta^l/\alpha^p]^2 = [T^{lp}]^2$$

(Achenbach 1984; Auld 1990; Graff 1991; Viktorov 1967).

Obviously, the guided wave scattering solutions in this study can be influenced by the inevitable error sources from the hybrid boundary element scheme: the approximated field variables, the numerical integration, the Gaussian quadrature with finite integration points, the discretized boundary with finite boundary nodes, and the finite number of propagating modes used in the mode superposition and the solution matching scheme between the finite scattering zone and the rest of an infinite waveguide. Consequently, to convince us of the accuracy of the technique, the error involved in the analysis can be estimated by the deviation of the total reflected and transmitted energies from the incident energy. Thus, we have the following expression for error:

$$\varepsilon = \left|\left[\bar{E}_p^I - \left(\sum_{j=1}^J \bar{E}_j^{BS} + \sum_{l=1}^L \bar{E}_j^{FS}\right)\right]\frac{1}{\bar{E}_p^I}\right| \times 100 = \left|1 - \left(\sum_{j=1}^J [R^{jp}]^2 + \sum_{l=1}^L [T^{lp}]^2\right)\right| \times 100,$$

where ε denotes the relative (percentage) error based on the energy conservation principle.

20.6 Sample Guided Wave Scattering Problems

20.6.1 Preliminaries

Figure 20-7 shows the flow chart of the computational steps necessary to implement the studies of guided wave scattering from geometries and/or defects. To perform parametric studies, the material properties listed here were used (along with the previously calculated dispersion curves for steel):

- L-wave velocity = 5.94 mm/μs;
- T-wave velocity = 3.2 mm/μs;
- density = 7.8 g/cm^3; and
- thickness = 1.0 mm.

Several computational aspects of the BEM–NMET approach are reviewed in the following paragraphs. The plate model will be treated as an infinite waveguide with various surface-breaking defects and various reflecting boundaries. The modeling is restricted to localized modeling zones within the infinite plate. Where the modeling zone intersects the left and right semi-infinite portions of the plate, we design special matching boundary conditions: rather than the typical elastic boundary conditions, displacements and tractions are specified by deriving the relations of the modal functions between displacements and tractions. This reduces the number of boundary equations required to represent intersections of the modeling zone with semi-infinite regions. See Figure 20-8.

We now present a comparative study of constant boundary displacements versus quadratic boundary displacements. The algebraic relations involved are

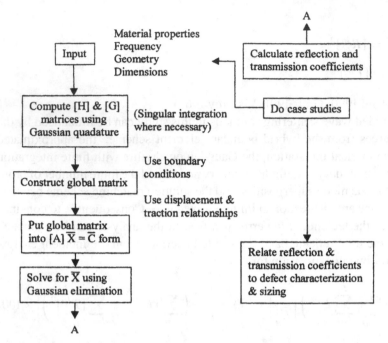

Figure 20-7. Flow chart showing computational steps necessary to implement the BEM–NMET study of guided wave interaction with scatterers.

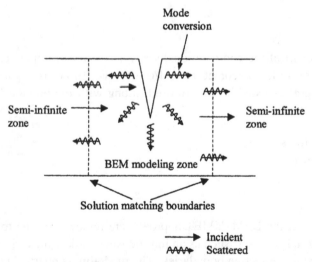

Figure 20-8. BEM–NMET modeling approach, showing semi-infinite and modeling zones.

$$\bar{u}_i = \text{constant} \quad \text{versus} \quad \bar{u}_i = a\bar{u}^a + b\bar{u}^b + c\bar{u}^c,$$

where a, b, c are the shape functions (defined as second-degree polynomials) to interpolate displacement functions inside a boundary element and where $\bar{u}^a, \bar{u}^b, \bar{u}^c$ are the discrete nodal displacements at nodes a, b, c. The shape functions are given by

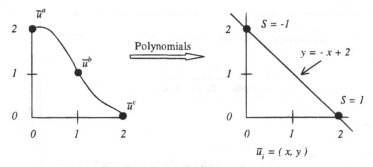

Figure 20-9. Example of the smoothing effect of polynomials on quadratic boundary elements.

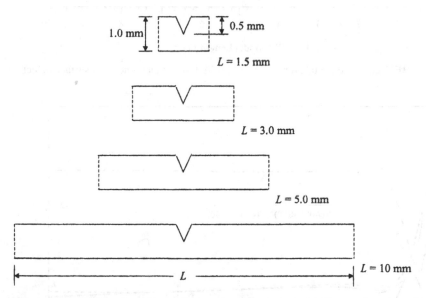

Figure 20-10. Illustration of models to evaluate numerical stability as a function of the near-field effects of a scattering center.

$$a = \tfrac{1}{2}S(S-1), \quad b = -(S-1)(S+1), \quad c = \tfrac{1}{2}S(S+1)$$

for $S \in [-1, 1]$. As S varies from -1 to 1 in a specific local coordinate system, $\bar{u}_i$ varies smoothly from $\bar{u}^a$ to $\bar{u}^c$.

A simple example is shown in Figure 20-9. Essentially, the polynomials act as weights to implement a smoothing function. In BEM analysis, these polynomials are called shape functions. After comparing constant versus quadratic elements, we can surmise that – for the studies to be discussed here – there are no significant differences between the two ways of dealing with displacements. Constant boundary elements will be used for the remainder of this section.

The distance between the boundaries of the modeling zone may be studied as they relate to the location of scattering centers (see Figure 20-10). The mode conversions and scattering near the scattering center may include evanescent modes with sufficient amplitude

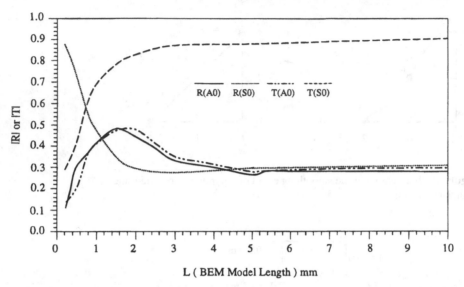

Figure 20-11. BEM convergence test for S0 ($fd = 1.0$ MHz-mm) incidence to a surface defect.

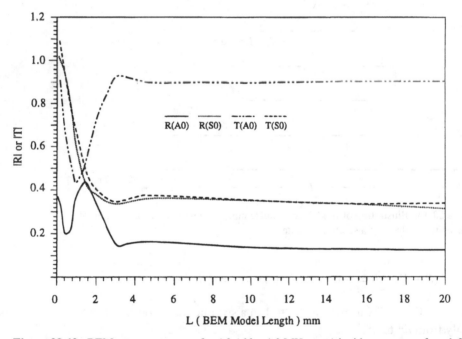

Figure 20-12. BEM convergence test for A0 ($fd = 1.0$ MHz-mm) incidence to a surface defect.

to cause near-field superposition effects. This is due to small values of x in the term $\exp[-k_{Im}x]$ (x is small near the scattering center). In order to avoid near-field effects, we should use a model length L equal to or greater than one half the wavelength of the incident mode; this will enable us to achieve stable numerical results. Convergence graphs for sample wave modes and fd values are shown in Figures 20-11, 20-12, and 20-13 as a function of model length.

Another concern, which can also be associated with evanescent modes, is the numerical accuracy of solutions near modal cutoff frequencies. In this case, improved numerical

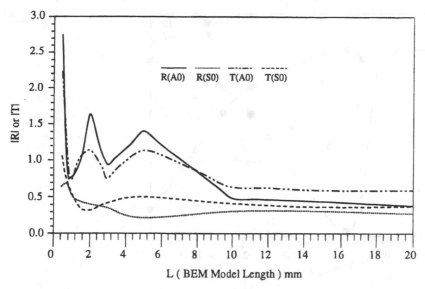

Figure 20-13. BEM convergence test for A0 ($fd = 1.5$ MHz-mm) incidence to a surface defect.

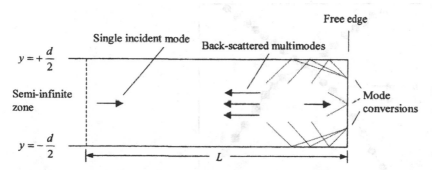

Figure 20-14. Guided wave scattering from a perpendicular free edge.

accuracy can be achieved by considering not only propagating modes but also evanescent modes with small attenuation.

We next examine the number of elements used to formulate the BEM boundaries as a function of modal frequency. Noting that the time-harmonic displacements are the waves (or wave packets), we will see that the higher the wave frequency, the greater the number of boundary element displacements (nodes) needed to accurately represent the wave. The reader should be cautioned that – in order to maintain computational efficiency and accuracy – trade-offs must be made between the number of elements necessary to (a) model expected modal frequencies and (b) avoid near-field scattering effects.

20.6.2 Reflections from a Free Edge

There is an extensive body of knowledge on the problem of scattering from a free edge; see Figure 20-14. To address issues associated with this problem, we present antisymmetric and symmetric incident modes and the resulting variations of the reflection coefficient.

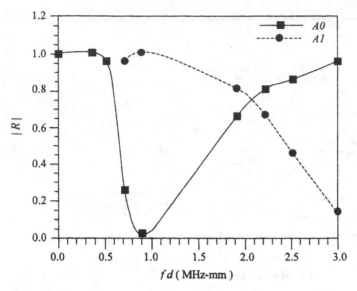

Figure 20-15. Variation of reflection coefficient as a function of modes (A0 and A1) and fd, for A0 incidence.

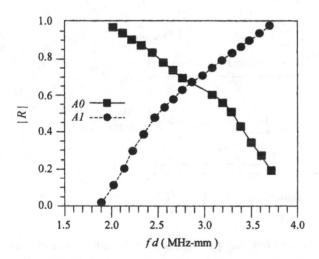

Figure 20-16. Variation of reflection coefficient as a function of modes (A0 and A1) and fd, for A1 incidence.

In turn, the reflection coefficient provides information about the mode conversions that can (and cannot) occur.

The antisymmetric cases include A0 and A1 mode incidence. For such cases, the absolute value of the reflection coefficient, $|R|$, as a function of the frequency–thickness product is used to study reflections and mode conversions. (Reflected symmetric modes were precluded by the symmetry of the waveguide models with respect to the midplane of the plate along the thickness direction.) Figure 20-15 illustrates the antisymmetric mode conversion of A0 to A1 and vice versa.

The range of variation shown in Figure 20-15 can be divided into three regions. The first region is between $fd = 0$ MHz-mm and $fd = 0.5$ MHz-mm, where the A0 mode maintains perfect reflection ($|R| = 1$) until it subsides and mode A1 becomes prominent.

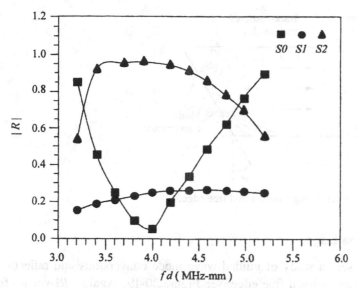

Figure 20-17. Variation of reflection coefficient as a function of modes (S0, S1, and S2) and fd, for S0 incidence.

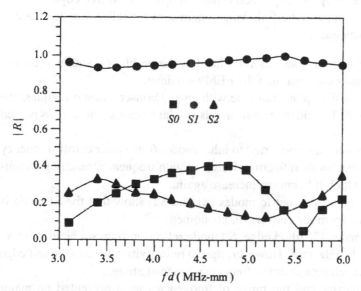

Figure 20-18. Variation of reflection coefficient as a function of modes (S0, S1, and S2) and fd, for S1 incidence.

The second region considered is between $fd = 0.9$ and 2.0 MHz-mm. In this region, the reflectivity of A0 and A1 modes approach each other, being equal at 2.0 MHz-mm. About 70% of the incident A0 mode amplitude will be reflected as A0 and A1 modes. The third region considered, $fd > 2.0$ MHz-mm, shows a decline in A1 reflected amplitude with frequency as the A0 mode becomes dominant again. Similar results were obtained using A1 mode incidence; see Figure 20-16.

We also studied the incident symmetric modes S0, S1, and S2. Sample results for S0 and S1 incident modes are shown in Figure 20-17 and Figure 20-18, respectively.

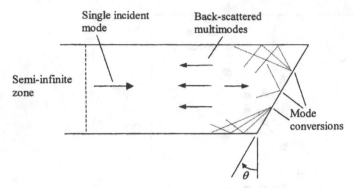

Figure 20-19. Guided wave scattering from a tilted free edge.

20.6.3 Inclined Free Edge

In this section we present a study of guided wave mode conversions and reflection coefficient variations for an inclined free edge; see Figure 20-19. Again, $|R|$-versus-fd diagrams are used for the analysis – in the same manner as for the perpendicular free edge. The principal conclusion is that mode conversions occur between antisymmetric and symmetric modes owing to the midplane asymmetry induced by the tilted free edge.

Experimental evidence supports the following conclusions for reflections from a free edge with guided wave incidence.

(1) The BEM–NMET approach successfully handles the problem – and in a manner more computationally efficient than the FEM technique.

(2) Edge reflections in a steel plate coincide with energy conservation principles; the cases of A0, A1, and S0 incidence show trends similar to cases where glass is used instead of steel.

(3) Overall, incident modes are converted to other modes (once their cutoff frequency is exceeded); incident mode reflection decreases with frequency, reaching a minimum value, after which it begins to increase again.

(4) Investigations of higher symmetric modes (S1 and S2) show that the S1 mode is virtually unaffected by mode conversion phenomena.

(5) For S0 incidence and a 45° tilted edge, S0 mode reflection shows a rapid increase in reflectivity at 1.4 MHz-mm. However, the S0 reflectivity with a 60° tilted edge showed virtually no change over the frequency range of study.

(6) Geometrical complexity and the range of frequencies used presented no major problems, suggesting that the BEM–NMET approach could handle arbitrary discontinuities.

20.6.4 Arbitrary Defects

In order to test assumptions about the utility of guided waves for defect detection, we require a BEM–NMET parametric investigation of the effects that defect sharpness has on guided wave reflection, transmission, and mode conversions. As Figure 20-20 shows, the defects modeled are represented as ellipses of varying eccentricities, where s and h denote half of the defect mouth and depth, respectively. The half-circle shown in the middle

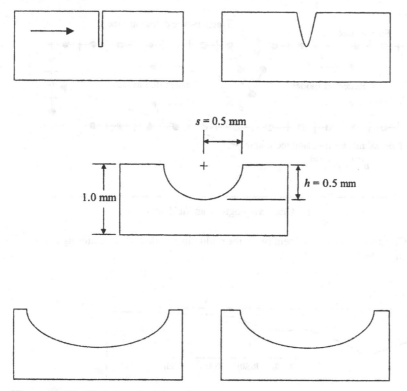

Figure 20-20. Variations in defect sharpness.

of the figure can be taken as a reference point. Depth is fixed as the mouth of the defect is varied above and below the eccentricity value of 1. The modes A0 and S0 were used as the incident modes from the left.

The goal of this exercise is identifying the most pronounced mode changes with respect to variations in defect sharpness. Such changes would lend themselves quite well to NDE. The effects can be studied using analysis and diagrams quite similar to those used for the free edge study, with the addition of the transmission coefficient $|T|$. An fd value of 1 MHz-mm is selected for A0 and S0 modes because of the large separation of the phase and group velocities of these modes from other modes. Also, both modes have nearly nondispersive group velocities. For this reason, the wave modes stably propagate under localized frequency sweeping around this fd value and hence can be experimentally obtained without any significant change of group velocity.

Boundary element discretization of the whole boundary, including the crack surface, is automatically established by a mesh generation routine inserted in the main hybrid BEM code. The defect surface is assumed as an elliptical contour with different aspect ratio, s/h. This contour can be mathematically defined by the equation of an ellipse and then divided into 60–70 constant boundary elements; flat boundaries have 150–200 elements.

The configuration of a constant boundary element mesh for a surface-breaking crack is given in Figure 20-21. The material properties for a steel plate used in the BEM simulation are the longitudinal wave velocity $c_L = 5.94$ mm/μs, the transverse wave velocity $c_T = 3.2$ mm/μs, and the density $\rho = 7.8$ g/cm^3. In Figure 20-22, the crack depth is fixed at 0.5, or 50% of the waveguide thickness; the width of the defect mouth varies for

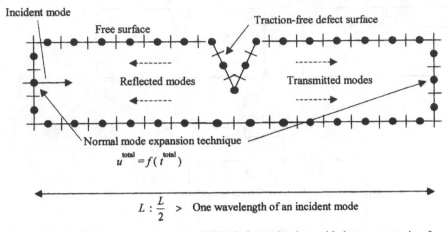

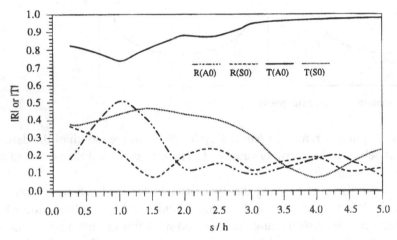

Figure 20-21. Constant boundary element mesh for evaluating guided wave scattering from a surface-breaking defect.

Figure 20-22. Variations in reflection and transmission versus aspect ratio of an elliptical surface defect for A0 incidence ($fd = 1.0$ MHz-mm; $h = 0.5$ or 50% of waveguide thickness).

different defect sharpnesses. (A sharp crack propagates much faster than a round surface "wastage" and so is considered more dangerous, even though the initial defect depth of both may be the same.) In order to evaluate differences in detection sensitivity, an incident mode (on the left cross-sectional boundary) is chosen and coefficients are plotted with respect to s/h ratios of a surface defect for both transmission T and reflection R. At $fd = 1.0$ MHz-mm, Figure 20-22 shows BEM results for an incident mode of A0; Figure 20-23 shows results for S0 incidence. Some of the curves exhibit a spline fit over approximately eleven evenly spaced points.

The incident energy, which is associated with two possible scattered modes (A0 and S0), can be scattered back and forward from a defect. The numerical calculation of T and R is carried out by changing the boundary mesh for different degrees of sharpness. As

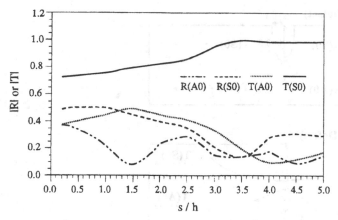

Figure 20-23. Same as Figure 20-22 but for S0 incidence.

shown in Figures 20-22 and 20-23, there is enough variation in R and T with respect to s/h change to enable a data acquisition scheme that might be used in classification analysis. In other words, this result convinces us that the guided wave technique can be used for defect classification purposes in separating a critical sharp crack from other types of round defects. For both A0 and S0 incidence, if $s/h > 4.0$ then the transmission factor of an incident mode is dominant (reaching unity), since mode conversion is substantially reduced owing to the increase in crack-tip bluntness. Mode conversion finally converges to a stable stage of a tapered waveguide, with smoothly varying thickness, when $s/h = 4.0$. At this point, defect classification (with A0 and S0 at $fd = 1.0$ MHz-mm) is no longer possible, according to the results graphed in the figures.

For both A0 and S0 incidence, the trend of T and R curves is similar: between $s/h = 1$ and 2, the reflection of the incident modes decreases monotonically with the increase in crack-tip bluntness. It is interesting that the reflection of the A0 mode for A0 incidence rapidly increases with the increase of s/h between 0.2 and 1.0, with the maximum at $s/h = 1.0$ when the defect shape becomes a half-circle. The S0 reflection for S0 incidence shows only a slight decrease in the same s/h range. Therefore, A0 mode incidence should make for better defect classification of sharp cracks than S0 incidence (if we monitor the incident mode reflection at $fd = 1.0$ MHz-mm). Conversely, when the S0 mode is chosen as incident, reflection of the A0 mode can be used (rather than the S0 reflection) because its variation is fairly steep up to about $s/h = 1.5$; this enables another possibility for defect classification.

Generally speaking, the reflection and transmission of an incident mode shows an overall inverse trend, indicating conservation of energy. This is because the major scattering still interacts with an incident mode even though other minor scattered modes can also be generated through mode conversion. It is also observed that the mode conversion behavior is proportional, which aids in detecting sharpness.

With respect to varying defect sharpness, the results so far show the potential of guided wave techniques for defect classification in practical NDE applications. Unlike the case of laboratory inspection, however, information may not be available on realistic defect shapes and size in the field. Hence a frequency-sweeping algorithm could be useful, particularly if source features and pattern recognition were employed. However, a more

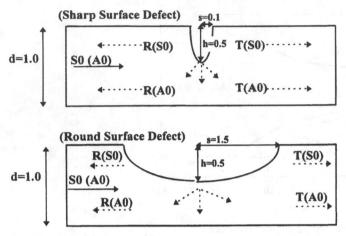

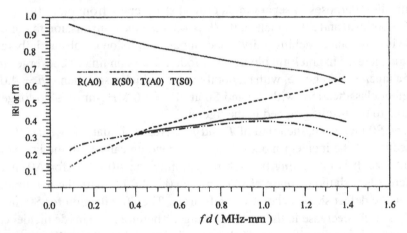

Figure 20-24. Guided wave sensitivity study with variation of incident frequency for different defects, with incident mode S0 or A0 between $fd = 0.1$ and 1.5 MHz-mm; $h = 0.5$ (50% of waveguide thickness).

Figure 20-25. Variation in reflection and transmission from a sharp surface crack ($s/h = 0.2$) versus S0 incident frequency.

realistic setup for BEM modeling of the guided wave scattering would be required, so we now report on guided wave sensitivity with variation in frequency.

Frequency sweeping and control are typical and realistic ways of obtaining features for defect classification. These techniques can be performed using various incident modes. Because each point on a dispersion curve provides a different wave structure, varying mode conversion is expected from sweeping frequency. In this section, incident frequency will be varied between $fd = 0.1$ and 1.5 MHz-mm, following two dispersion curves of the incident A0 and S0 modes for a sharp crack and a round wastage ($s/h = 0.2$ and 3.0, resp.). The problem statement is illustrated in Figure 20-24.

The meshing scheme is applied to discretize both boundaries of the waveguide and the defect surface. The results of reflection and transmission factors associated with S0 and A0 incidence are plotted in Figures 20-25–20-28 for the problem laid out in Figure 20-24.

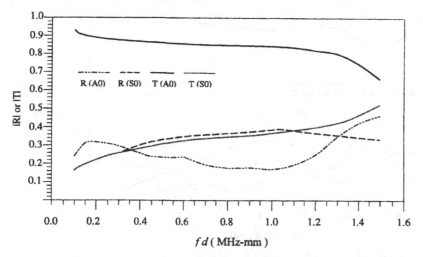

Figure 20-26. Same as Figure 20-25 but for A0 incidence.

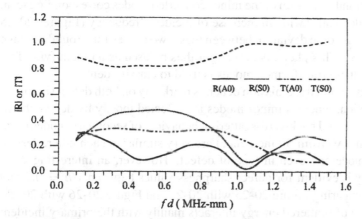

Figure 20-27. Variation in reflection and transmission from a rounded surface wastage ($s/h = 3.0$) versus S0 incident frequency.

As seen in the four graphs, there is a significant difference between the two defect types in variation of reflection and transmission factors with an increase of incident frequency. For a crack ($s/h = 0.2$) subjected to S0 incidence, the S0 mode transmission and reflection coefficients monotonically decrease and increase (resp.) with an fd increase, as shown in Figure 20-25, whereas A0 incidences yields only slight changes (Figure 20-26). This trend suggests that to monitor the scattering of S0 rather than A0 incidence would result in better sensitivity to defect classification of IGA (intergranular corrosion attack) cracks.

For a round wastage ($s/h = 3.0$), the S0 mode transmission factor dominates the entire frequency-sweeping range in Figure 20-27 (S0 incidence). However, the trend is quite different from the case of an IGA crack (Figure 20-25). Figure 20-27 shows: (i) transmission and reflection in S0 mode with (resp.) minimum and maximum sensitivities at about $fd = 0.6$ MHz-mm; and (ii) the parabolic variation from 0.6 to 1.0 MHz-mm. Considering the peak at about $fd = 0.3$ MHz-mm shown in Figure 20-28, we ought to use the A0 reflection of A0 incidence for a round wastage classification.

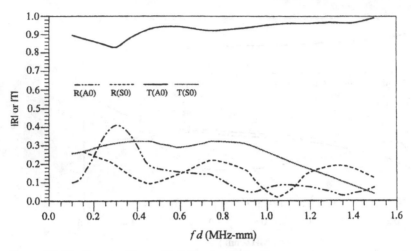

Figure 20-28. Same as Figure 20-27 but for A0 incidence.

Note that transmission and reflection of the minor converted modes corresponding to an IGA crack vary almost identically with an increase of incident frequency (Figures 20-25 and 20-26), whereas there is some deviation between these two curves for a round wastage (Figures 20-27 and 20-28). This phenomenon might thus be used as an identifying feature when monitoring of these transformed modes is used to classify defects.

The variation in reflection and transmission depends markedly on both defect sharpness and incident mode selection, whereas minor modes are affected mostly by defect sharpness. For example, forward- and backward-scattering behaviors of secondary modes (e.g., S0 from A0 incidence and A0 from S0 incidence) are very similar to each other, despite differences in incident mode and even in type of defect. However, an interesting deviation from this similarity is evident for defect sharpness values between $s/h = 0.2$ and 3.0, as can be seen by comparing Figure 20-25 with 20-27 and Figure 20-26 with 20-28. Overall, the major portion of scattered energy interacts mainly with the primary incident modes.

We conclude this section by outlining a sample study on guided wave sensitivity with variation of defect depth. It is clear that the guided wave technique has sufficient potential in flaw classification analysis, especially for distinguishing critical from noncritical defects. Once a guideline for defect classification is established, a sizing study for a critical crack would be the next task in the nondestructive testing procedure for obtaining more quantitative data.

Our main emphasis now is thus to explore the feasibility of guided wave techniques for sizing defect depth. For this study, we choose a sharp IGA crack ($s = 0.05$) subjected to A0 or S0 incidence at $fd = 1.0$ MHz-mm. The depth of an IGA crack varies between $h/d = 0.1$ and 0.9, as shown in Figure 20-29.

Figures 20-30 and 20-31 show the reflection and transmission for both primary (incident) and secondary (converted) modes under S0 and A0 incidence, respectively. As we can see in the figures, reflection and transmission for a secondary mode increase monotonically with defect depth, because the deeper reflector surface creates more mode conversion. However, after reaching their maximum, the secondary mode scatterings show a tendency to decrease with respect to increasing defect depth, since the problem converges

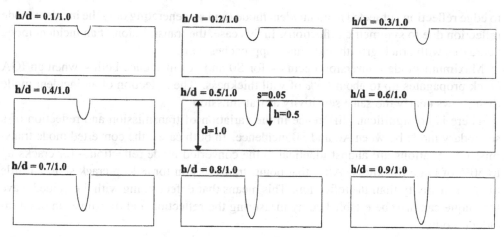

Figure 20-29. Range of IGA crack profiles used in study of guided wave sensitivity to variation in defect depth.

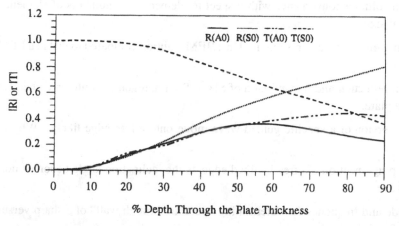

Figure 20-30. Reflection and transmission versus percentage depth of a sharp surface crack ($s = 0.05$, $d = 1.0$) for S0 incidence.

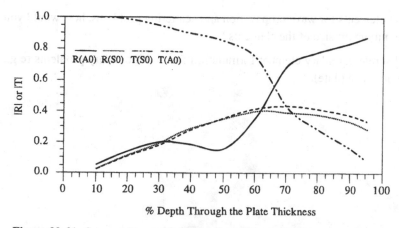

Figure 20-31. Same as Figure 20-30 but for A0 incidence.

to edge reflection subjected to an incident mode (and so generating only the incident mode reflection due to symmetric reflection). In this case, the transmission of an incident mode decreases with crack growth and finally approaches zero.

Maximum mode conversion occurs – for S0 and A0 incidence both – when an IGA crack propagates up to about 65% of wall thickness. The reflection of an incident mode then shows nearly the same sensitivity as transmission.

There is no significant difference in the variation of transmission and reflection of a secondary mode between A0 and S0 incidence. In both cases, the converted mode transmission variations are almost identical to the converted mode reflection – for cracks up to 50% of plate thickness. After that point, transmission for a deep crack shows a little higher sensitivity than the reflection. This means that defect sizing with the guided wave technique can also be established by measuring the reflection and transmission of an incident mode.

20.7 Exercises

1. Comment on solution convergence with respect to element size, numbers of elements, and structural size.

2. Comment on number of elements required in a BEM solution compared to that in a FEM solution.

3. Comment on reflection and transmission of SH or SV impingement onto an embedded cylinder in a plate.

4. Is mode conversion of a specific guided wave mode onto a free edge likely? Why or why not?

5. Describe impingement onto a free inclined edge as the inclination angle becomes normal to the wave vector.

6. Select a mode and frequency for determining "percent through wall" of a sharp versus a smooth defect in a plate.

7. Select a possible mode and frequency that merely differentiates a sharp from a smooth defect in a plate.

8. Using constant elements for wave propagation and scattering problems, how would you determine the maximum size of the elements?

9. Express the discrete boundary element formulation for multiboundary problems (e.g., an embedded hole in a plate).

APPENDIX A

Ultrasonic Nondestructive Testing Principles, Analysis, and Display Technology

A.1 Some Physical Principles

It will be useful to review some widely used basic concepts in ultrasonic nondestructive evaluation as a complement to the more detailed aspects of the mechanics and mathematics of wave propagation and ultrasonic nondestructive evaluation (NDE). Of first concern will be defining such fundamental ultrasonic field parameters as near field and angle of divergence. These will be followed by elements of instrumentation and display technology, along with aspects of axial and lateral resolution of an ultrasonic transducer.

Wave velocity, one of the key parameters of wave propagation study, is the velocity at which a disturbance propagates in some specified material. Its value depends on material, structure, and form of excitation. Many different formulas for wave velocity are presented. The most widely used wave velocity value used in ultrasonic NDE is the bulk longitudinal wave velocity, generally thought of as directly proportional to the square root of the elastic modulus over density. Another common velocity is the bulk shear wave velocity, which is proportional to the square root of the shear modulus over density. These velocities are called bulk velocities. Bulk waves do not require a boundary for support. Guided waves, on the other hand, require a boundary for propagation. Many tables of wave velocity values for different materials are available in the literature.

The *particle velocity* in a material is also a useful wave propagation parameter. Note that, for a given material, the maximum wave velocity can be a thousand times greater than a maximum particle velocity. Particle velocity initiates motion once the wave reaches the specific particle being considered.

There are two basic modes of particle velocity motion with respect to the wave type generated within the material. For a longitudinal wave, the particle velocity direction is in the same direction as the wave vector or the wave velocity vector. For transverse or shear wave propagation, the particle velocity vector is at 90° to the direction of the wave vector. The transverse particle velocity vector can sometimes be called a vertical shear wave or a horizontal shear wave, depending on the coordinates used in the study. It should be pointed out that all kinds of wave motion (e.g., a guided wave consisting of a Lamb-type wave, a surface wave, an interface wave) are actually made up of a superposition of longitudinal and shear wave particle velocity components.

We now outline some additional key parameters of wave propagation. *Attenuation* can come about from internal friction or energy absorption in a material, or it can also be geometric as in the case of spherical or cylindrical wave propagation. In general, attenuation

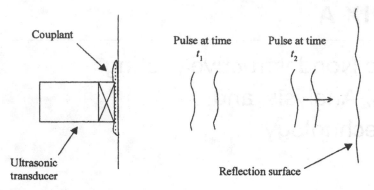

Figure A-1. Basic ultrasonic test principle.

produces a reduction in magnitude even if pulse duration is constant (nondispersive media). In special cases, however, where attenuation can be considered a function of frequency (dispersive media), pulse spreading can be observed, which obviously leads to magnitude reduction as well.

Scattering can produce both magnitude reductions and pulse spreading as a result of wave transmission from a transducer or from wave interaction with a small obstacle or flaw. Many approximations of scattering analysis are used in ultrasonic analysis. The simplest scattering theory utilizes Huygens's principle and spherical scattering integrated over the entire surface area of the obstacle. Attenuation of the $1/r$ type is considered at a point near the receiving transducer. More advanced theories include mode conversion and energy partitioning to shear, as well as phase and magnitude reduction considerations.

A variety of *interface waves* can be produced in certain materials. For example, imagine an angle beam impinging on an air–solid interface (surface wave), a fluid–solid interface (Sholte) wave, or a solid–solid interface (Stonely) wave. Note that the depth of penetration of a surface wave equals approximately half to twice the wavelength. Surface waves can follow a curved surface, provided the radius of curvature is roughly greater than one half the wavelength. These topics are covered in Chapters 7 and 9.

Ultrasonic waves can be generated inside a material by placing a piezoelectric element in contact with the surface and then pulsing the element with an appropriate voltage-versus-time profile. The piezoelectric element converts electrical energy into mechanical energy by what is commonly known as the piezoelectric effect. Application of a suitable couplant material between the piezoelectric element and the material in question allows ultrasonic waves to propagate efficiently from the transducer element into the test material. The basic test principle is illustrated for a bulk wave in Figure A-1.

Good sound transmission is important in order to perform appropriate signal analysis. Amplification of the echoes recorded on a test instrument can occur either before the sound leaves the transducer or after it is received by the transducer. Ultrasonic wave frequencies are normally used in the range between 40 kHz and 500 MHz, although recent applications in concrete testing go even lower than 40 kHz and applications in acoustic microscopy often go beyond 500 MHz. Note that the audible region starts at close to 20 kHz.

Wavelength in a material can be calculated as a function of the input frequency f and wave velocity c, which is a characteristic of the material and structure in question:

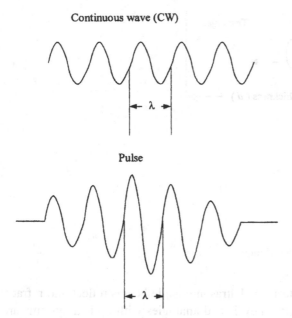

Continuous wave (CW)

Pulse

Figure A-2. Basic ultrasonic waveforms.

$$c = f\lambda. \tag{A.1}$$

For example, here is the wavelength calculation for a 1.5-MHz transducer in water with $c = 1,500$ m/s:

$$\lambda = \frac{c}{f} = \frac{1,500 \text{ m/s}}{1.5 \times 10^6 \text{ cycles/s}}$$
$$= 1,000 \times 10^{-6} \text{ m/cycle}$$
$$= .001 \text{ m/cycle}$$
$$= 1 \text{ mm/cycle}.$$

In steel, $c = 6,000$ m/s and so $\lambda = 4$ mm/cycle for the same 1.5-MHz transducer.

Pulse-echo ultrasonic techniques are often employed in NDE. In general, input waves may be either continuous wave (CW) or of pulse mode, as shown in Figure A-2. Note that λ in the pulse case is approximate because of the frequency bandwidth of the pulse. The wavelength λ agrees more closely with the formula $\lambda = c/f$ if the pulse is of narrowband frequency content and hence of many cycles (20 or more). For a highly damped pulse of (say) $1\frac{1}{2}$ to 3 cycles, the pulse is very broadbanded in frequency content and hence the λ value changes across the pulse, with the average value being quite close to that obtained with the formula if the frequency is close to the center frequency of the pulse.

A simple example of a pulse-echo experiment is illustrated in Figure A-3. First, take a look at the pulse-echo contact ultrasonic measurement as shown in the figure. Results are displayed on a cathode ray tube (CRT), giving us a plot of amplitude versus time. Time can be measured from the oscilloscope display as t_1. The distance or thickness d of the test object can then be computed from the following formula: $d = ct_1/2$.

Several concepts related to these principles may be outlined as follows. Reflection factor analysis is essential in ultrasonic physics. A plane wave encountering an interface between two materials is divided into two components, one of which is reflected and

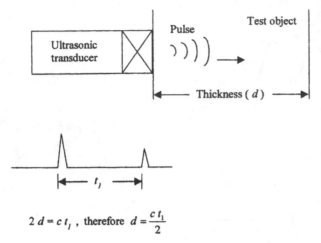

$2\,d = c\,t_l$, therefore $d = \dfrac{c\,t_l}{2}$

Figure A-3. Basic ultrasonic pulse-echo technique.

one that is transmitted beyond the interface. Ultrasonic angle beams reflect and refract, or bend, at an interface. Critical angle analysis and analogies with optical systems are useful, along with a detailed explanation of mode conversion into longitudinal and shear waves.

An understanding of ultrasonic field analysis is critical in both nondestructive testing (NDT) and diagnostic ultrasound. Such topics – which include near-field and far-field analysis, ultrasonic beam angle of divergence, and pressure variations through the ultrasonic field – are useful in transducer design analysis, signal interpretation work, signal processing analysis, and also in understanding reflector and material classification. Although a detailed understanding of the analytical tools required in ultrasonic field analysis are not necessary for carrying out successful work in either NDT or diagnostic analysis, the basic philosophy on which these calculations are based is useful for understanding and appreciating the work tasks associated with advanced field analysis, transducer and instrument selection, signal display techniques, and problems associated with resolution, image improvement, and reflector classification. An understanding of ultrasonic field analysis also allows us to perform and modify tests on an interactive basis, thereby providing us with an effective feedback process for improving data acquisition, signal interpretation, and so forth.

Let's now review Huygens's principle and examine its role in ultrasonic field analysis and basic scattering theory. Scattering theory is associated with the calculation process of pressure or stress waves being reflected from a reflector.

Essentially, a plane wavefront could be approximated as an infinite number of spherical waves propagating from a source distribution along a straight line. Huygens's principle states that a finite scatterer in a material (considered in a fluid, initially) could be replaced by an infinite number of point sources placed over the surface area of that object. It is now possible to examine the wave propagation characteristics reflecting from the surface by mathematically adding together the contributions of all point sources at some particular desired point in the ultrasonic field. The summing process could take place by integration and consideration of a differential pressure function emanating from a differential element on the surface of the scatterer. An alternate approach exists where the scatterer could be replaced by a finite number of point sources and summing point by point to obtain the total

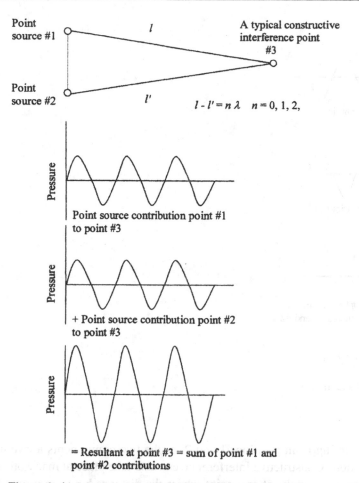

Figure A-4. A sample constructive interference problem.

solution. This approach is acceptable if the wavelength selected for the analysis is greater than the largest length of the finite element segment established on the surface of the scatterer. Aspects of Huygens's principle can be extended to include wave motion analysis in solid materials. The problem becomes difficult, however, because of the mode conversion and energy partitioning of both longitudinal and shear waves reflecting from the surface of an arbitrarily shaped scatterer. Much research has been focused on this subject; see, for example, Hueter and Bolt (1955), Kinsler et al. (1982), Pain (1993), and Pierce (1989).

A.2 Wave Interference

An understanding of wave interference is the key to understanding most aspects of ultrasonic wave propagation. Here we outline a brief explanation of the superposition process of waveforms that lead to constructive and destructive interference. First, consider the constructive interference problem illustrated in Figure A-4. If two point-source ultrasonic wave generators are stationed at positions 1 and 2, variation of ultrasonic pressure throughout the ultrasonic field will take place as a result of interference phenomena of the two wave fields as they propagate toward some point in question. In the constructive interference problem shown, the final solution at point 3 in space consists of the direct

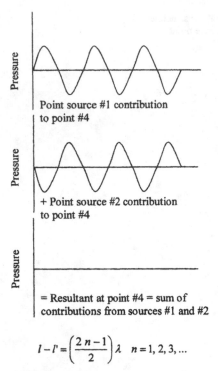

Point source #1 contribution
to point #4

+ Point source #2 contribution
to point #4

= Resultant at point #4 = sum of
contributions from sources #1 and #2

$$l - l' = \left(\frac{2n-1}{2} \right) \lambda \quad n = 1, 2, 3, \ldots$$

Figure A-5. A sample destructive
interference problem.

sum of the components coming from positions 1 and 2, as the two components arrive in
phase at the point in question. Constructive interference can also take place at many other
points in the ultrasonic field – namely, at any point where the distance between points 1
and 3 and points 2 and 3 differ by an integral number of wavelengths.

A destructive interference problem is shown in Figure A-5. In this case, the two components arrive at point 4 out of phase by half a wavelength. The net result in this case
is total destructive interference. An intermediate interference problem for an arbitrary
point 5 in space is illustrated in Figure A-6.

Throughout this textbook, the interference process (based on waveform superposition)
is used repeatedly to explain various physical concepts of great value in ultrasonic analysis and in diagnostic ultrasound. Superposition phenomena are useful in understanding
principles of frequency analysis and beam focusing as well as in multi-element transducer
design.

A.3 Computational Model for a Single Point Source

A first-order approximation of the ultrasonic field in solid media is based on the computational model of a single point source in a fluid. Pressure variations resulting from
excitation of a piezoelectric element (or from ultrasonic wave interaction with a small reflector) can be calculated by considering the well-known point-source solution in a fluid
in combination with Huygens's principle. A point-source pressure excitation in a fluid
produces a spherical pressure field of the form

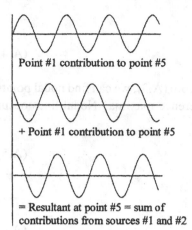

Point #1 contribution to point #5

+ Point #1 contribution to point #5

= Resultant at point #5 = sum of
contributions from sources #1 and #2

$$l - l' \neq n\lambda \quad n = 0, 1, 2,$$

$$l - l' \neq \left(\frac{2n-1}{2}\right)\lambda \quad n = 1, 2, 3, \ldots$$

Figure A-6. A sample intermediate
interference problem.

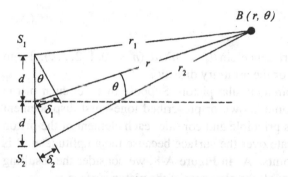

Figure A-7. The far-field interference from two equal sources.

$$p(r, t) = \frac{A_0}{r} e^{i(kr - \omega t)}, \quad \text{(A.2)}$$

where $A_0 = \rho c U_0$, $\omega = 2\pi f$, U_0 is amplitude of the outgoing wave, c is the wave velocity, and ρ is density. Consider the pressure at the appropriate point B in Figure A-7, where r_1 and r_2 are much larger compared with d ($r_1 \gg d$, $r_2 \gg d$). This allows us to make the approximation $\delta_1 \approx \delta_2 = \delta$ and $(d \sin\theta)r_2 - r = r - r_1 = \delta$. The resulting pressure at the point B is the superposition from points S_1 and S_2:

$$p_B = \frac{A_0}{r_1} e^{i(kr - k\delta - \omega t)} + \frac{A_0}{r_2} e^{i(kr + k\delta - \omega t)}$$

$$\approx \frac{A_0}{r} e^{i(kr - \omega t)} [e^{ik\delta} + e^{-ik\delta}]$$

$$= \frac{2A_0}{r} e^{i(kr - \omega t)} \cos[kd \sin\theta], \quad \text{(A.3)}$$

where $r \gg d$ and

$$k = \frac{2\pi}{\lambda}. \tag{A.4}$$

The parameter δ governs the interference pattern. From (A.3), we can find nodal points where the pressure is almost zero – hence, an interference maxima. Nodal lines occur where

$$\cos(kd \sin \theta) = \cos\left(\frac{2\pi}{\lambda}(d \sin \theta)\right) = 0. \tag{A.5}$$

This leads to the relationship

$$\frac{2\pi}{\lambda}(d \sin \theta) = (2n+1)\frac{\pi}{2} \quad (n = 0, 1, 2, \ldots). \tag{A.6}$$

From (A.6) we obtain that

$$\sin \theta_n = \left(n + \frac{1}{2}\right)\frac{\lambda}{2d} \quad \text{(nodal lines)}. \tag{A.7}$$

For maximum value of interference, equation (A.5) leads to

$$\cos\left(\frac{2\pi}{\lambda}(d \sin \theta)\right) = 1; \tag{A.8}$$

this gives directions for maxima interference $\sin \theta_n = n\lambda/d$ $(n = 0, 1, 2, \ldots)$. From (A.3), we can calculate an amplitude for the arbitrary direction.

We will now visualize radiation from a circular piston. Suppose that a circular piston of radius a vibrates uniformly with some known or prescribed tone-burst displacement function. We can start with Huygens's principle and consider each element of the piston as a single source; we can then integrate over the surface because the amplitude at B is the result of superposition from all points. As in Figure A-8, we consider the resulting pressure at some point B obtained from all the elements of the piston surface.

The field due to a reflecting source can then be derived by integrating the point-source solution over the face of the source, which is actually a superposition process of all point-source contributions over the face of the ultrasonic transducer. The geometry of such a situation is shown in Figure A-8 for a cylindrical element. A numerical approximation procedure exists for obtaining a solution to the ultrasonic field variations from a piezo-electric source or reflecting element. A reflecting element could be divided into a finite number of elements of approximately equal area, as illustrated in Figure A-9.

The pressure variations generated from the arbitrarily shaped transducer element or reflecting element can then be calculated by summing the solutions (at a desired point in space) resulting from a finite number of individual solutions from each finite element on the reflecting surface. The magnitude variations can be calculated from the well-known point-source fluid solution. The phase variations, however, must be considered from a three-dimensional geometric point of view. Note that, as the size of the finite elements diminish, the resulting pressure function approaches the exact result that would be obtained by an exact mathematical integration process. Reasonable results and convergence

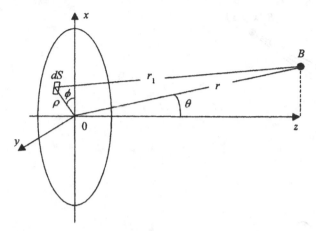

Figure A-8. Coordinate system used to derive characteristics of a flat cylindrical piston.

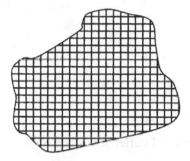

Figure A-9. Finite element grid for field computation.

to the true solution are found by selecting element sizes with dimensions less than the wavelength being considered in the field analysis problem. In fact, a general numerical approximation rule calls for wavelength λ to be greater than l, the largest linear dimension contained within the finite element. For wavelengths larger than the maximum transducer diameter, the resulting ultrasonic field solution is simply the well-known point-source spherical wave solution in the fluid.

In order to demonstrate this numerical solution process, let us briefly consider the example presented in Figure A-10. If a solution is desired at point B in three-dimensional space from the piezoelectric element shown (containing just four elements), the calculation process would proceed as follows. The resulting solution would consist of a superposition process of four spherical wave solutions. Distances traveled from the piezoelectric element to point B must be calculated by trigonometry, resulting in values for x_1, x_2, x_3, and x_4. Appropriate attenuation could then be calculated by letting the B values in the point-source solutions be equal to x_1, x_2, x_3, and x_4. Note that the resulting solution at point B would consist of summing four solutions, each solution having slightly different arrival times and amplitudes: $p(x, y, z) = p_1 + p_2 + p_3 + p_4$. The resulting solution is obtained by adding the individual solutions together for each time position. Although the solution presented is for a continuous wave excitation case, provisions for extensions into

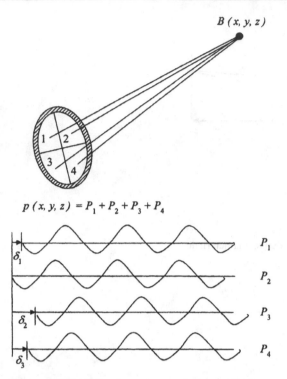

$B\,(x,\,y,\,z)$

$p\,(x,\,y,\,z) = P_1 + P_2 + P_3 + P_4$

P_1

P_2

P_3

P_4

Figure A-10. Finite element numerical procedure.

a pulse-type excitation can be made rather easily by way of Fourier series and transform analysis.

A.4　Directivity Function for a Cylindrical Element

A consideration of the coordinate system used to derive characteristics of a flat cylindrical piston (illustrated in Figure A-8) allows us to consider a more precise mathematical closed-form solution of the field emanating from a cylindrical element. The first approximation in this solution involves the representation of the actual distance between an arbitrary point on the surface of the disc source and an arbitrary point in the field.

Consider when point B is far from the piston. In this case, piston radius b is much smaller than r and r_1 ($r \gg b$, $r_1 \gg b$). The difference ($r - r_1$) approximately equals the projection value of vector $\bar{\rho}$ on the vector $\overline{OB}$. As a result, we conclude that

$$r - r_1 = \rho \sin\theta \cos\phi. \tag{A.9}$$

The pressure in the field point B (see Figure A-8) can be obtained by dividing the surface of the piston into an infinite number of elements dS. The pressure at the point B produced by element dS can be written as

$$dp = \frac{A_0\,dS}{r_1}e^{i(kr_1 - \omega t)}, \tag{A.10}$$

where r and r_1 are large compared to the piston radius. By substituting (A.9) into (A.10), we obtain the following approximation (in polar coordinates):

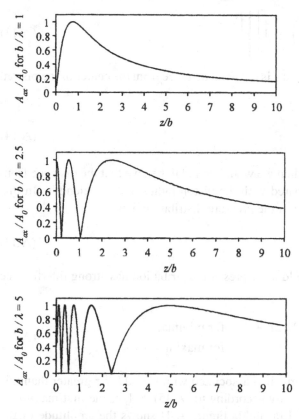

Figure A-11. Profile of the presssure distribution along the axial axis, where b is piston radius, λ is wavelength, and z is the axial distance from the piston.

$$dp = \frac{A_0}{r} e^{i(kr-\omega t)} e^{-ik\rho \sin\theta \cos\phi} \rho \, d\rho \, d\phi. \tag{A.11}$$

The total pressure is:

$$p = \frac{A_0}{r} e^{i(kr-\omega t)} \int_0^b \int_0^{2\pi} e^{-ik\rho \sin\theta \cos\phi} \rho \, d\rho \, d\phi$$

$$= 2\pi \frac{A_0}{r} e^{i(kr-\omega t)} \int_0^b J_0(k\rho \sin\theta) \rho \, d\rho$$

$$= \pi b^2 \frac{A_0}{r} e^{i(kr-\omega t)} \left[\frac{2J_1(kb \sin\theta)}{kb \sin\theta} \right]. \tag{A.12}$$

Some numerical results of the far-field pressure distribution are graphed in Figure A-11.

A detailed study of this approximation and resulting errors in field analysis is included in Rose (1975). Directivity functions for other multiple point or line source oscillators can be found in the literature; see, for example, Hueter and Bolt (1955).

The amplitude of the axial pressure distribution in the near field of the piston can be described (Kinsler et al. 1982; Pierce 1989) as:

$$A_{ax} = 2A_0 \left| \sin\left\{ \frac{1}{2}kz\left[\sqrt{1 + \left(\frac{b}{z}\right)^2} - 1 \right] \right\} \right|, \tag{A.13}$$

where A_{ax} denotes axial amplitude and z is the axial distance from the center of the piston. For $z/b \gg 1$,

$$\sqrt{1 + \left(\frac{b}{z}\right)^2} \approx 1 + \frac{1}{2}\left(\frac{b}{z}\right)^2. \tag{A.14}$$

In equation (A.11), if $z/b \gg kb$ then we would be led to believe that the distance from the observation point is large compared with the piston radius and wavelength. In this case, the asymptotic axial amplitude of the pressure distribution is

$$A_{ax} = \frac{1}{4}A_0 \cdot \frac{b}{z} \cdot kb. \tag{A.15}$$

As follows from (A.13), the near-field axial pressure distribution has strong interference phenomena. This effect occurs for

$$\pi \frac{z}{\lambda}\left[\sqrt{1 - \left(\frac{b}{z}\right)^2} - 1 \right] = \begin{cases} (\pi/2)(2n+1) & \text{for minima,} \\ \pi n & \text{for maxima,} \end{cases} \tag{A.16}$$

where $n = 0, 1, 2, \ldots$. At the first maxima associated with $n = 0$ for z greater than this value, pressure decreases monotonically according to (A.15) as $1/r$; the first maxima is therefore a border between near and far field. Figure A-11 shows the amplitude of the near-field pressure distribution calculated from equation (A.13).

A.5 Ultrasonic Field Presentations

Pressure variations in the ultrasonic field can be presented in a variety of ways. Although the distribution is three-dimensional, the practical presentation techniques are two-dimensional. Although many modified techniques are used and illustrated in the literature, two principle ones are most beneficial in the field analysis problem.

(1) *Axial pressure profile* – plots the maximum pressure of an ultrasonic waveform as a function of the axial coordinate emanating from the center line of a transducer element. The maximum pressure value is extracted as a peak-to-peak magnitude feature of the entire amplitude-versus-time profile as it passes the coordinate axis z. Sample curves are shown in Figure A-11.

(2) *Polar coordinate diffraction-type presentation* – plots maximum pressure against an angle θ. As we examine rays extending from the coordinate center point at this arbitrary angle θ, the maximum pressure value occurring at that angle can be measured along the radial coordinate, as shown in Figure A-12. This kind of presentation often produces side lobes of pressure energy that occur naturally because of constructive and destructive interference phenomena occurring in the superposition process of ultrasonic waveforms.

The characteristics of an ultrasonic transducer are caused by the interference of the point sources. For continuous wave (CW) operation, there is a zone near the source that

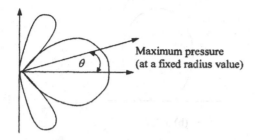

Figure A-12. Polar coordinate diffraction-type presentation.

is characterized by large variations in field intensity (near field). Beyond this, the field intensity decreases smoothly (far field). Ultrasonic field data can be presented in a variety of ways. Both axial and transverse pressure profiles were presented here in an attempt to explain parameters for the near field and the angle of divergence field as well as aspects of computation based on the two profiles.

The polar coordinate presentation is an alternate and often useful format. This diffraction pattern provides us with useful information on beam angle of divergence. Even more importantly, the side lobe information and possible zero interaction pressures with scatterers at some off-angle position can easily be seen. In the plot of Figure A-12, the center line pressure (obtained at the coordinates $z = z_0$, $\theta = 0$) is arbitrarily set to unity. From this plot, angle-of-divergence values can easily be measured to (say) a zero pressure point, a 6-dB-down pressure point, and so forth. Note that, as the frequency and transducer radius b change, the diffraction pattern changes. As a rule of thumb, the following observation can be made: increasing f or b decreases the angle of divergence and increases the number of side lobes (see Figure A-13).

A.6 Near-Field Calculations

The near-field distance of a CW-type transducer is usually defined as that point on the axis of the transducer separating a region of intense oscillations from a region of a smooth intensity decay. The location of this point is the last of several local maxima. The farthest maximum for CW excitation occurs when

$$z = \frac{b^2}{\lambda} - \frac{\lambda}{4} = \frac{D^2 - \lambda^2}{4\lambda}, \tag{A.17}$$

as illustrated earlier in this appendix. The last maximum is easily observed in Figure A-11.

It is generally assumed that λ^2 can be neglected compared to D^2, in which case the near-field location is given by

$$N = \frac{D^2}{4\lambda}. \tag{A.18}$$

A.7 Angle-of-Divergence Calculations

One commonly used characteristic of the ultrasonic field is the beam angle of divergence. This characteristic is used as an indication of the relative intensity distribution of the beam. Basic textbooks show the ultrasonic field – confined to a cylinder having the

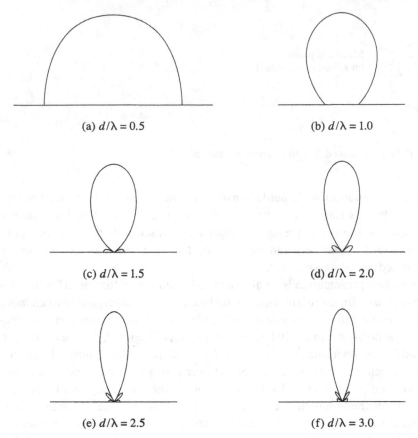

Figure A-13. Polar coordinate beam pressure profiles for circular pistons of various d/λ ratios, where $d = $ diameter and $\lambda = $ wavelength.

same diameter as the transducer – extending from the transducer face to an axial distance of one near field. From this point, the field is generally considered (from a qualitative viewpoint) as being conical, with the apex at the center of the transducer face. From a quantitative viewpoint, the half-angle of divergence to a zero pressure point, based on the calculation of the first zero of the Bessel function for CW far-field work, yields

$$\sin\alpha = \frac{0.6\lambda}{b}, \tag{A.19}$$

where b is the transducer radius.

A.8   Ultrasonic Beam Control

Ultrasonic field variations are controlled by varying transducer geometry, frequency, and size. Advanced topics on transducer design and beam control are presented in many texts. For now, basic elements of beam control can be evaluated by studying the near-field formula

$$N = \frac{D^2}{4\lambda} = \frac{D^2 f}{4c}$$

and the formula (A.19) for angle of divergence α: $\sin\alpha = 0.6\lambda/b$.

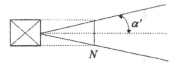

Figure A-14. Ultrasonic beam control.

A small angle of divergence is usually desirable. Examining (A.19) shows that this can be obtained by using a small wavelength (which is the same as high frequency) and a transducer with large radius b. However, the problem with this approach is that the near field increases tremendously. In order to decrease the near field, it is desirable in most applications to avoid the confusion zone of constructive and destructive interference; hence it is necessary to *decrease* transducer diameter D (radius b) as well as frequency. Thus, beam control calls for a compromise in the selection of transducers for specific applications.

A more qualitative understanding of the relationship between near field and beam angle of divergence α can be obtained by studying the simplified diagram in Figure A-14. Clearly, as N increases, α' decreases (where α' is simply the angle in the box shown from the transducer diameter to the near-field point). Note that α' and α are different: α is used in a quantitative sense and α' in a qualitative sense.

A.9 A Note on Ultrasonic Field Solution Techniques

We shall now detail a comparison of two solution techniques as an exercise to illustrate the value and utility of a directivity function. The comparison – between approximate and exact (finite element) formulations – was conducted to determine type and magnitude of the errors introduced. A transverse field distribution was generated for each technique for CW b/λ ratios of 10 and 40; these distributions were computed at several axial distances. The results, shown in Figure A-15, indicate there are two types of errors that must be considered: signal amplitude variations along the center line z, and the spatial distribution profile at particular values of z.

The amplitude error does not become negligible until a distance of approximately "two near fields" is reached. (By the way, identical results were obtained for the case of $b/\lambda = 40$ and $b = 0.259$ inches, except that the relative pressure amplitude was four times less than the $b/\lambda = 10$ case.) If one of the signals being considered is within two near fields and the other is far from the transducer, a considerable error could be introduced by using the approximate equations. If the transverse distributions are considered then we will find that – although the error in magnitude of the axial intensity does not decrease to an acceptable level until a distance of two near fields is reached – the shapes of the distribution converge within a distance of one near field. The approximate solution is not at all reasonable with respect to axial distances less than the near field of the transducer.

A.10 Pulsed Ultrasonic Field Effects

The discussion has thus far been restricted to continuous wave sources and the fields they generate. Most realistic applications of ultrasonics, however, deal with pulsed transducers. In many cases, the expected beam angle of divergence or near-field distance for

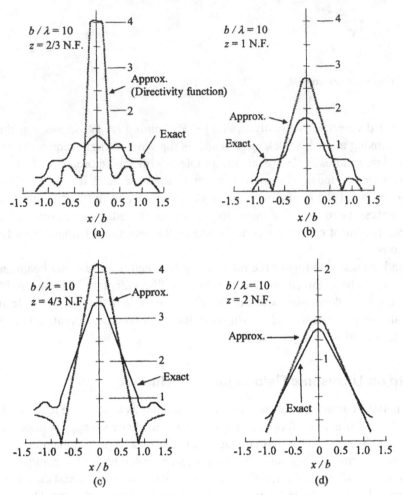

Figure A-15. Comparison of the exact and approximate pressure field distributions (N.F. denotes near-field distance).

a pulsed transducer is calculated using formulations based on a CW source. Yet this approach is flawed insofar as ultrasonic field distributions are a function of pulse shape.

Fourier analysis can enable a mathematical computation of the true beam angle of divergence. First, the Fourier components of the pulse are computed and then a traditional ultrasonic field is generated for each of the continuous wave components. The intensity distribution of the pulse is the summation of the component distributions. It must be mentioned, however, that each of the components will have a slightly different pressure profile. Although all have the same magnitude when normalized on the transducer axis, the intensity of the components having higher D/λ values will increase more rapidly as the field point is moved away from the transducer axis. This implies that the pulse shape actually changes, since the spectral composition of the pulse is changing. The extent of this change can only be determined once the spectral content of the transmitted pulse is known.

Many other differences in ultrasonic field parameters exist for broadband transducers as compared with narrowband ones, and this should be carefully considered when doing ultrasonic field work. A qualitative comparison of spatial distribution results (transverse

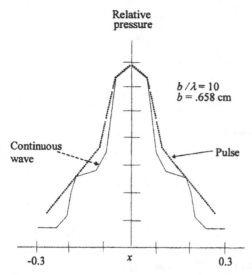

Figure A-16. Comparison of spatial distribution results for a pulse and continuous wave response function.

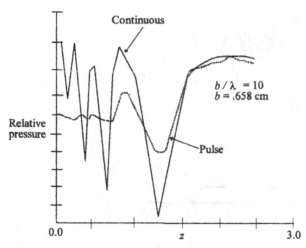

Figure A-17. Comparison of axial pressure results for a pulse and continuous wave response function.

pressure profiles) for a continuous waveform versus a typical ultrasonic pulse form is shown in Figure A-16. A relatively smooth profile is obtained for the pulse waveform. Figure A-17 compares two axial pressure profiles, where the CW response function shows severe oscillations at values less than 3.0 inches. Once again, a smoother distribution occurs for the pulse waveform. In both cases, however, the near field is at 2.75 inches, which is close to the theoretical values using center-line frequency (from the frequency profile) as the nominal frequency in the calculations.

An ultrasonic field for an arbitrarily shaped ultrasonic waveform (a pulse) can be evaluated by considering the superposition of many different ultrasonic fields, one for each individual frequency component. As indicated earlier, irregularities associated with the

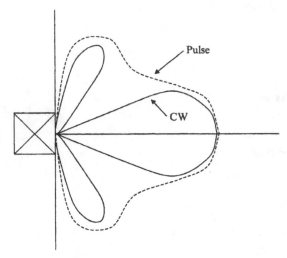

Figure A-18. Pulse effects on a diffraction pattern.

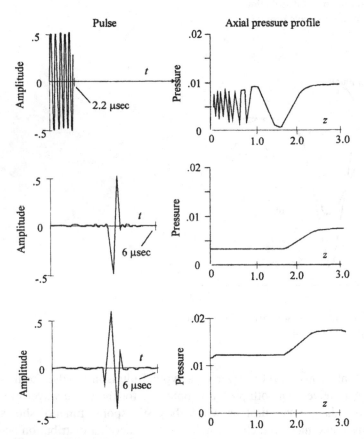

Figure A-19A. Pulse effects on axial pressure profile.

axial and transverse profiles are generally reduced substantially by a smoothing effect. A diffraction pattern is likewise smoothed, as illustrated in Figure A-18.

We have calculated axial pressure profiles for six different ultrasonic response functions; the results are shown in Figures A-19A and A-19B. In each case, D/λ was selected

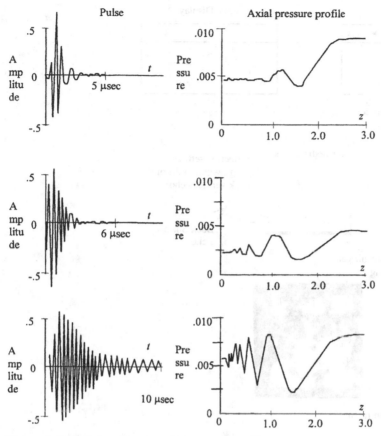

Figure A-19B. Pulse effects on axial pressure profile.

close to 20 with a center-line nominal frequency of 2.25 MHz. As the ringing decreases, oscillations in the near field decrease; as the ringing increases, results approach those of the continuous wave case. Additional details on the mathematical solution process associated with Fourier transform analysis are presented in many papers and textbooks.

A.11   Introduction to Display Technology

Here we present a brief outline of display technology as an aid to understanding the basic elements of ultrasonic analysis. The four principal scanning types may be listed as follows.

(1) *A-scan*. This is an amplitude-versus-time display, samples of which are shown in Figure A-16.

(2) *B-scan*. This represents the brightness mode. An image of a cross-section can be presented on a cathode ray tube (CRT) by keeping track of the amplitudes reflected from various portions of the cross-section (see Figure A-20). Time gain compensation (TGC) is often used to present an improved image, whereby echoes appearing from greater distances are displayed as being greater in magnitude. Experience allows us to develop a nonlinear amplification function enabling improved image resolution. This technique is particularly useful in diagnostic ultrasound.

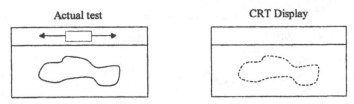

Figure A-20. B-scan concept.

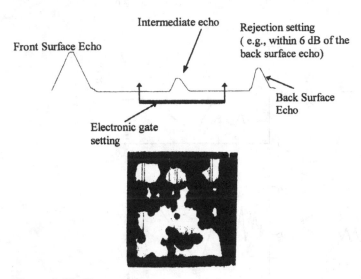

Figure A-21. C-scan testing.

(3) *M-scan*. This is a time position scan that allows us to observe moving objects. A popular example involves motion within a body, as in scanning a moving cardiac wall by plotting echo position versus time.

(4) *C-scan*. This is a planar view at a particular depth range of the defects in that plane.

A sample C-scan result is presented in Figure A-21. Two Plexiglas plates were bonded together; the white areas indicate good bonding whereas black indicates no-bond areas. In this case, the intermediate echo is rather substantial in size. Note that C-scan control maps with gray scaling can be obtained by rerunning a C-scan at different reject levels – say, at 4 dB, 6 dB, and 8 dB.

A.12   Amplitude Reduction of an Ultrasonic Waveform

We will now list the ways in which an ultrasonic waveform can be reduced in magnitude.

(1) The *reflection factor* at an interface between two materials can reduce waveform magnitude.

(2) Ultrasonic waveform magnitudes are reduced by *dispersion* due to geometric effects induced by cylindrical or plate-type structures or layered media. Such effects are actually a summation of reflection factor effects.

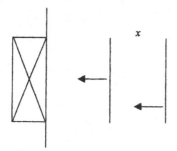

Figure A-22. Axial resolution definition.

(3) Ultrasonic *beam spreading* causes a reduction in magnitude (see Section A.5 for additional details). Scattering from a reflector is also included in beam-spread amplitude reduction, since additional beam spreading occurs as an ultrasonic wave is reflected from a scatterer or from a foreign object or structural defect inside a material.

(4) *Attenuation* due to either spherical or cylindrical wave propagation can cause reduction in amplitude. This attenuation is nonabsorbing and nondispersive; it is due to geometric decay only.

(5) Attenuation due to *absorption* mechanisms includes internal friction and internal scattering, which also reduce wave magnitudes.

(6) Electronic *processing* (e.g., filtering and amplification) can also cause amplitude reduction.

A.13 Resolution and Penetration Principles

The resolution and penetrating power of an ultrasonic wave depends on the wavelength of excitation inside the material in question. Greater wavelengths or lower frequencies generally penetrate much further into a material and result in less absorption. Higher-frequency ultrasonic excitations with smaller wavelengths generally decay more rapidly inside a material, but resolution capability is improved. Now, we will address some basic definitions with respect to axial and lateral resolution of an ultrasonic transducer.

A.13.1 Axial Resolution

The smallest axial distance that can be resolved between two axially located reflections is called the *axial resolution* of the transducer; see Figure A-22. For example, axial resolution is the smallest distance before superposition of the incident and reflected echoes makes an axial separation measurement impossible. A formula for axial resolution can be derived as a function of pulse duration. Axial resolution is defined as the capability of separating two point reflectors in an axial direction from the ultrasonic transducer. The best possible axial resolution can therefore come about by using as short a pulse as possible. This is illustrated in Figure A-23, where the best axial resolution is achieved for a high-frequency and highly damped ultrasonic transducer.

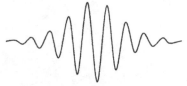

(a) Low frequency, poorly damped

(b) High frequency, poorly damped

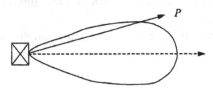

(c) High frequency, highly damped

Figure A-23. Axial resolution improvement.

(a) Poor lateral resolution potential

(b) Good lateral resolution potential

Figure A-24. Lateral resolution improvement.

A.13.2   Lateral Resolution

We now move to lateral resolution improvement. *Lateral resolution* is defined as the ability to resolve two point reflectors in a lateral direction at some specific point inside the structure. It is clear that the narrowest possible ultrasonic beam should be used. Two polar coordinate profiles at a specified radius showing pressure versus angle are illustrated in Figure A-24. The narrow beam can be achieved by using a transducer of larger diameter or a higher frequency, or by using a focused transducer. Lateral resolution will vary with depth because of ultrasonic beam divergence. The best lateral resolution will be achieved at an axial distance that is equal to either the near-field point or the design focal point of the ultrasonic transducer.

A.14 Exercises

1. Illustrate schematically the approximate qualitative change in a typical waveform as it travels from one position to another, showing (a) dispersion due only to structure for a typical $c(f)$ curve, (b) attenuation as a function of frequency only, and (c) both dispersion and attenuation.

2. In a wave-scattering Huygens's principle exercise, how could you tell the difference between a flat reflector and a volumetric reflector? Also, how might you determine the size of each?

3. Calculate wavelength in steel and in water for a 1-inch transducer at 2.25 MHz.

4. Compute the near field and angle of divergence for a 1-inch-diameter transducer of 2.25 MHz in steel and in water.

5. How can you focus an ultrasonic transducer inside a fluid or solid material? Describe the principle of focusing and several ways of achieving focus.

6. When using a finite element grid for field computation, how would you determine the mesh size?

7. Describe how to compute or find directivity functions in acoustics for transducers shaped other than as a circular disk.

8. Explain the axial pressure profile smoothing process as a function of pulse frequency bandwidth.

9. Calculate axial resolution as a function of pulse duration.

10. Show how the zeros of the Bessel function could be used to estimate beam angle of divergence for a cylindrical piston source.

11. From the literature, list some directivity functions for different shaped transducers.

APPENDIX B

Basic Formulas and Concepts in the Theory of Elasticity

B.1 Introduction

Elements of elasticity theory are fundamental to the studies of wave propagation in solid media. A brief review of the formulas and concepts are presented in this appendix for reference purposes. Many textbooks are available that cover the theory of elasticity, including such classic texts as Chou and Pagano (1992), Sokolnikoff (1956), and Timoshenko and Goodier (1987).

B.2 Nomenclature

Principal formulas and definitions may be listed as follows.

A = transformation matrix, E = Young's modulus,

F = body force, K = bulk modulus,

T = traction force, U = strain energy/volume,

u_i = displacement component, $\mu = G$ = shear modulus,

ε_{ij} = strain tensor, σ_{ij} or τ_{ij} = stress tensor,

ϕ = strain invariant (dilatation), θ = stress invariant,

ρ = density, τ = stress matrix,

v = Poisson's ratio, λ = Lamé constant = $\dfrac{2\mu v}{1 - 2v}$,

δ_{ij} = Kronecker delta = $\begin{cases} 1 & \text{if } i = j, \\ 0 & \text{if } i \neq j; \end{cases}$

u_i is the deformation in the x_i direction.

For the alternating tensor,

$$\epsilon_{ijk} = \begin{cases} +1 & \text{all unequal, in cyclic order,} \\ -1 & \text{unequal, not in cyclic order,} \\ 0 & \text{any two indices equal.} \end{cases}$$

The strain tensor is

$$\varepsilon_{ij} = \frac{1}{2}\left(\frac{\partial u_i}{\partial x_j} + \frac{\partial u_j}{\partial x_i}\right) = \varepsilon_{ji},$$

and the rotation tensor is

$$-\omega_k = \omega_{ij} = -\omega_{ji} = \frac{1}{2}\left(\frac{\partial u_i}{\partial x_j} - \frac{\partial u_j}{\partial x_i}\right);$$

the dilatation is

$$\phi = \varepsilon_{ii} = \frac{\partial u_i}{\partial x_i} = \varepsilon_{11} + \varepsilon_{22} + \varepsilon_{33}.$$

Cartesian coordinate vector operators are

$$\nabla = \frac{\partial}{\partial x}\hat{i} + \frac{\partial}{\partial y}\hat{j} + \frac{\partial}{\partial z}\hat{k} \quad \text{(gradient)},$$

$$\nabla^2 = \frac{\partial^2}{\partial^2 x} + \frac{\partial^2}{\partial^2 y} + \frac{\partial^2}{\partial^2 z} \quad \text{(Laplacian)};$$

cylindrical coordinate vector operators are

$$\nabla = \frac{\partial}{\partial r}\hat{e}_r + \frac{1}{r}\frac{\partial}{\partial \theta}\hat{e}_\theta + \frac{\partial}{\partial z}\hat{e}_z \quad \text{(gradient)},$$

$$\nabla^2 = \frac{\partial^2}{\partial^2 r} + \frac{1}{r}\frac{\partial}{\partial r} + \frac{1}{r^2}\frac{\partial^2}{\partial \theta^2} + \frac{\partial^2}{\partial z^2} \quad \text{(Laplacian)}.$$

In cylindrical coordinates, u_r, u_θ, and u_z are deformations in the r, θ, and z directions (respectively), and strain tensor components are

$$\varepsilon_{rr} = \frac{\partial u_r}{\partial r},$$

$$\varepsilon_{\theta\theta} = \frac{1}{r}\frac{\partial u_\theta}{\partial \theta} + \frac{u_r}{r},$$

$$\varepsilon_{zz} = \frac{\partial u_z}{\partial z},$$

$$\varepsilon_{r\theta} = \frac{1}{2}\left(\frac{1}{r}\frac{\partial u_r}{\partial \theta} + \frac{\partial u_\theta}{\partial \theta} - \frac{u_\theta}{r}\right),$$

$$\varepsilon_{rz} = \frac{1}{2}\left(\frac{\partial u_z}{\partial r} + \frac{\partial u_r}{\partial z}\right),$$

$$\varepsilon_{\theta z} = \frac{1}{2}\left(\frac{\partial u_\theta}{\partial z} + \frac{1}{r}\frac{\partial u_z}{\partial \theta}\right).$$

The stress tensor τ_{ij} is a symmetric tensor that can be written as

$$\tau_{ij} = \tau_{ji} = \begin{bmatrix} \tau_{11} & \tau_{12} & \tau_{13} \\ \tau_{21} & \tau_{22} & \tau_{23} \\ \tau_{31} & \tau_{32} & \tau_{33} \end{bmatrix}$$

(see Figure B-1). Note also that, in Cartesian coordinates, $\tau_{11} = \tau_{xx}$ and so forth (i.e., indices 1, 2, 3 map to axes x, y, z, resp.). The strain tensor ε_{ij} is also symmetric and can be written as

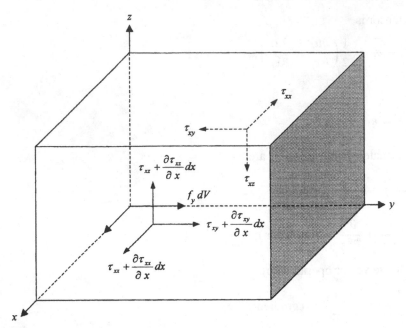

Figure B-1. Differential three-dimensional element. (A surface is denoted by the axis to which it is perpendicular. The stresses shown are on the positive surfaces. On the opposite or negative surfaces, the stresses are in the opposite directions.)

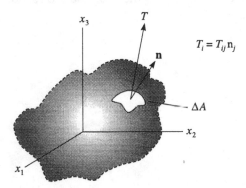

$$T_i = T_{ij}\,\mathrm{n}_j$$

Figure B-2. Differential surface area element (**n** is the normal unit vector to the area ΔA; T is not in the direction of **n**).

$$\varepsilon_{ij} = \begin{bmatrix} \varepsilon_{11} & \varepsilon_{12} & \varepsilon_{13} \\ \varepsilon_{21} & \varepsilon_{22} & \varepsilon_{23} \\ \varepsilon_{31} & \varepsilon_{32} & \varepsilon_{33} \end{bmatrix}.$$

Finally, the dot product $\bar{a} \bullet \bar{b} = a_i b_i$; the cross product $\bar{a} \times \bar{b} = \bar{c}$ ($C_i = \epsilon_{ijk} a_j b_k$).

B.3   Stress, Strain, and Constitutive Equations

Consider the application of stress on a differential element, as in Figure B-1. The relationship between traction and stress is $T_i = \tau_{ij} n_j$ (see Figure B-2). A general stress–strain relationship for anisotropic media is

$$\tau_{ij} = \lambda \delta_{ij} \phi + 2\mu \varepsilon_{ij} = \frac{E\nu\phi}{(1+\nu)(1-2\nu)} \delta_{ij} + \frac{E}{1+\nu} \varepsilon_{ij},$$

where $\tau_{ij} = C_{ijkl}\varepsilon_{kl}$ (Hooke's law) and dilatation $\phi = \varepsilon_{11} + \varepsilon_{22} + \varepsilon_{33}$. In terms of strain we have

$$\varepsilon_{ij} = -\frac{\lambda}{2\mu(3\lambda + 2\mu)} \delta_{ij}\theta + \frac{1}{2\mu} \tau_{ij} = \frac{1+\nu}{E} \tau_{ij} - \frac{\nu}{E} \theta \delta_{ij},$$

where $\theta = \tau_{11} + \tau_{22} + \tau_{33}$ is the stress invariant and $\tau_{ik} = 2G\varepsilon_{ik} + \lambda\theta\delta_{ik}$. The equilibrium equations of elasticity can be written as

$$\frac{\partial \tau_{ij}}{\partial x_j} + F_i = 0.$$

B.4   Elastic Constant Relationships

Also note the following elastic constant relationships for an isotropic material:

$$E = \frac{\mu(3\lambda + 2\mu)}{\lambda + \mu} = \frac{\lambda(1+\nu)(1-2\nu)}{\nu},$$

$$\mu = \frac{E}{2(1+\nu)}, \quad \lambda = \frac{\nu E}{(1+\nu)(1-2\nu)}, \quad \nu = \frac{\lambda}{2(\lambda+\mu)}.$$

The strain energy is

$$U = \tfrac{1}{2}(\lambda + 2\mu)[\varepsilon_{xx}^2 + \varepsilon_{yy}^2 + \varepsilon_{zz}^2]$$
$$+ 2\mu[\varepsilon_{yx}^2 + \varepsilon_{xz}^2 + \varepsilon_{yz}^2] + \lambda[\varepsilon_{xx}\varepsilon_{yy} + \varepsilon_{zz}\varepsilon_{xx} + \varepsilon_{zz}\varepsilon_{yy}].$$

In terms of stress,

$$U = \frac{1}{2E}(\tau_{xx}^2 + \tau_{yy}^2 + \tau_{zz}^2) - \frac{\nu}{E}(\tau_{xx}\tau_{yy} + \tau_{yy}\tau_{zz} + \tau_{xx}\tau_{zz}) + \frac{1}{2\mu}(\tau_{xy}^2 + \tau_{xz}^2 + \tau_{yz}^2).$$

B.5   Vector and Tensor Transformation

We now review briefly the particulars of vector and tensor transformation for rotating coordinate systems. Consider A as an orthogonal matrix of direction cosines between two different coordinate systems, primed and unprimed.

For a first-order tensor, $x_i' = a_{ij}x_j$, where $a_{ij} = \cos(x_i', x_j)$. Note that i is a free index and that the js appearing twice on the RHS of the equation indicate summation according to the Einstein summation rule. In vector notation, $\bar{x}' = A\bar{x}$, where

$$A = \begin{array}{c} \\ x' \\ y' \\ z' \end{array} \begin{array}{ccc} x & y & z \\ \left[\begin{array}{ccc} a_{11} & a_{12} & a_{13} \\ a_{21} & a_{22} & a_{23} \\ a_{31} & a_{32} & a_{33} \end{array}\right] \end{array}$$

and also (if needed) $x_i = a_{ji}x_j'$.

For a second-order tensor, $\tau_{ik}' = a_{ij}a_{kl}\tau_{jl}$ or $\tau' = A\tau A^{-1}$ for matrices in a similarity transformation.

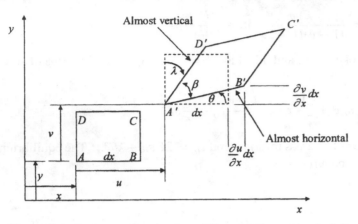

Figure B-3. Strain consideration for a differential element in elasticity.

B.6 Principal Stresses and Strains

We now consider the eigenvalue problem for principal stress and principal strain computation. The principal stresses are given by

$$|\tau_{ij} - \tau\delta_{ij}| = 0$$

and the vectors A_j by

$$|\tau_{ij} - \tau\delta_{ij}|[A_j] = 0.$$

The principal strains are given by the determinant equation

$$|\varepsilon_{ij} - \varepsilon\delta_{ij}| = 0,$$

and the eigenvectors A_j are given by

$$|\varepsilon_{ij} - \varepsilon\delta_{ij}|[A_j] = 0.$$

B.7 The Strain Displacement Equations

In this section we examine the derivation of the strain displacement equations

$$\varepsilon = [\underbrace{\varepsilon_x, \varepsilon_y, \varepsilon_z}_{\substack{\text{normal} \\ \text{strains}}}, \underbrace{\gamma_{yz}, \gamma_{xz}, \gamma_{xy}}_{\substack{\text{engineering} \\ \text{shear strain}}}]^T.$$

For infinitesimally small deformations, consider translation and deformation of a two-dimensional element, as shown in Figure B-3. Note that

$$\varepsilon_x = \frac{A'B' - AB}{AB} = \frac{A'B' - dx}{dx}, \tag{$*$}$$

$$\varepsilon_y = \frac{A'D' - AD}{AD} = \frac{A'D' - dy}{dy},$$

$$\gamma_{xy} = \frac{\pi}{2} - \beta = \theta - \lambda \quad \text{(since counterclockwise is positive)}.$$

Point B is displaced by

$$u + \frac{\partial u}{\partial x} dx, \qquad v + \frac{\partial v}{\partial x} dx;$$

point D is displaced by

$$u + \frac{\partial u}{\partial y} dy, \qquad v + \frac{\partial v}{\partial y} dy.$$

Therefore, in trying to solve for ε_x, we need $A'B'$. From $(*)$ we have

$$(A'B')^2 = [dx(1 + \varepsilon_x)]^2.$$

Separately,

$$(A'B')^2 = \left(dx + \frac{\partial u}{dx} dx \right)^2 + \left(\frac{\partial v}{dx} dx \right)^2.$$

Therefore, $(A'B')^2 = (A'B')^2$ implies that

$$\varepsilon_x^2 + 2\varepsilon_x + 1 = 1 + 2\frac{\partial u}{\partial x} + \left(\frac{\partial u}{\partial x} \right)^2 + \left(\frac{\partial v}{\partial x} \right)^2.$$

For small strain and displacement, note that

$$\left(\frac{\partial u}{\partial x} \right)^2 \to 0 \quad \text{and} \quad \left(\frac{\partial v}{\partial x} \right)^2 \to 0$$

and that second-order effects are zero. Hence

$$\varepsilon_x = \frac{\partial u}{\partial x}$$

and, similarly,

$$\varepsilon_y = \frac{\partial v}{\partial y}, \qquad \varepsilon_z = \frac{\partial w}{\partial z}.$$

Now consider engineering shear strain:

$$\tan \theta \approx \theta = \frac{\dfrac{\partial v}{\partial x} dx}{dx + \left(\dfrac{\partial u}{\partial x} \right) dx}$$

$$= \frac{\dfrac{\partial v}{\partial x} dx}{dx \left(1 + \dfrac{\partial u}{\partial x} \right)},$$

where $\partial u/\partial x$ is small when compared with unity. Therefore,

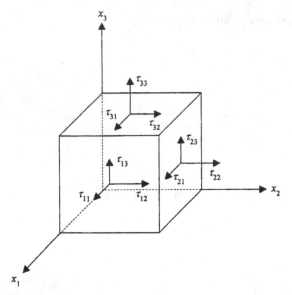

Figure B-4. Differential element used to derive the equilibrium equation and the governing three-dimensional wave equation.

$$\theta = \frac{\partial v}{\partial x}.$$

Similarly,

$$\lambda = \frac{-\partial u}{\partial y} = -\tan \lambda = \frac{\dfrac{\partial u}{\partial y}\, dy}{dy \left(1 + \dfrac{\partial v}{\partial y}\right)}$$

and so

$$\gamma_{xy} = \theta - \lambda = \frac{\partial u}{\partial y} + \frac{\partial v}{\partial x}.$$

These are the well-known strain displacement equations from elasticity. Note the linear approximations and small strain assumptions.

B.8 Derivation of the Governing Wave Equation

In light of the aspects of elasticity discussed so far, we may now consider how to derive Navier's governing wave equation. Stresses are shown on each face in Figure B-4, which depicts a cubic solid (six sides formed by three sets of parallel planes). The sum of the stresses is zero for the static case in (x, y, z)-coordinates, or equal to ma (from Newton's second law). Looking at the x direction, we see that $\tau_{xy} = \tau_{yx}$, $\tau_{yz} = \tau_{zy}$, and so forth. Note that $\sigma_x = \tau_{xx}$. For the static case we have

$$\sigma_x + \frac{\partial \sigma_x}{\partial x}\, dx - \sigma_x - \tau_{yx} + \tau_{yx} + \frac{\partial \tau_{yx}}{\partial y}\, dy - \tau_{zx} + \tau_{zx} + \frac{\partial \tau_{zx}}{\partial z}\, dz + f_x\, dV = 0.$$

Therefore,

$$\frac{\partial \sigma_x}{\partial x} + \frac{\partial \tau_{xy}}{\partial y} + \frac{\partial \tau_{xz}}{\partial z} + f_x = 0 \quad \{x, y, z\}$$

(here, $\{x, y, z\}$ signifies the cyclic permutations required to obtain three equations). These are the equilibrium equations of elasticity.

Assuming a free index i and summing over j, we may write

$$\sigma_{ij,j} = 0, \qquad \sigma_{11,1} + \sigma_{12,2} + \sigma_{13,3} = 0$$

for $i = 1, 2, 3$. We can also let 1, 2, 3 become x, y, z as follows:

$$\partial \frac{\tau_{xx}}{\partial x} + \partial \frac{\tau_{xy}}{\partial y} + \partial \frac{\tau_{xz}}{\partial z} = 0.$$

Given $F = ma$ for motion, it follows that $\sigma_{ij,j} = \rho \ddot{u}_i$ for $i, j = 1, 2, 3$.

B.9   Anisotropic Elastic Constants

Consider the following general linearly elastic constitutive equation for anisotropic elastic constants: $\sigma_{ij} = C_{ijkl} \varepsilon_{kl}$. The term C_{ijkl} is a rank-4 tensor with $3^4 = 81$ components; however, because of symmetry, only 21 coefficients are needed to describe a general anisotropic body. Stress and strain are also symmetric. The four subscripts of C_{ijkl} can be reduced to two as follows.

For the tensor notation: 11 22 33 23, 32 31, 13 12, 21

Use the abbreviated matrix notation: 1   2   3   4   5   6

This allows us to consider the simple matrix multiplication $\tau = C\varepsilon$, where τ is a 6×1 matrix, C is 6×6, and ε is 6×1:

$$\begin{bmatrix} \tau_{11} \\ \tau_{22} \\ \tau_{33} \\ \tau_{12} \\ \tau_{23} \\ \tau_{13} \end{bmatrix} = \begin{bmatrix} C_{11} & C_{12} & \cdot & \cdot & \cdot & C_{16} \\ C_{21} & C_{22} & & & & \\ \cdot & & \cdot & & & \\ \cdot & & & \cdot & & \\ \cdot & & & & \cdot & \\ C_{61} & & & & & C_{66} \end{bmatrix} \begin{bmatrix} \varepsilon_{11} \\ \varepsilon_{22} \\ \varepsilon_{33} \\ \varepsilon_{12} \\ \varepsilon_{23} \\ \varepsilon_{13} \end{bmatrix}$$

The number of constants for various types of anisotropic materials may be listed as follows:

- 21 constants for triclinic materials;
- 13 constants for monoclinic materials;
- 9 constants for orthorhombic or orthotropic materials;
- 7 or 6 constants for tetragonal materials;
- 7 constants for trigonal materials;
- 5 constants for hexagonal or transversely isotropic materials;
- 3 constants for cubic materials;
- 2 constants for isotropic materials.

We now summarize the elastic constant stiffness coefficient matrices for a few selected anisotropic materials. For more detail on anisotropy, see Auld (1990) or Pollard (1977).

Triclinic: 21 constants

$$
\begin{matrix}
C_{11} & C_{12} & C_{13} & C_{14} & C_{15} & C_{16} \\
 & C_{22} & C_{23} & C_{24} & C_{25} & C_{26} \\
 & & C_{33} & C_{34} & C_{35} & C_{36} \\
 & & & C_{44} & C_{45} & C_{46} \\
 & & & & C_{55} & C_{56} \\
 & & & & & C_{66}
\end{matrix}
$$

Monoclinic: 13 constants (standard orientation)

$$
\begin{matrix}
C_{11} & C_{12} & C_{13} & 0 & C_{15} & 0 \\
 & C_{22} & C_{23} & 0 & C_{25} & 0 \\
 & & C_{33} & 0 & C_{35} & 0 \\
 & & & C_{44} & 0 & C_{46} \\
 & & & & C_{55} & 0 \\
 & & & & & C_{66}
\end{matrix}
$$

Orthorhombic: 9 constants

$$
\begin{matrix}
C_{11} & C_{12} & C_{13} & 0 & 0 & 0 \\
 & C_{22} & C_{23} & 0 & 0 & 0 \\
 & & C_{33} & 0 & 0 & 0 \\
 & & & C_{44} & 0 & 0 \\
 & & & & C_{55} & 0 \\
 & & & & & C_{66}
\end{matrix}
$$

Tetragonal: 7 constants   6 constants

$$
\begin{matrix}
C_{11} & C_{12} & C_{13} & 0 & 0 & C_{16} \\
 & C_{11} & C_{13} & 0 & 0 & -C_{16} \\
 & & C_{33} & 0 & 0 & 0 \\
 & & & C_{44} & 0 & 0 \\
 & & & & C_{55} & 0 \\
 & & & & & C_{66}
\end{matrix}
\qquad
\begin{matrix}
C_{11} & C_{12} & C_{13} & 0 & 0 & 0 \\
 & C_{11} & C_{13} & 0 & 0 & 0 \\
 & & C_{33} & 0 & 0 & 0 \\
 & & & C_{44} & 0 & 0 \\
 & & & & C_{55} & 0 \\
 & & & & & C_{66}
\end{matrix}
$$

Trigonal: 7 constants

$$
\begin{matrix}
C_{11} & C_{12} & C_{13} & C_{14} & -C_{25} & 0 \\
 & C_{11} & C_{13} & -C_{14} & C_{25} & 0 \\
 & & C_{33} & 0 & 0 & 0 \\
 & & & C_{44} & 0 & C_{25} \\
 & & & & C_{44} & C_{14} \\
 & & & & & \tfrac{1}{2}(C_{11} - C_{12})
\end{matrix}
$$

Hexagonal: 5 constants

$$
\begin{array}{cccccc}
C_{11} & C_{12} & C_{13} & 0 & 0 & 0 \\
 & C_{11} & C_{13} & 0 & 0 & 0 \\
 & & C_{33} & 0 & 0 & 0 \\
 & & & C_{44} & 0 & 0 \\
 & & & & C_{44} & 0 \\
 & & & & & \frac{1}{2}(C_{11} - C_{12})
\end{array}
$$

Cubic: 3 constants

$$
\begin{array}{cccccc}
C_{11} & C_{12} & C_{12} & 0 & 0 & 0 \\
 & C_{11} & C_{12} & 0 & 0 & 0 \\
 & & C_{11} & 0 & 0 & 0 \\
 & & & C_{44} & 0 & 0 \\
 & & & & C_{44} & 0 \\
 & & & & & C_{44}
\end{array}
$$

Isotropic: 2 constants

$$
\begin{array}{cccccc}
C_{11} & C_{12} & C_{12} & 0 & 0 & 0 \\
 & C_{11} & C_{12} & 0 & 0 & 0 \\
 & & C_{11} & 0 & 0 & 0 \\
 & & & \frac{1}{2}(C_{11} - C_{12}) & 0 & 0 \\
 & & & & \frac{1}{2}(C_{11} - C_{12}) & 0 \\
 & & & & & \frac{1}{2}(C_{11} - C_{12})
\end{array}
$$

The anisotropic factor A is commonly used to evaluate the level of anisotropy:

$$
A = \frac{2C_{44}}{C_{11} - C_{12}};
$$

if the material is isotropic then $A = 1$.

A number of effective modulus theories are available to reduce an inhomogeneous multilayered composite material to a single homogeneous anisotropic layer for wave propagation and strength considerations. See Chapter 17 for more detail.

APPENDIX C

Complex Variables

C.1 Introduction

Topics from advanced mathematics are essential if one is to master the subject of stress waves in solids. The subject of complex variables is particularly important. As a result, a summary of key concepts is outlined in this appendix; many excellent reference textbooks are also available.

Certain solutions of algebraic equations led to the creation of imaginary numbers. Consider, for example, the following equation:

$$2x^2 + 5x - 4 = 0.$$

It is well known that the solution of this equation can be written down directly by use of the quadratic formula

$$x = \frac{-b \pm \sqrt{b^2 - 4ac}}{2a},$$

which gives the solution:

$$x = \frac{5}{4} \pm \frac{\sqrt{-7}}{4}.$$

The problem then arises of how to handle the -7 under the radical sign. The artifice of the imaginary number was created for such purposes. An imaginary number is defined as a number i such that $i^2 = -1$. Using this definition, it follows that:

$$\sqrt{-7} = \sqrt{i^2 7} = i\sqrt{7},$$

and so the solution to the previous equation could be rewritten as

$$x = \frac{5}{4} \pm i\frac{\sqrt{7}}{4}.$$

When the imaginary number was combined with real numbers (to create, e.g., such numbers as $5i$ and $-2i$), it was found that such a system of numbers was closed under multiplication and division but not under addition, subtraction, or the extraction of roots. *Closed* means that the same type of numbers should result from a mathematical operation; in the case of mixing real and imaginary numbers, an operation should result in a number that is either real or imaginary.

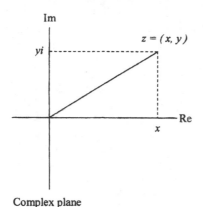

Complex plane

Figure C-1. Rectangular representation
of the complex number $z = x + iy$.

It was easily seen that multiplication and division resulted in either real or imaginary
numbers. For instance,

$$5 \times 6 = 30, \quad 2i \times -3 = -6i, \quad 4i \times 2i = 8i^2 = -8,$$

and so on; division is likewise closed. But when addition or subtraction were attempted,
the result was neither real nor imaginary. For example, $3 + 4i = ?$. Thus, the complex
number was created to enable a number system that was closed under all operations.

C.2 Complex Numbers

A complex number is defined as an ordered pair (x, y) of two real numbers x and y,
where the x is the "real" part and the y is the "imaginary" part. A complex number is
typically denoted by the letter z; $z = (x, y)$. The operations of addition, subtraction, multiplication, division, and root extraction are defined such that this system of numbers is
closed (result in other complex numbers) under all these operations, which will be defined
shortly.

The complex number z can be represented as $z = x + iy$. This is the rectangular representation (other representations are described in Section C.2.3). With this representation,
a plane – called the complex plane or z-plane – is defined as shown in Figure C-1. In the
figure, "Im" means imaginary and denotes the imaginary axis; "Re" means real and des-
ignates the real axis. These symbols are also used in an operational sense such that

$$\text{Re}\,z = x \quad \text{and} \quad \text{Im}\,z = y.$$

The operation Re (or Im) becomes particularly important for physical interpretation of
wave propagation problems formulated in terms of complex variables.

The reflection of a complex number about the real axis is called its *complex conjugate*
and is typically denoted z^* (a bar over z is also sometimes used). See Figure C-2. The
conjugate is obtained simply by reversing the sign of the imaginary part of a complex
number: if $z = x + iy$ then $z^* = x - iy$.

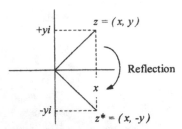

Figure C-2. Illustration of the complex conjugate z^* of a complex number z.

C.2.1 Elementary Operations

For two complex numbers z_1 and z_2 with $z_1 = x_1 + iy_1$ and $z_2 = x_2 + iy_2$, the basic operations are defined as follows.

Addition:

$$z_1 + z_2 = (x_1 + x_2) + i(y_1 + y_2).$$

Subtraction:

$$z_1 - z_2 = (x_1 - x_2) + i(y_1 - y_2).$$

Multiplication:

$$z_1 z_2 = (x_1 + iy_1)(x_2 + iy_2)$$
$$= x_1 x_2 + x_1 iy_2 + iy_1 x_2 + iy_1 iy_2$$
$$= (x_1 x_2 - y_1 y_2) + i(x_1 y_2 + y_1 x_2).$$

Division:

$$\frac{z_1}{z_2} = \frac{z_1 z_2^*}{z_2 z_2^*} = \frac{(x_1 + iy_1)(x_2 - iy_2)}{(x_2 + iy_2)(x_2 - iy_2)} = \frac{(x_1 x_2 + y_1 y_2)}{(x_2^2 + y_2^2)} + i\frac{(y_1 x_2 - x_1 y_2)}{(x_2^2 + y_2^2)},$$

provided $(x_2^2 + y_2^2)$ is not 0.

C.2.2 Important Properties

Inequalities: $z_1 > z_2$ or $z_1 < z_2$ have no meaning unless both z_1 and z_2 are real.

Signs: There are no positive or negative complex numbers.

Absolute value: For $z = x + iy$, the absolute value of z (denoted by $|z|$) is given by

$$|z| = \sqrt{x^2 + y^2};$$

$|z|$ is also known as the *modulus* or the *magnitude* of z. Since $zz^* = x^2 + y^2$, this means $zz^* = |z|^2$ or, equivalently, $|z| = \sqrt{zz^*}$. Thus we have

$$z + z^* = 2\,\mathrm{Re}\,z, \qquad z - z^* = 2\,\mathrm{Im}\,z,$$
$$(z_1 + z_2)^* = z_1^* + z_2^*, \qquad (z_1 z_2)^* = z_1^* z_2^*.$$

Geometry: Complex numbers obey the algebra of two-dimensional vector algebra; $|z_1 - z_2|$ is the distance between z_1 and z_2. The *dot product* is

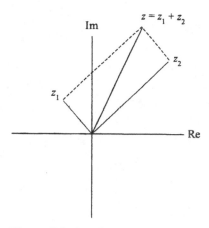

Figure C-3. Parallelogram law for complex numbers.

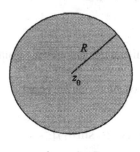

$$|z - z_0| < R$$

Figure C-4. R-neighborhood of z_0.

$$(z_1 \bullet z_2) = \mathrm{Re}(z_1^* z_2) = \mathrm{Re}(z_1 z_2^*),$$

and the *cross product* is

$$[z_1 \times z_2] = \mathrm{Im}(z_1^* z_2) = -\mathrm{Im}(z_1 z_2^*).$$

As shown in Figure C-3, $z = z_1 + z_2$ obeys the *parallelogram law*. Finally, we have the *triangle inequalities:*

$$|z_1 + z_2| \le |z_1| + |z_2| \quad \text{and} \quad |z_1 - z_2| \ge ||z_1| - |z_2||.$$

When dealing with complex numbers, the expression $|z - z_0| < R$ is often encountered. This expression defines a circular area of radius R centered at the point z_0. In Figure C-4, the shaded area is called the *R-neighborhood* of z_0. As an example, $|f(z) - w_0| < \delta$ would mean $f(z)$ is in the δ-neighborhood of w_0.

C.2.3 Other Representations

There are two other commonly used representations of complex numbers, the trigonometric and the polar representation. We can easily identify the basis for the *trigonometric representation* by referring to Figure C-5. The complex number z can be written as:

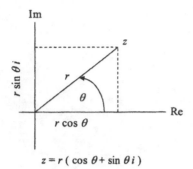

$$z = r(\cos\theta + \sin\theta\, i)$$

Figure C-5. Trigonometric representation
of a complex number z.

$$z = x + iy = r\cos\theta + r\sin\theta i = r(\cos\theta + i\sin\theta).$$

The variables r and θ are related to the real variables x and y as follows:

$$r = \sqrt{x^2 + y^2}, \qquad \theta = \tan^{-1}(y/x).$$

Note that $r = |z|$. The angle θ is called the *argument* of z (although the terms "polar angle" and "phase" are also used).

In this representation, when two complex numbers $z_1 = r_1(\cos\theta_1 + i\sin\theta_1)$ and $z_2 = r_2(\cos\theta_2 + i\sin\theta_2)$ are multiplied together, the result is

$$z_1 z_2 = r_1 r_2[(\cos\theta_1\cos\theta_2 - \sin\theta_1\sin\theta_2) + i(\cos\theta_1\sin\theta_2 + \cos\theta_2\sin\theta_1)].$$

Using a well-known trigonometric identity, this becomes

$$z_1 z_2 = r_1 r_2[\cos(\theta_1 + \theta_2) + i\sin(\theta_1 + \theta_2)].$$

To ensure that the resulting product is uniquely defined, the angle $\phi = (\theta_1 + \theta_2)$ is either reduced by 2π or increased by 2π so that $-\pi < \phi \le \pi$. If z_1 and z_2 were, say, equal to z (i.e., if $z = r(\cos\theta + i\sin\theta)$, then z^2 would be $z^2 = r^2(\cos 2\theta + i\sin 2\theta)$. By induction, this result is easily generalized to

$$z^n = R(\cos\phi + i\sin\phi),$$

where $R = r^n$ and $\phi = n\theta \pm 2\pi k$ (k being selected such that $-\pi < \phi \le \pi$). This last condition is necessary because $\cos(\theta \pm 2\pi) = \cos\theta$ and $\sin(\theta \pm 2\pi) = \sin\theta$ yet a unique value for z^n is desired. For example, if $\theta = 87°$ and $n = 5$ then $n\theta$ would be $435°$. Hence "minus" would be selected and the value of $k = 1$ used; then ϕ would be $(435° - 360°) = 75°$. This result is called De Moivre's formula.

With De Moivre's formula, it is possible to calculate the roots of complex numbers. If $w = r^{1/n}(\cos\theta/n + i\sin\theta/n)$ then w is the nth root of $z = r(\cos\theta + i\sin\theta)$ since, using De Moivre's formula, $w^n = z$. However, there are additional roots due to the periodicity of the sine and cosine functions. By using an integer k, all the n roots of a complex number z can be written as

$$w_k = r^{1/n}(\cos(\theta + 2\pi k)/n + i\sin(\theta + 2\pi k)/n) \quad \text{for } k = 0, 1, 2, \ldots, (n-1).$$

When $k = 0$, the root w_0 is called the *principal root* of z.

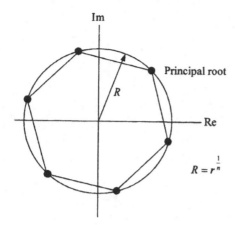

Figure C-6. Complex roots of $z^{1/n}$ ($n = 6$).

The roots of a complex number z can always be geometrically represented as a regular polygon of n sides inscribed in a circle of radius $R = r^{1/n}$; see Figure C-6. As an example, consider the mapping of the point $z = 1 + 2i$ into the w-plane under $w = z^{1/3}$. In the trigonometric representation,

$$z = 5^{1/2}(\cos 63.4° + i \sin 63.4°).$$

Since $r = (1^2 + 2^2)^{1/2}$ and $\theta = \tan^{-1}(2/1)$, the roots are

$$w_k = 5^{1/6}\{\cos[(63.4° + 2\pi k)/3] + i \sin[(63.4° + 2\pi k)/3]\} \quad \text{for } k = 0, 1, 2.$$

The cube roots of $w = (1 + 2i)^{1/3}$ are:

$$w_0 = 1.220 + 0.472i \quad (k = 0),$$
$$w_1 = -1.018 + 0.821i \quad (k = 1),$$
$$w_2 = -0.202 - 1.292i \quad (k = 2).$$

In order to derive the *polar representation* of a complex number, we must first consider the power series expansion of e^x with $x = i\theta$ as follows:

$$e^x = 1 + x + \frac{x^2}{2!} + \frac{x^3}{3!} + \cdots,$$

$$e^{i\theta} = 1 + (i\theta) + \frac{(i\theta)^2}{2!} + \frac{(i\theta)^3}{3!} + \cdots.$$

Using $i^2 = -1$ and separating terms into real and imaginary parts yields

$$e^{i\theta} = \left(1 - \frac{\theta^2}{2!} + \frac{\theta^4}{4!} - \cdots\right) + i\left(\theta - \frac{\theta^3}{3!} + \frac{\theta^5}{5!} + \cdots\right).$$

These power series are readily identified as the cosine and sine. Therefore,

$$e^{i\theta} = \cos\theta + i \sin\theta.$$

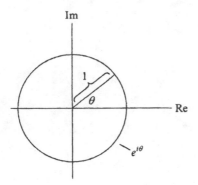

Figure C-7. The unit circle in the complex plane.

This relationship is called Euler's formula. Because $|e^{i\theta}| = \cos^2\theta + \sin^2\theta = 1$, $e^{i\theta}$ is often called the *unit circle* on the complex plane. See Figure C-7.

When a magnitude r is adjoined to $e^{i\theta}$, we finally have the polar representation:

$$z = (x + iy) = r(\cos\theta + \sin\theta) = re^{i\theta},$$

which is the most compact notation for a complex number z. The polar representation is the one used most often when studying wave propagation. As an example, let

$$w(t) = A\cos(\omega t - \theta),$$

which represents a sinusoidally varying wave with amplitude A, angular frequency ω, phase angle θ, and time t. Now let $h(t)$ be a function such that $h(t) = Be^{-i\omega t}$, where $B = Ae^{i\theta}$. Then

$$h(t) = Ae^{i\theta}e^{-i\omega t} = Ae^{-i(\omega t - \theta)} = A[\cos(\omega t - \theta) - i\sin(\omega t - \theta)]$$

or, equivalently,

$$w(t) = \operatorname{Re} h(t) = A\cos(\omega t - \theta).$$

The term B contains the amplitude and phase information about the wave, while $e^{-i\omega t}$ contains the frequency information. The form $Be^{-i\omega t}$ is used because of its notational simplicity and because of the ease with which calculus operations (such as differentiation and integration) are effected with the exponential function. The real part, Re, is always assumed to be used (although seldom explicitly displayed) with any results because that part has physical meaning (i.e., $w(t) = \operatorname{Re} Be^{-i\omega t}$ has physical meaning). The imaginary part can also be used, but it is not common practice to do so.

C.3   Multivalued Functions

Functions of a complex variable are usually written as $f(z)$. Because the system of complex numbers is closed, $f(z)$ for $z = x + iy$ can be written as $f(z) = u(x, y) + iv(x, y)$, or simply as $f(z) = u + iv$.

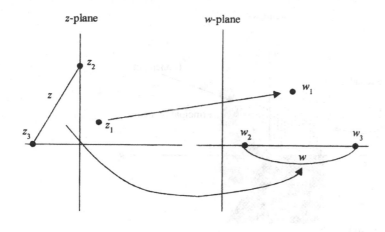

$$w(z) = 2z^2 + 3$$

Figure C-8. Example of a mapping from the z-plane to the w-plane.

When speaking of complex functions, the function $f(\cdot)$ is said to map from the z-plane to the w-plane (or to the (u, v)-plane); see Figure C-8. Note that u and v are two real independent functions of two real independent variables x and y. For example, Figure C-8 depicts the function $w(z) = 2z^2 + 3$, where the u and v functions are

$$u(x, y) = 2(x^2 - y^2) + 3 \quad \text{and} \quad v(x, y) = 4xy,$$

respectively. In the figure, the line between z_2 and z_3 is mapped onto the parabolic segment between w_2 and w_3, and the point z_1 is mapped to the point w_1.

Some functions of a complex variable map onto more than one point on the w-plane; these are called *multivalued* functions. Consider the function $w = z^{1/3}$, which maps the entire z-plane into three distinct regions of the w-plane. To see this, note that three single-valued functions are contained in the original function. They are (for $z = re^{i\theta}$):

$$w(z)_0 = r^{1/3}[\cos(\theta/3) + i \sin(\theta/3)],$$
$$w(z)_1 = r^{1/3}[\cos(\theta/3 + 120°) + i \sin(\theta/3 + 120°)],$$
$$w(z)_2 = r^{1/3}[\cos(\theta/3 + 240°) + i \sin(\theta/3 + 240°)].$$

Each of these functions is called a *branch*; $w(z)_0$ is the principal branch, $w(z)_1$ is the second branch, and so on. Figure C-9 shows how the z-plane is mapped to the w-plane. Note that θ is restricted to $-180° < \theta \leq 180°$ in the z-plane. Letting $\phi_0 = \theta/3$ be the argument of $w(z)_0$, we can see that $-60° < \phi_0 \leq 60°$ in the w-plane. In a similar fashion, $60° < \phi_1 \leq 180°$ and $180° < \phi_2 \leq 300°$ (this is easily obtained by adding $120°$ and $240°$, resp., to the inequality for ϕ_0).

The rays $\theta = 60°$, $180°$, and $300°$ are called *branch cuts*. A branch cut is a curve of singular points that separate branches (where by a "singular point" we mean that a function is discontinuous at the point and/or its derivatives do not exist at the point). The point $z = 0$, common to all the branches, is called a *branch point*. A branch point is a *singularity* of all the branches.

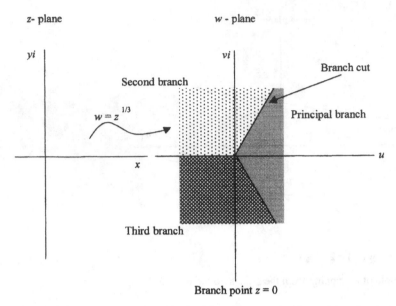

z- plane

w - plane

Figure C-9. Illustration of multivalued function mapping from the z-plane to the w-plane.

An important multivalued function is $\log z$. By using the polar representation $z = re^{i\theta}$, it follows that

$$\log z = \log re^{i\theta} = \log r + i\theta \quad \text{for } r > 0.$$

Since $e^{i\theta} = e^{i(\theta \pm 2\pi k)}$ $(k = 0, 1, 2, \ldots)$, $\log z$ can be written as

$$\log z = \log r + i\phi, \quad \text{where } \phi = (\theta \pm 2\pi k) \text{ and } k = 0, 1, 2, \ldots.$$

Here, k and the sign of $\phi = (\theta \pm 2\pi k)$ are selected so that $-\pi < \phi \leq \pi$. When $k = 0$, the value of $\log z$ is called the *principal value*. There are an infinite number of branch cuts, each corresponding to the value and sign of k chosen. Note that $z = 0$ is the branch point of $\log z$.

C.4 Analytic Functions

A function of a complex variable, $f(z)$, is said to be *continuous* at a point z_0 if $|f(z) - w_0| < \delta$ (δ arbitrarily small) whenever $|z - z_0| < \varepsilon$, where ε is a very small number. In words, a function is continuous if, whenever z is in the neighborhood of z_0, $f(z)$ is in the neighborhood of w_0 (i.e., $f(z_0) = w_0$; see Figure C-4).

The derivative of complex function $f(z)$ is given by

$$f(z) = \lim_{\Delta z \to 0} \frac{f(z + \Delta z) - f(z)}{\Delta z}.$$

For the function $f(z) = u(x, y) + iv(x, y)$, this amounts to

$$f'(z) = \lim_{\substack{\Delta x \to 0 \\ \Delta y \to 0}} \frac{[u(x + \Delta x, y + \Delta y) - u(x, y)] + i[v(x + \Delta x, y + \Delta y) - v(x, y)]}{\Delta x + i\Delta y}.$$

The limits can be approached separately. If Δy is set to zero and the limit is approached from the real axis, then

$$f'(z) = \lim_{\Delta x \to 0} \frac{[u(x + \Delta x, y) - u(x, y)] + i[v(x + \Delta x, y) - v(x, y)]}{\Delta x}$$

$$= \frac{\partial u}{\partial x} + i \frac{\partial v}{\partial x}.$$

If Δx is set to zero and the limit is approached from the imaginary axis, then

$$f'(z) = \lim_{\Delta y \to 0} \frac{[u(x, y + \Delta y) - u(x, y)] + i[v(x, y + \Delta y) - v(x, y)]}{i \Delta y}$$

$$= -i \frac{\partial u}{\partial y} + i \frac{\partial v}{\partial y}.$$

Since $f'(z)$ should be independent of the manner of approach of the limits, it follows that

$$\frac{\partial u}{\partial x} + i \frac{\partial v}{\partial y} = -i \frac{\partial u}{\partial y} + \frac{\partial v}{\partial y}$$

and hence that

$$\frac{\partial u}{\partial x} = \frac{\partial v}{\partial y}, \qquad \frac{\partial u}{\partial y} = -\frac{\partial v}{\partial x}.$$

These last two equalities are called the *Cauchy–Riemann conditions*.

If a function has continuous first partial derivatives that satisfy the Cauchy–Riemann conditions in some neighborhood of a point z, then $f(z) = u + iv$ is said to be *analytic* at z (otherwise known as "regular" or "holomorphic"). As an example, consider the function $f(z) = 1/z$:

$$\frac{1}{z} = \frac{1}{x + iy} = \frac{x}{x^2 + y^2} - \frac{iy}{x^2 + y^2},$$

$$\frac{\partial u}{\partial x} = \frac{\partial v}{\partial y} = \frac{y^2 - x^2}{(x^2 + y^2)^2},$$

$$\frac{\partial u}{\partial y} = -\frac{\partial v}{\partial x} = \frac{-2xy}{(x^2 + y^2)^2}.$$

Obviously, the Cauchy–Riemann conditions are satisfied, but the first partials are not continuous at $z = 0$ (i.e., $x = 0$ and $y = 0$). Thus $1/z$ is analytic everywhere *except* at $z = 0$. Note that

$$f'(z) = \frac{\partial u}{\partial x} + i \frac{\partial v}{\partial x} = \frac{\partial v}{\partial y} - i \frac{\partial u}{\partial y}.$$

The derivatives of analytic functions obey the same operational properties as do the derivatives of real continuous functions. For instance,

$$\frac{d(z^n)}{dz} = nz^{n-1},$$

$$\frac{d(f_1 f_2)}{dz} = f_1 \frac{df_2}{dz} + f_2 \frac{df_1}{dz},$$

and so on. By taking second partial derivatives, it can be demonstrated that all analytic functions satisfy Laplace's equation. For example:

$$\frac{\partial u}{\partial x} = \frac{\partial v}{\partial y}, \qquad \frac{\partial u}{\partial x} = -\frac{\partial u}{\partial y},$$

$$\frac{\partial^2 u}{\partial x^2} = \frac{\partial^2 v}{\partial x \, \partial y}, \qquad \frac{\partial^2 v}{\partial x \, \partial y} = -\frac{\partial^2 u}{\partial y^2},$$

$$\frac{\partial^2 u}{\partial x^2} + \frac{\partial^2 u}{\partial y^2} = 0 \quad (\text{i.e., } \nabla^2 u = 0).$$

Through a similar calculation, v is also found to satisfy Laplace's equation. Functions that satisfy Laplace's equation are called *harmonic* functions. Since both the real and imaginary parts of an analytic function satisfy Laplace's equation, u and v are called conjugate harmonics of each other. It is important to note that if a function is analytic then both its real part and its imaginary part will satisfy Laplace's equation.

In the trigonometric or polar representations, the Cauchy–Riemann conditions are given by

$$\frac{\partial u}{\partial r} = \frac{1}{r}\frac{\partial v}{\partial \theta}, \qquad \frac{1}{r}\frac{\partial u}{\partial \theta} = -\frac{\partial v}{\partial r}.$$

For

$$z = r(\cos\theta + i\sin\theta), \qquad z = re^{i\theta},$$

the derivatives are

$$f'(z) = (\cos\theta - i\sin\theta)\left(\frac{\partial u}{r} + i\frac{\partial v}{\partial r}\right)$$

$$= e^{-i\theta}\left(\frac{\partial u}{r} + i\frac{\partial v}{\partial r}\right).$$

Both u and v satisfy Laplace's equation in polar coordinates:

$$r^2\frac{\partial^2 u}{\partial r^2} + \frac{1}{r}\frac{\partial u}{\partial r} + \frac{\partial^2 u}{\partial\theta^2} = 0.$$

C.5 Integrals

Before introducing complex integration, definitions of various types of curves must be available. Most complex integration is performed parametrically. This will be apparent after the few examples given here are reviewed.

An *arc* is defined as a continuous set of points (x, y) that are parametrically defined by

$$x = g(t), \quad y = h(t) \quad (a \le t \le b),$$

where g and h are continuous functions of t. If x and y are unique for each value of t, then the arc is called a *Jordan* arc. If $g(b) = g(a)$ and $h(b) = h(a)$, but otherwise each

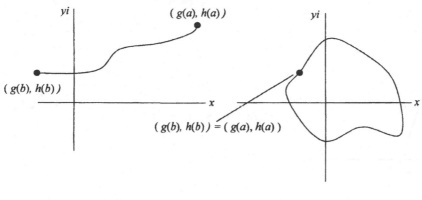

Jordan arc   Jordan curve

Figure C-10. Examples of a Jordan arc and a Jordan curve.

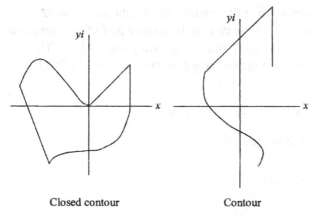

Closed contour   Contour

Figure C-11. Examples of a contour and a closed contour.

(x, y) corresponding to t is unique, then the arc is called a Jordan curve or a *simple closed curve.* See Figure C-10.

When the functions $g(t)$ and $h(t)$ have continuous derivatives and are not zero simultaneously, the arc or curve they represent is called a *smooth* curve. If an arc or curve is smooth, then its length L is given by:

$$L = \int_a^b \sqrt{[g(t)']^2 + [h(t)']^2}\, dt.$$

A *contour* is a continuous connection of smooth Jordan arcs. Figure C-11 shows examples of contours.

When a function of a complex variable, $f(z)$, is integrated along a contour, the integral is called a *contour* (or *line*) *integral.* Contour integrals are written as $I = \int_C f(z)\, dz$, where C is the contour upon which the integral is evaluated. Typically, contour integrals are evaluated by recasting the integral in parametric form – that is, by rewriting the integrand $f(z)$ and the variable of integration dz as functions of a single parameter and integrating over the applicable range of that parameter. A few examples will make this clear.

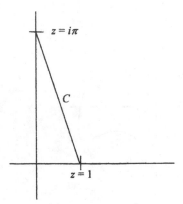

Figure C-12. Contour for integration
of the function e^z.

Consider integrating the function e^z over the contour (a straight line) $z = i\pi$ to $z = 1$. This contour is shown in Figure C-12. Note that the imaginary part of the contour can be written as a function of x, $y = \pi(1-x)$, and that x varies from 0 to 1. Then $z = x + i\pi(1-x)$ on the contour. Rearranging, we have $z = x(1-i\pi) + i\pi$. From this, it follows that

$$e^z = e^{x(1-i\pi)+i\pi} = -e^{x(1-i\pi)} \quad \text{and} \quad dz = (1-i\pi)\,dx.$$

The integral now becomes a function of the parameter x:

$$\int_C f(z)\,dz = -\int_0^1 e^{x(1-i\pi)}(1-i\pi)\,dx.$$

Performing the integration yields

$$\int_C f(z)\,dz = -\int_0^1 e^{x(1-i\pi)}(1-i\pi)\,dx$$

$$= -(1-i\pi)\left[\frac{e^{x(1-i\pi)}}{(1-i\pi)}\right]_0^1$$

$$= -[-e-1]$$

$$= 1+e.$$

As a second example, let $f(z) = (z+2)/z$ with the contour of integration the semi-circle $z = 2e^{i\theta}$, where $\theta = [0, \pi]$. This contour is shown in Figure C-13. In terms of θ, $f(z)$ can be written as $(2e^{i\theta} + 2)/2e^{i\theta}$ or $1 + e^{-i\theta}$. Since $z = 2e^{i\theta}$ on the contour, $dz = 2ie^{i\theta}\,d\theta$ as θ ranges from 0 to π. The integral now becomes

$$2i\int_0^\pi (e^{i\theta} + 1)\,d\theta = 2i\left[\frac{e^{i\theta}}{i}\right]_0^\pi + 2\pi i = -4 + 2\pi i.$$

The Cauchy–Goursat theorem and the Cauchy integral formula are two very important consequences of contour integration. The Cauchy–Goursat theorem states that the contour integral of a function interior to and on a closed curve is equal to zero:

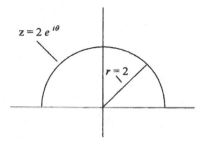

Figure C-13. Contour for integration of the function $(z + 2)/z$.

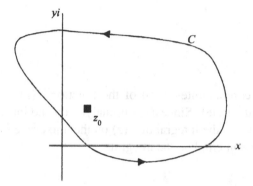

Figure C-14. Typical geometry involved in the application of the Cauchy–Goursat theorem and the Cauchy integral formula.

$$\int_C f(z)\, dz = 0.$$

The Cauchy integral formula states that if a function $f(z)$ is analytic on and interior to a closed contour and if z_0 is any point interior to the contour, then

$$f(z_0) = \frac{1}{2\pi i} \int_C \frac{f(z)}{(z - z_0)}\, dz.$$

For both the Cauchy–Goursat theorem and the Cauchy integral formula, the integration is performed by traveling around the contour in the counterclockwise (positive) direction. Figure C-14 illustrates the geometry.

An important aspect of analytic functions is manifest in the Cauchy integral formula: *organic character*. This means that if a function is analytic on the boundary of a region then the function's values in the region within the boundary are determined by its values on the boundary. Conversely, its values within the interior of the boundary determine its values on the boundary; if the function changes value within the interior, these changes are reflected by changes in its boundary values. For derivatives, the Cauchy integral formula may be written as

$$f^{(n)}(z_0) = \frac{n!}{2\pi i} \int_C \frac{f(z)}{(z - z_0)^{n+1}}\, dz.$$

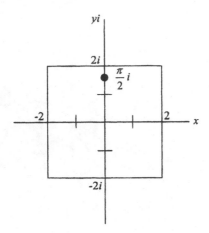

Figure C-15. Contour for integration of $e^{-z}/(z - i\pi/2)$.

Here are some examples. Consider the contour integration of the function $f(z) = z^2/(z - 3)$ around the contour $|z| = 1$ (the unit circle). Since $f(z)$ is analytic on and interior to the unit circle ($z = 3$ is outside of $|z| = 1$), the integral of $f(z)$ on the unit circle is zero, by the Cauchy–Goursat theorem. Next contemplate the integral

$$\int_C \frac{e^{-z}}{z - \pi i/2} \, dz$$

evaluated on the contour shown in Figure C-15, where $f(z)$ is identified as e^{-z} and z_0 is identified as $\pi i/2$. Since the point $z = \pi i/2$ is within the contour of Figure C-15, we have

$$e^{-\pi i/2} = -i = \frac{1}{2\pi i} \int_C \frac{e^{-z}}{z - \pi i/2} \, dz \quad \text{or} \quad \int_C \frac{e^{-z}}{z - \pi i/2} \, dz = 2\pi.$$

As another example, we will use the derivative version of Cauchy's formula to evaluate the integral

$$\int_C \frac{z^3 + 2z}{(z - z_0)^3} \, dz,$$

where C encloses z_0. Notice that the power 3 of the denominator can be identified as the $n + 1$ of the Cauchy formula for derivatives at z_0:

$$f^{(n)}(z_0) = \frac{n!}{2\pi i} \int_C \frac{f(z)}{(z - z_0)^{n+1}} \, dz.$$

Therefore, $n = 2 f(z)$ can be taken as $z^3 + 2z$, which means that $f''(z) = 6z$ and $f''(z_0) = 6z_0$. It follows that

$$6z_0 = \frac{2!}{2\pi i} \int_C \frac{z^3 + 2z}{(z - z_0)^3} \, dz \quad \text{or} \quad \int_C \frac{z^3 + 2z}{(z - z_0)^3} \, dz = 6\pi i z_0.$$

This last example shows the organic property in that, for arbitrary contour C (bounding the region containing z_0), the integral is determined by the selection of the point z_0.

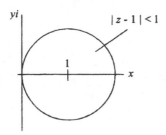

Figure C-16. Circle of convergence for Taylor series expansion of $1/z$ about the point $z = 1$.

Two important theorems that can be derived using the Cauchy integral formula are the maximum modulus theorem and the Liouville theorem.

The *maximum modulus theorem* states that if a function is analytic in a bounded region, is not constant, and is continuous on the bounding contour, then $|f(z)| < M$, where M is the maximum value of $|f(z)|$ on the boundary. In simple terms, the maximum modulus of an analytic function never occurs in the interior of a region but rather attains its maximum value on the enclosing boundary.

Before stating Liouville's theorem, we must define the term "entire." A function is said to be *entire* if it is analytic at every point of the complex plane (has derivatives everywhere). The *Liouville theorem* states that if the modulus of a function, $|f(z)|$, is bounded for all z in the complex plane and if $f(z)$ is entire, then $f(z)$ is a constant.

Indefinite complex integrals are evaluated in the same manner as indefinite real integrals. For example, if $F'(z) = f(z)$ then $F(z) = \int f(z)\,dz$. Definite integrals are also evaluated in a manner similar to real definite integrals. For instance, $\int_a^b f(z)\,dz = F(b) - F(a)$ with $F'(z) = f(z)$.

C.6 Power Series

The *Taylor series expansion* of a function $f(z)$ about the point z_0 is given by

$$f(z) = f(z_0) + f'(z_0)(z - z_0) + \frac{f''(z_0)}{2!}(z - z_0)^2 + \cdots + \frac{f^{(n)}(z_0)}{n!}(z - z_0)^n + \cdots .$$

This series converges to $f(z)$ if $f(z)$ is analytic at all points within a circle around z_0. If r is the radius of the circle in which $f(z)$ is analytic, then r is called the *radius of convergence* of the series. If the point z_0 is taken as zero, then the series is called a *Maclaurin series* and can be written as

$$f(z) = f(0) + f'(0)z + \frac{f''(0)}{2!}z^2 + \cdots + \frac{f^{(n)}(0)}{n!}z^n + \cdots .$$

As an example of a Taylor series expansion, consider the function $1/z$ expanded about the point $z = 1$. Note that $f(z)$ is not analytic at $z = 0$ and so the circle in which $f(z)$ is analytic must exclude $z = 0$. Such a circle is $|z - 1| < 1$, shown in Figure C-16. The nth derivatives of $1/z$ are $(-1)^n n!\, z^{-n-1}$; at $z = 1$, these become $(-1)^n n!$. Using these values, the Taylor series is

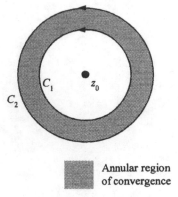

Annular region
of convergence

Figure C-17. The annular region of
convergence used for Laurent series
expansion about z_0.

$$\frac{1}{z} = \sum_{n=0}^{\infty} (-1)^n (z-1)^n .$$

The most common example of a Maclaurin series is the expansion of e^z about $z = 0$ because, for $f(z) = e^z$, $f^{(n)}(z) = e^z$, and $e^0 = 1$, the Maclaurin series is

$$e^z = 1 + \sum_{n=1}^{\infty} \frac{z^n}{n!}.$$

Since e^z is analytic everywhere on the complex plane, the circle of convergence is the entire complex plane.

The *Laurent series* is a series in both positive and negative powers of z. It is used when the region of convergence is limited to the annular region between two circles with a common center at the point z_0. Figure C-17 shows the region of convergence for a Laurent series. The Laurent series may be written as

$$f(z) = \sum_{n=0}^{\infty} a_n (z-z_0)^n + \sum_{n=1}^{\infty} \frac{b_n}{(z-z_0)^n},$$

where

$$a_n = \frac{1}{2\pi i} \int_{C_2} \frac{f(z)}{(z-z_0)^{n+1}} \, dz \quad (n = 0, 1, 2, \ldots),$$

$$b_n = \frac{1}{2\pi i} \int_{C_1} \frac{f(z)}{(z-z_0)^{-n+1}} \, dz \quad (n = 1, 2, \ldots).$$

The contour integrals are evaluated in the counterclockwise direction. The part of the series containing positive powers of z is called the *regular* part of the series, and the part containing the negative part of the series is called the *principal* part of the series.

These formulas can be used to find the Laurent series of a function, but this is rarely done. Usually the Taylor, Maclaurin, or other types of series can be associated with the function and with the applicable annulus of convergence. As an example, think of the function $1/z(1 - z)$. Obvious regions of convergence are $0 < |z| < 1$ and $1 < |z| < \infty$, since either of these contain $z = 0$ or $z = 1$ where the function is not analytic. The function can be rewritten as

$$\frac{1}{z(1 - z)} = \frac{1}{z} + \frac{1}{(1 - z)}.$$

Since $0 < |z| < 1$, the second term on the right is actually the sum of a geometric series $1 + z + z^2 + z^3 + \cdots = 1/(1 - z)$. Therefore, the Laurent series for $1/z(1 - z)$ within the annulus $0 < |z| < 1$ can be written as:

$$\frac{1}{z(1 - z)} = \frac{1}{z} + 1 + z + z^2 + z^3 + \cdots \quad (0 < |z| < 1).$$

For the annulus $1 < |z| < \infty$, the following recasting of $1/z(1 - z)$ is necessary:

$$\frac{1}{z(1 - z)} = \frac{1}{z} - \frac{1}{z(1 - 1/z)}.$$

Since $0 < |1/z| < 1$, the geometric series can be used again. When z is replaced by $1/z$, we have

$$\frac{1}{(1 - 1/z)} = 1 + \frac{1}{z} + \frac{1}{z^2} + \frac{1}{z^3} + \cdots.$$

It follows that

$$\frac{1}{z(1 - z)} = -\frac{1}{z^2} - \frac{1}{z^3} - \frac{1}{z^4} - \cdots \quad (1 < |z| < \infty).$$

C.7 Poles and Zeros

If $f(z_0) = 0$ then z_0 is called a *zero* of $f(z)$. If the derivatives

$$f^{(1)}(z_0) = f^{(2)}(z_0) = \cdots = f^{(m-1)}(z_0) = 0$$

but $f^{(m)}(z_0) \neq 0$, then z_0 is called a *zero of order m*. If $m = 1$ then the zero is called a *simple* zero.

A *pole* is defined with reference to the principal part of the Laurent series representation of a function:

$$f(z) = \underbrace{\sum_{n=0}^{\infty} a_n(z - z_0)^n}_{\text{regular part}} + \underbrace{\sum_{n=1}^{\infty} \frac{b_n}{(z - z_0)^n}}_{\text{principal part}}.$$

If the number of terms in the principal part is finite (if, e.g., there are only m terms) then we have

$$f(z) = \frac{b_1}{(z - z_0)} + \frac{b_2}{(z - z_0)^2} + \cdots + \frac{b_m}{(z - z_0)^m} + \sum_{n=0}^{\infty} a_n(z - z_0)^n;$$

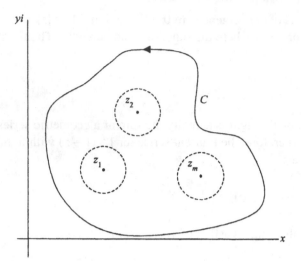

Figure C-18. Illustration of a contour used for evaluating the integral of a function $f(z)$, with m singularities, using the residue theorem.

in this case, z_0 is said to be a pole of order m. It can be shown that, as z approaches a pole, $|f(z)|$ becomes infinite. A pole of order 1 is called a *simple* pole.

If a function $f(z)$ is analytic in the neighborhood of a point z_0 but not at z_0, then z_0 is called an *isolated singularity* of $f(z)$. If $f(z)$ can be made analytic by choosing an appropriate value for $f(z_0)$, then z_0 is called a *removable* singularity. For example, the function $(\sin z)/z$ can be made analytic at $z = 0$ by assigning the value 1 to $(\sin z)/z$ at $z = 0$. If a function has poles of order m, then its isolated singularies are removable. If a function $f(z)$ has a Laurent series, expanded about a point z_0, that requires an infinite number of terms in its principal part, then z_0 is said to be an *essential* singularity: it is not removable. A good example is $\log z$, which has essential singularities at $z = 0$ and the entire negative real axis.

The first coefficient, b_1, of the principal part of a Laurent series of a function is called the *residue* of the function at the isolated singularity z_0. From the integral formula for b_1 of a Laurent series, we have

$$\text{Res } f(z) \text{ at } z = z_0 = b_1 = \frac{1}{2\pi i} \int_C f(z)\, dz,$$

where C encloses the point z_0 and the integral is taken in the counterclockwise direction around C. The residue theorem states that if a function has a finite number of singularities (or e.g. m poles) then the integral of $f(z)$ around a contour that encloses the singularities is equal to the sum of the residues of the singularities. See Figure C-18. Letting R_i denote the residues of the singularities $z_1, z_2, \ldots, z_m$ for $i = 1, 2, \ldots, m$, the residue theorem is written as

$$\int_C f(z)\, dz = 2\pi i (R_1 + R_2 + \cdots + R_m).$$

The residue(s) of a function can be found using the integral representation, but other methods are more convenient. Some examples will make this clear.

One way to find the residue of a function is to expand the function as a Laurent series and identify the coefficient of the $(z - z_0)^{-1}$ term. Suppose we want to find the residue of $e^{-z}/(z - 1)^2$. The isolated singularity is $z_0 = 1$, so we can use the annulus $0 < |z - 1| < \infty$. Note that e^{-z} can be written as $e^{-(z-1)}e^{-1}$. The Taylor expansion of $e^{-(z-1)}$ is then

$$e^{-(z-1)} = 1 - (z - 1) + \frac{(z - 1)^2}{2!} - \frac{(z - 1)^3}{3!} + \cdots .$$

From this, we obtain

$$\frac{e^{-z}}{(z - 1)^2} = \frac{e^{-1}e^{-(z-1)}}{(z - 1)^2} = \frac{e^{-1}}{(z - 1)^2} - \frac{e^{-1}}{(z - 1)} + e^{-1}\left[\frac{1}{2!} - \frac{(z - 1)}{3!} + \cdots\right].$$

Since the coefficient of the $(z - 1)^{-1}$ term is $-e^{-1}$, the residue is $-1/e$.

There is also a formula for residues when a function has m poles:

$$\text{Res at } z_0 = \frac{1}{(m - 1)!} \frac{d^{m-1}}{dz^{m-1}}[(z - z_0)^m f(z)]|_{z=z_0}.$$

For example, e^z/z^4 has a fourth-order pole at $z = 0$. Using $m = 4$,

$$\text{Res at } z_0 = \frac{1}{3!} \frac{d^3}{dz^3}\left(z^4 \frac{e^z}{z^4}\right)\bigg|_{z=0} = \frac{1}{6}.$$

If $m = 1$, a simple pole, then the formula reduces to

$$\text{Res at } z_0 = (z - z_0)f(z)|_{z=z_0}.$$

For instance, the function $(7z - 8)/z(z - 1)$ has simple poles at $z = 0$ and $z = 1$. Using the simple-pole residue formula, we obtain

$$\text{Res at } 0 = z\frac{7z - 8}{z(z - 1)}\bigg|_{z=0} = 8,$$

$$\text{Res at } 1 = (z - 1)\frac{7z - 8}{z(z - 1)}\bigg|_{z=1} = -1.$$

If these functions were the integrands of contour integrals with appropriate contours bounding the regions containing the poles, we would have:

$$\int_C \frac{e^{-z}}{(z - 1)^2} \, dz = 2\pi i(-e^{-1}) = -\frac{2\pi i}{e},$$

$$\int_C \frac{e^z}{z^4} \, dz = 2\pi i\left(\frac{1}{6}\right) = \frac{\pi i}{3},$$

$$\int_C \frac{7z - 8}{z(z - 1)} \, dz = 2\pi i(8 + -1) = 14\pi i.$$

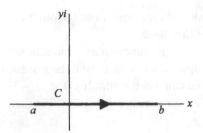

Figure C-19. Contour and direction of integration for evaluating real integrals via complex variables.

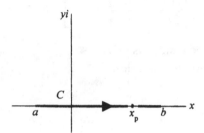

Figure C-20. Contour with singularity on the real axis.

C.8 Evaluating Real Integrals via Complex Variables

Real integrals of the form $\int_a^b f(x)\,dx$ can be evaluated as complex contour integrals (thus allowing use of the Cauchy integral formula and the residue theorem) $\int_C f(z)\,dz$ when the contour C is evaluated with the inclusion of the part of the real or x-axis between the points a and b. Figure C-19 shows an example of such a contour.

If the real integral has a singularity (or singularities) on the real axis then, in a strict sense, the integral does not exist. This problem can be circumvented by defining a value of the integral with*out* the singularity or singularities included. Consider the integral $\int_a^b g(x)\,dx$ with a singularity at x_p. Figure C-20 depicts the contour with a singularity, using $f(x) = (x - x_p)g(x)$. The integral can be recast as $\int_a^b f(x)/(x - x_p)\,dx$ or, in the complex plane, as $\int_C f(z)/(z - x_p)\,dz$ with the contour as shown in Figure C-20. The Cauchy principal value of such an integral is defined as

$$P \int_a^b \frac{f(x)}{x - x_p}\,dx = \lim_{r \to 0}\left[\int_a^{x_p - r} \frac{f(x)}{x - x_p}\,dx + \int_{x_p + r}^b \frac{f(x)}{x - x_p}\,dx\right];$$

see Figure C-21.

If the integral is cast in complex form then, for a simple pole at $z = x_p$, the Laurent series of $f(z)$ would be

$$f(z) = \sum_{n=0}^\infty a_n(z - x_p)^n + \frac{\text{Res}}{z - x_p}.$$

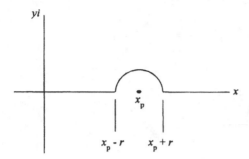

Figure C-21. Contour for Cauchy principal value
of a real integral with a singularity on the x-axis.

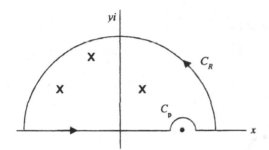

Figure C-22. Completion of contour when additional
poles are introduced.

As z approaches x_p, $f(z)$ approaches $\mathrm{Res}(z - x_p)^{-1}$, since the regular part approaches
zero and the principal part dominates. To integrate around the semicircular contour shown
in Figure C-22, let $z - x_p = re^{i\theta}$ and note that dz becomes $ire^{i\theta}\,d\theta$ at $\theta = \pi$ ($z = x_p - r$)
and at $\theta = 0$ ($z = x_p + r$). The integral can then be written as

$$\int_C \frac{f(z)}{z - x_p}\,dz = \int_\pi^0 \frac{\mathrm{Res}}{re^{i\theta}} ire^{i\theta}\,d\theta = -i \int_0^\pi d\theta\,\mathrm{Res} = -i\pi\,\mathrm{Res}$$

as z approaches x_p (note reversal of the integration limits).

If additional poles are introduced when a real integral is reformulated as a complex
contour integral, then a closed contour can be constructed as shown in Figure C-22. The
evaluation of such an integral then becomes

$$\int_C f(z)\,dz = \underbrace{P\int f(z)\,dz}_{\substack{\text{Cauchy} \\ \text{principal} \\ \text{value}}} + \underbrace{(-i\pi\,\mathrm{Res})}_{\substack{\text{semicircular} \\ \text{contribution}}} + \underbrace{\int_{C_R} f(z)\,dz}_{\substack{\text{large} \\ \text{semicircular} \\ \text{contribution}}} = 2\pi i\left[\sum_{\text{inside } C} \mathrm{Res}\right],$$

which yields

$$P\int f(z)\,dz = 2\pi i\left[\sum_{\text{inside } C} \mathrm{Res} + \frac{1}{2}\mathrm{Res}\,x_p\right] - \int_{C_R} f(z)\,dz.$$

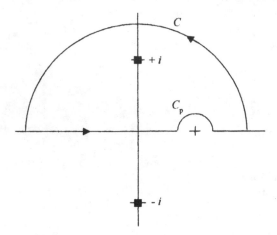

Figure C-23. Possible contour for integration of the
function $(1 + z^2)^{-1}(1 - z)^{-1}$.

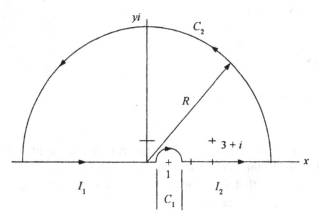

Figure C-24. Contour for integration of a real integral with a
pole on the real axis.

As an example, consider the following integrand in complex formulation:

$$\frac{1}{(1 + x^2)(1 - x)} \rightarrow \frac{1}{(1 + z^2)(1 - z)}.$$

Note that the pole at $z = 1$ remains, but two new poles at $-i$ and $+i$ have been introduced.
To use either the Cauchy integral formula or the residue theorem, the contour shown in
Figure C-23 can be used. Now consider the integral

$$\int_{-\infty}^{\infty} \frac{1}{(x - 1)(x^2 - 6x + 10)} \, dx.$$

The integrand has poles at $x = 1$ (on the real axis) and complex poles at $x = 3 + i$ and
$x = 3 - i$. The contour shown in Figure C-24 can be used.

Referring to the figure, we can see that the Cauchy principal value of the integral will be the sum of the integrals I_1 and I_2. To evaluate the integral, the residue theorem can be used as follows:

$$\int_C f(z)\, dz = 2\pi i \sum_{\text{inside } C} \text{Res}.$$

This yields

$$\int_C f(z)\, dz = I_1 + C_1 + I_2 + C_2 = 2\pi i\, \text{Res at } (3+i).$$

The value of the integral around the contour C_1 is simply $-i\pi$ Res at $z = 1$. Using Res $= (z - z_0) f(z)|_{z=z_0}$, we have

$$(z - 1)\frac{1}{(z-1)(z^2 - 6z + 10)}\bigg|_{z=1} = \frac{1}{5}$$

and so the integral around C_1 is $-i\pi/5$. To evaluate the integral around the contour C_2, consider that $z = Re^{i\theta}$ on this contour and (to satisfy the limits of the real integral) that R will approach infinity; θ will vary from 0 to π. Since $dz = iRe^{i\theta}\, d\theta$, it follows that

$$\lim_{R\to\infty} \int_0^\pi \frac{iRe^{i\theta}}{(Re^{i\theta} - 1)(R^2 e^{i2\theta} - 6Re^{i\theta} + 10)}\, d\theta = 0,$$

because the integrand varies as $1/R^2$ for large R. Thus,

$$I_1 + -i\pi/5 + I_2 + 0 = 2\pi i\, \text{Res at } (3+i) \quad \text{or}$$

$$I_1 + I_2 = 2\pi i\, \text{Res at } (3+i) + i\pi/5.$$

Using the residue formula once more, for the residue at $(3 + i)$ we obtain

$$(z - (3+i))\frac{1}{(z-1)(z-(3-i))(z-(3+i))}\bigg|_{z=(3+i)} = \frac{1}{(2+i)(2i)}.$$

Replacing $I_1 + I_2$ yields

$$\int_{-\infty}^\infty \frac{1}{(x-1)(x^2 - 6x + 10)}\, dx = \frac{\pi}{(2+i)} + \frac{i\pi}{5} = \frac{2\pi}{5}.$$

If a function has an essential singularity and/or branches (i.e., if it is multivalued), then care must be taken in constructing the contour. A good example is the integral

$$\int_0^\infty \frac{1}{1 + x^3}\, dx.$$

Because the integrand is odd, this integral cannot be extended to $-\infty$ (if a large semicircle of radius R were to be considered). One way to avoid this problem is by introducing the multivalued function $\ln z$ into the integrand – noting that, on the x-axis, the principal branch of $\ln z$ is $\ln x$ and the first branch is $\ln x + 2\pi i$. (Remember that $\ln z = \ln|z| + i(\theta + 2\pi k)$ for $k = 0, 1, 2, \ldots$.)

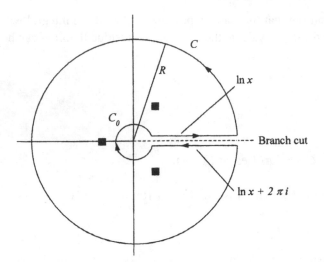

Figure C-25. Contour illustrating use of a branch cut.

With this change, we have

$$\int_0^\infty \frac{\ln z}{1 + z^3} \, dz,$$

and the contour of Figure C-25 is now available. The figure shows the poles of $1 + z^3$, and the essential singularity of $\ln z$ at $z = 0$ is circumscribed with a contour C_0 of radius r. The branch cut "cuts" the principal branch and the first branch of $\ln z$. The large circular contour C has radius R, which can be assumed to approach ∞ and so satisfy the upper limit of the real integral. The contour integral is

$$\int_C f(z) \, dz = \text{principal branch} + C + \text{first branch} + C_0 = 2\pi i \sum_{\text{inside } C} \text{Res}.$$

This may be rewritten as

$$\int_0^\infty \frac{\ln x}{1 + x^3} \, dx + C + \int_\infty^0 \frac{\ln x}{1 + x^3} \, dx + 2\pi i \int_\infty^0 \frac{1}{1 + x^3} \, dx + C_0 = 2\pi i \sum_{\text{inside } C} \text{Res}.$$

The integrals around C_0 and C can easily vanish, and the reversal of integration limits reverses the sign of an integral. Thus we have

$$-2\pi i \int_0^\infty \frac{1}{1 + x^3} \, dx = 2\pi i \sum_{\text{inside} C} \text{Res} \quad \text{or} \quad \int_0^\infty \frac{1}{1 + x^3} \, dx = -\sum_{\text{inside } C} \text{Res}.$$

The poles of the integrand are the cube roots of -1: $e^{i\pi/3}$, $e^{i\pi}$, and $e^{i5\pi/3}$. The residues at these poles can be evaluated using an alternative formula for residues – valid for the ratio of analytic functions – given by

$$(z - z_0) f(z)|_{z=z_0} = \frac{N(z_0)}{D'(z_0)} \quad \text{when} \quad f(z) = \frac{N(z)}{D(z)}$$

(see Section C.9). Since $f(z) = (\ln z)/(1 + z^3)$,

$$(z - z_0)f(z)|_{z=z_0} = \frac{\ln z_0}{3z_0^2}.$$

Evaluating and summing the residues yields

$$\frac{i\pi}{3} \frac{1}{3e^{i2\pi/3}} + i\pi \frac{1}{3e^{i2\pi}} + \frac{i5\pi}{3} \frac{1}{3e^{i10\pi/3}} \quad \text{or}$$

$$\frac{i\pi}{9}(e^{-i2\pi/3} + 3e^{-i2\pi} + 5e^{-i10\pi/3}),$$

which is

$$\frac{i\pi}{9}\{[\cos(120°) + 3\cos(360°) + 5\cos(600°)]$$
$$-i[\sin(120°) + 3\sin(360°) + 5\sin(600°)]\}.$$

This can be evaluated as $\sum_{\text{inside } C} \text{Res} = -2\pi(\sqrt{3}/9)$. The value of the integral then becomes

$$\int_0^\infty \frac{1}{1 + x^3}\, dx = -\sum_{\text{inside } C} \text{Res} = 2\pi \frac{\sqrt{3}}{9}.$$

C.9 Alternative Residue Formula

If the functions involved in $f(z) = N(z)/D(z)$ are analytic, then the numerator and the denominator can each be expanded in a Taylor series about the point z_0. Since z_0 is assumed to be a pole of $f(z)$, it follows that $D(z_0) = 0$. Expanding, we have

$$f(z) = \frac{N(z_0) + N'(z_0)(z - z_0) + N''(z_0)(z - z_0)^2/2! + \cdots}{D(z_0) + D'(z_0)(z - z_0) + D''(z_0)(z - z_0)^2/2! + \cdots}.$$

Since $D(z_0) = 0$,

$$f(z) = \frac{N(z_0) + N'(z_0)(z - z_0) + N''(z_0)(z - z_0)^2/2! + \cdots}{D'(z_0)(z - z_0) + D''(z_0)(z - z_0)^2/2! + \cdots}.$$

We now multiply by $(z - z_0)$ to obtain

$$(z - z_0)f(z) = \frac{N(z_0) + N'(z_0)(z - z_0) + N''(z_0)(z - z_0)^2/2! + \cdots}{D'(z_0) + D''(z_0)(z - z_0)/2! + \cdots};$$

with $z = z_0$, the $(z - z_0)$ terms vanish, leaving

$$(z - z_0)f(z)|_{z=z_0} = \frac{N(z_0)}{D'(z_0)}$$

as an alternative residue formula.

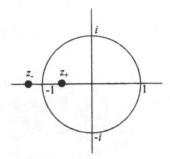

Figure C-26. Location of roots
with respect to the unit circle.

C.10 Trigonometric Integrals

Integrals of the form $\int_0^{2\pi} f(\cos\theta, \sin\theta)\, d\theta$, where $f(\cos\theta, \sin\theta)$ is the ratio of polynomials in $\cos\theta$ and/or $\sin\theta$, can be evaluated by considering θ to vary over the unit circle. With this assumption, the following relations hold:

$$z = e^{i\theta},$$

$$\cos\theta = \frac{e^{i\theta} + e^{-i\theta}}{2} = \frac{z + 1/z}{2}, \qquad \sin\theta = \frac{e^{i\theta} - e^{-i\theta}}{2i} = \frac{z - 1/z}{2i},$$

$$dz = ie^{i\theta}\, d\theta, \qquad dz = iz\, d\theta,$$

$$d\theta = -i\frac{dz}{z}.$$

Using these relationships, we can effect the following transformation:

$$\int_0^{2\pi} f(\cos\theta, \sin\theta)\, d\theta \;\rightarrow\; -i\int_C g(z)\, dz,$$

where $g(z)$ is a polynomial in z and C is the unit circle.

For example, consider the integral $\int_0^{2\pi} d\theta/(1 + k\cos\theta)$. First, note that if $k^2 < 1$ then there will be no zeros in the denominator, since 1 will always be greater than $|k\cos\theta|$ for either positive or negative values of k satisfying the inequality. Using the θ-to-z transformations, the integral may be rewritten as

$$-i\int_C \frac{dz}{z[1 + k(z + 1/z)/2]}$$

or (by algebraic manipulation) as

$$-2i\int_C \frac{dz}{kz^2 + 2z + k}.$$

The roots of the polynomial are

$$z_+ = -1 + \sqrt{1 - k^2} \quad \text{and} \quad z_- = -1 - \sqrt{1 - k^2}.$$

The location of these roots with respect to the unit circle is shown in Figure C-26.

Because $k^2 < 1$, both roots lie on the real axis. Because the z_+ root is within the unit circle (considered here as the integration contour), the residue theorem may be applied. The residue of $g(z) = (kz^2 + 2z + k)^{-1}$ at z_+ is $1/2\sqrt{1-k^2}$, since

$$-2i \int_C \frac{dz}{kz^2 + 2z + k} = -2i \frac{2\pi i}{2\sqrt{1-k^2}}.$$

The original integral then becomes

$$\int_0^{2\pi} \frac{d\theta}{1 + k\cos\theta} = \frac{2\pi}{\sqrt{1-k^2}} \quad \text{for } k^2 < 1.$$

APPENDIX D

Schlieren Imaging and Dynamic Photoelasticity

D.1 Discussion

Techniques for visualizing wave propagation can be very useful with regard to both the educational and research aspects of studying various wave propagation phenomena. Schlieren imaging techniques were used decades ago to study wave propagation in the fluids surrounding an immersed solid structure. One could, by way of the density gradients in the fluid material, photograph ultrasonic waves and pulses as they entered (and departed) a material. One could thereby estimate various wave velocities in the solid material as well as reflection and refraction angles. Both continuous wave and pulse-type propagation have been studied with this method; see Barnes and Burton (1949), Chinnery, Humphrey, and Beckett (1997), and Neubauer (1973).

More recently, waves in glass materials have been photographed using dynamic photoelasticity concepts together with an appropriate light source, camera, polarizer, analyzer, and quarter-wave plates. As polarized light enters a photoelastic test specimen, it divides into components that travel with different speeds along the respective principal axes. Thus, by the "photoelastic effect" for birefringent materials, the resultant magnitude of the light vector produces a colored fringe pattern that is proportional to the difference in principal strains at the particular point being considered in the material. The effect permits the visual study of stress wave propagation in solid photoelastic materials. Several applications of

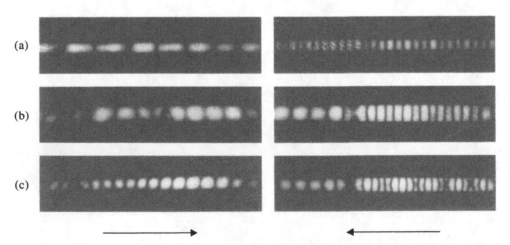

Figure D-1. Reflection of incident A1 modes: (a) $fd = 2.2$ MHz-mm; (b) $fd = 3.23$ MHz-mm; (c) $fd = 4.25$ MHz-mm. (Copyright 1988 © The American Society for Nondestructive Testing, Inc. Reprinted with permission from *Materials Evaluation*.)

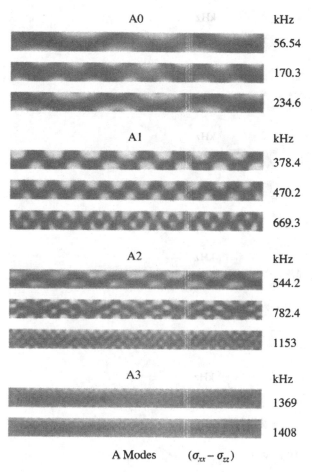

A0   kHz

56.54

170.3

234.6

A1   kHz

378.4

470.2

669.3

A2   kHz

544.2

782.4

1153

A3   kHz

1369

1408

A Modes $(\sigma_{xx} - \sigma_{zz})$

Figure D-2. Visualized resonant Lamb wave patterns of A and S modes for various frequencies. The bright spots correspond to the compressive component of the difference of normal stress, $\sigma_{xx} - \sigma_{zz}$, in the plate. (Reprinted from *Ultrasonics*, Vol. 32, No. 4, Hyo Ung Li and Katsuo Negishi, "Visualization of Lamb mode patterns in a glass plate," pg. 243, 1993, with permission from Elsevier Science.)

dynamic photoelasticity in flaw detection analysis are presented by Rose, Mortimer, and Chou (1972) for a shock-type loading. This concept is extended to the visualization of ultrasonic waves in Zhang, Shen, and Ying (1988) and Li and Negishi (1994).

We shall now consider two dynamic photoelastic photographs of waves traveling in a glass plate. First, the reflection of a Lamb wave by a free plate edge is studied by Zhang et al. (1988). In an illustration taken from their paper, the mode conversion process of a guided wave striking the end of a plate is clearly illustrated in Figure D-1. Notice that, for an incident A1 mode at three different fd values, the multiple modes that are reflected – especially for fd values of 3.23 and 4.25 MHz-mm – each travel at different group velocities and so make visualization easy (A0 and A1 modes are reflected).

Second, Figure D-2 presents another interesting result taken from Li and Negishi (1994). A variety of symmetric and antisymmetric modes in a plate are shown for different frequencies. Bright spots from the resonant mode patterns (as frequency is tuned in the

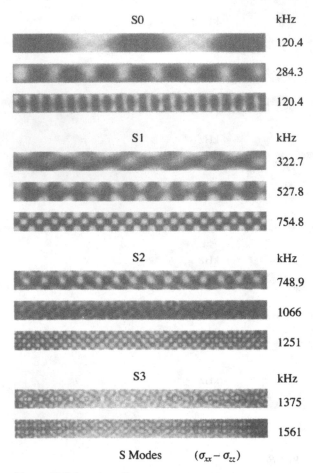

Figure D-2 *(continued)*.

120-mm-long, 10-mm-thick plate) allow us to compute wavelength and subsequent phase velocity. Frequency tuning is used to obtain well-organized results, as shown in the figure. Excellent agreement between theory and experiment is obtained in identifying modes and frequencies on the dispersion curves for glass. Note, for example, that the S0 mode at 120.4 kHz has 2.5 wavelengths in the 120-mm-long glass plate. Wavelength is therefore 48 mm, corresponding to a phase velocity of 5,779 m/s. Similar calculations can be performed for all of the other photos shown in the figure.

Work on numerical simulation and visualization of elastic waves has also received considerable attention. For pioneering efforts on the subject, see: K. Harumi (1986, "Computer simulation of ultrasonics in a solid," *Nondestructive Testing International* 19: 315–32), who uses a mass–spring lattice model; and H. Yamawaki and T. Saito (1992, "Numerical calculation of surface waves using new nodal equations," *Nondestructive Testing and Evaluation* 8/9: 379–89), who use a finite difference algorithm.

D.2 Exercises

1. Dynamic photoelasticity can be useful in studying wave propagation. Give a specific example.

2. Make a sketch showing all waveform possibilities of a bulk plane longitudinal wave impinging onto (a) a cylindrical cavity and (b) a penny-shaped crack tip.

3. How could the Schlieren imaging of ultrasonic waves be used to detect a defect in a structure? Be specific.

4. How could dynamic photoelasticity be used to study guided waves for thickness degradation due to corrosion attack?

5. For a given input guided wave mode at a particular frequency, describe what modes and frequencies might be reflected from a free end or a defect.

6. From the data presented in Figure D-2, plot a phase velocity dispersion curve for glass.

7. If there are any disagreements between theory and experiment for the results shown in Figure D-2, what might the causes be?

APPENDIX E

Key Wave Propagation Experiments

In this appendix we outline a few basic experiments in ultrasonic wave propagation. Of the hundreds of possibilities, seven key experiments are introduced to allow the student to become familiar with some basic instrumentation and ultrasonic transducer concepts and test procedures.

Experiment 1 concerns wave velocity measurement, whether it be a bulk longitudinal or shear wave or some type of guided wave. Wave refraction – and the use of angle beam transducers for generation and reception – is the subject of Experiment 2. In Experiment 3, skew angle is measured in anisotropic media. It is quite intriguing to see a wave travel in a direction other than where you are trying to send it. Of course, this phenomenon is a natural consequence of the physics and mechanics of waves in anisotropic media. Experiments 4, 5, and 6 are guided wave experiments in plates, rods, and tubes, respectively. Some interesting aspects of wave propagation, wave structure, and energy leakage are presented along with some practical aspects of ultrasonic NDE. Finally, Experiment 7 is on horizontal shear wave propagation. This interesting waveform also has some unusual and useful characteristics.

E.1   Experiment 1: Wave Velocity Measurements

E.1.1   Overview

Our main goal is to learn about procedures for measuring longitudinal wave velocity, shear wave velocity, and surface wave velocity. We shall employ the following equipment:

(1) ultrasonic pulser/receiver;
(2) digital oscilloscope;
(3) appropriate normal beam longitudinal wave and shear wave transducers;
(4) surface wave transducer or angle beam transducer and mediator;
(5) standard calibration block;
(6) test specimens (steel, aluminum, brass, etc.); and
(7) couplant – mineral oil for longitudinal waves, honey for shear waves.

Figure E.1-1 shows the equipment setup.

E.1.2   Procedure

Longitudinal Waves and Shear Waves

(1) Place the transducer on the specimen as shown in Figure E.1-1. Obtain the multiple echoes on the CRT screen.

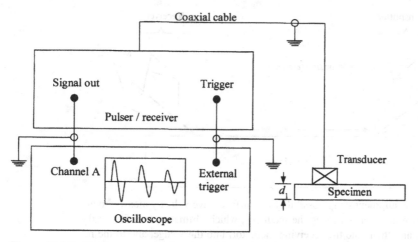

Figure E.1-1. Equipment setup for normal beam longitudinal wave and shear wave velocity measurement.

Table E.1-1. *Experimental results*

Specimen number and type	d_1 (mm)	T_L (μs)	c_L (mm/μs)	T_T (μs)	c_T (mm/μs)	Δd (mm)	Time (μs)	c_R (mm/μs)
1 Steel	49.53	8.5	5.82	16.8	2.95	20	8.3	2.41
2 Brass	38.49	8.6	4.48	16.6	2.32	20	9.9	2.02
3 Aluminum	50.18	7.9	6.27	16.1	3.14	20	7.1	2.82
4 Plexiglas	51.36	18.8	2.74	36.9	1.39	20	16.1	1.24
5 Graphite epoxy	16.51	5.3	3.12	8.9	1.86	20	14.1	1.42
6 Ceramic	17.88	3.1	5.75	5.5	3.25	20	6.5	3.08
7 Sintered powder metal	19.59	3.7	5.26	6.7	2.93	20	8.4	2.38

(2) Read the arrival time difference (Δt) between the first and second backwall echoes using the oscilloscope controls (or between second and third backwall echoes for more accuracy).

(3) Compute the longitudinal wave velocity and the shear wave velocity in the medium as follows:

$$c = 2d/\Delta t \text{ mm}/\mu s.$$

(4) Tabulate the readings in a chart similar to that in Table E.1-1 for the various materials.

Surface Waves

(1) Place the angle beam transducer and mediator on the specimen, as shown in Figure E.1-2. Obtain the signal for a distance d_2.

(2) Obtain a signal for a greater distance, say d_3.

(3) Using the oscilloscope controls, read the arrival time difference (Δt) between the first and second signals.

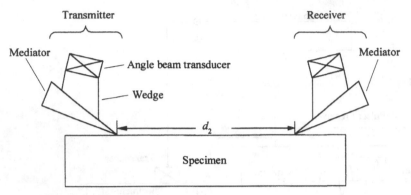

Figure E.1-2. Equipment setup possibility for surface wave velocity measurement. (A surface wave is generated along the mediator, which ultimately transfers to the test specimen and then onto the receiving mediator, into the wedge, and to the receiving transducer.)

(4) Compute the surface wave velocity in the medium as follows:

$$c_R = \Delta d / \Delta t \text{ mm}/\mu s, \quad \text{where} \quad \Delta d = d_3 - d_2.$$

(5) Tabulate the readings as in Table E.1-1 for the various materials.

E.1.3   Results and Comments

See Table E.1-1 for the experimental results. See also the sample waveform shown in Figure E.1-3.

Many ultrasonic references include a table of wave velocities for different materials; however, they are generally limited to only typical materials. It is therefore often necessary to obtain the wave velocities for a particular material before tackling a specific inspection problem.

There are many different ways of measuring arrival time of a waveform. When having excellent signal to noise ratio (S/N), the problem is straightforward. For reduced S/N and improved accuracy, one could use cross-correlation between two signals: a Hilbert transform for envelope profiling followed by a dB down-threshold value from the peak of the Hilbert transform, counting the specific number of zero crossings after achieving a specific threshold value.

The angle beam transducer should be placed on the mediator at the third critical angle, which (according to Snell's law) is

$$\theta_1 = \sin^{-1} \frac{c_{\text{wedge}}}{c_R}.$$

The term c_{wedge} denotes the longitudinal wave velocity in the wedge material; c_R is the approximate surface wave velocity in the test specimen. Note that a commercial surface wave transducer consists of a normal beam transducer and a wedge set at an appropriate angle for a specified material.

E.1.4   Exercises

1. Why would measuring the difference between a first and second backwall echo be more accurate than between the main bang and the first backwall echo?

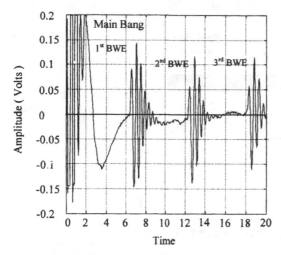

a. Longitudinal wave velocity RF waveforms for brass

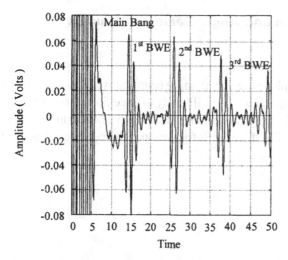

b. Shear wave velocity RF waveforms for brass

Figure E.1-3. Sample waveforms.

2. Why would measuring the difference between a second and third backwall echo sometimes be more acurate than between the first and second backwall echo?

3. At what angle should the angle beam transducer be placed on the mediator for best results?

4. Why is a mediator required to generate a surface wave in (a) an epoxy, (b) a composite, and (c) brass? Demonstrate why an ordinary surface wave angle beam transducer might not work.

5. Why does the polarizing axis of a shear wave transducer produce a different wave velocity in a unidirectional graphite epoxy – but *not* in a steel – test specimen as it is rotated by 0°, 45°, and 90°?

6. How do you measure a shear wave velocity without having a shear wave transducer?

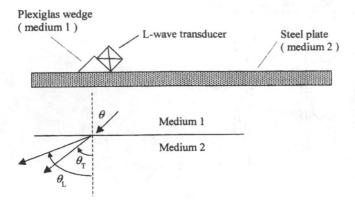

For the Plexiglas-Steel Boundary: $\theta_{cr1} = 28.15°$, $\theta_{cr2} = 58.4°$.

Figure E.2-1. Angle beam refraction.

E.2   Experiment 2: Refraction and Angle Beam Transducers

E.2.1   Overview

The goal of this experiment is to explore the use of angle beam transducers and refraction in ultrasonic wave analysis and inspection. We will study beam exit-point calibration using the International Institute of Welding (IIW) block, angle verification using the IIW block, and distance calibration using skip distance information. The equipment needed may be listed as follows:

(1) ultrasonic pulser–receiver;
(2) digital oscilloscope;
(3) IIW block;
(4) couplant (mineral oil);
(5) test specimens; and
(6) several angle beam transducers.

E.2.2   Procedure

A typical transducer setup is shown in Figure E.2-1. An L-wave transducer is placed on a wedge to direct the beam at an angle. The transducer and wedge together are called the angle beam transducer. An L wave is thus incident at an angle θ, and the refracted wave consists of both longitudinal and shear wave components.

The first critical angle is the angle at which θ_L becomes equal to 90°. If θ is greater than the first critical angle, θ_{cr1}, then the refracted wave in medium 2 consists of only a shear wave. For reasons of simplicity, this is the method used for most measurements with angle beam transducers. If θ is greater than the second critical angle, θ_{cr2}, then only a surface wave can propagate. This method is used to generate surface waves in materials (e.g., steel and aluminum) with high velocities.

The calibration block most widely used in angle beam inspection is the IIW block (see Figure E.2-2), which was developed by the International Institute of Welding for

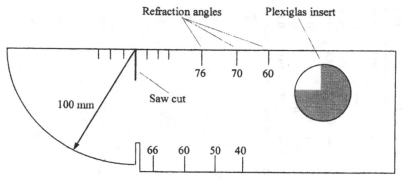

Figure E.2-2. IIW calibration block.

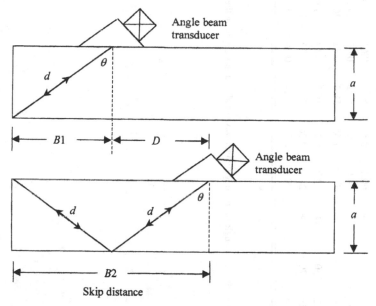

Figure E.2-3. Distance calibration procedure utilizing skip distance.

calibration of angle beam transducers. The radius shown is exactly 100 mm for the saw cut on top. This radius reflects ultrasonic waves for all transducer angles, with maximum reflection when the beam exit point is directly at the center of the radius.

The beam exit point is critical for locating flaws, and it is found by maximizing the reflected signal and marking the point where the saw cut intersects the transducer. Once the beam exit point has been marked, the beam angle can be found by moving the transducer along the IIW block, maximizing the reflection from the circular Plexiglas insert, and then determining the refraction angle as marked.

The angle beam transducer must be calibrated for a particular specimen before we can measure any distances with it. The transducer is placed on the specimen on one edge and moved in the x direction until the echo from the lower corner or edge B_1 is maximized, as shown in Figure E.2-3. (For this example, we'll assume that only shear waves are present in the test specimen.) This point is marked on the specimen and also on the time scale of the oscilloscope (say, t_1). The transducer is then moved farther down along the x direction

Table E.2-1. *Experimental results*[a]

Transducer rating[b]	Incident angle	Material															
		Steel				Aluminum				Plexiglas				Brass			
		θ_L	D	θ_T	D	θ_L	D	θ_T	D	θ_L	D	θ_T	D	θ_L	D	θ_T	D
55 ST	44.25	—	—	55.0	0.71	—	—	51.6	0.63	44.2	0.49	21.2	0.19	—	—	30.9	0.30
65 ST	50.54	—	—	65.0	1.07	—	—	60.1	0.87	50.5	0.61	23.6	0.22	—	—	34.6	0.34
70 ST	53.18	—	—	70.0	1.37	—	—	64.0	1.03	53.2	0.67	24.5	0.23	—	—	36.1	0.36
90 ST	58.42	—	—	90.0	—	—	—	73.1	1.65	58.4	0.81	26.2	0.25	—	—	38.8	0.40
45 L	19.49	45.0	0.50	23.1	0.21	50.1	0.60	22.0	0.20	19.5	0.18	9.9	0.09	31.1	0.30	14.2	0.13
60 L	24.12	60.0	0.87	28.7	0.27	70.1	1.38	27.3	0.26	24.1	0.22	12.2	0.11	39.3	0.41	17.5	0.16
70 L	26.32	70.0	1.37	31.7	0.30	—	—	29.8	0.29	26.3	0.25	13.3	0.12	43.4	0.47	19.0	0.17

Note: D is one half the skip distance (in inches).
[a] All plates 0.5" thick.
[b] ST = angle of the shear wave produced in steel using the given wedge. L = angle of the longitudinal wave produced in steel using the given wedge.

until the echo from the upper corner or edge at B_2 is maximized, as shown in the figure. This point is also marked on the specimen and the time t_2 noted. The distance traveled by the first signal down and back, $2d$, can be calculated by using $d^2 = B_1^2 + a^2$; the second signal travels twice this distance. Therefore, the difference between t_1 and t_2 gives the time the wave takes to travel distance $2d$. The wave velocity is calculated by $c_T = 2d/(t_2 - t_1)$.

Calibration allows us to locate the acoustic zero, which is usually embedded inside or close to the main bang echo, and also gives us the shear wave velocity in the medium. The actual refracted angle θ can also be found. This is important from an experimental standpoint, since θ varies from material to material.

The experimental results are displayed in Table E.2-1.

E.2.3   Exercises

1. Why does refraction occur?

2. If an angle beam transducer were labeled as $60°$ shear in steel, what refraction angle would be produced in aluminum?

3. Check some of the results in Table E.2-1.

E.3   Experiment 3: Skew Angle Measurement

E.3.1   Overview

Our goal is to understand the concept of skew angle as associated with wave propagation in anisotropic media. Toward this end, we use the following equipment:

(1) glass fiber polyester (or other thick composite material) test specimen;
(2) two broadband transducers with angle wedges; and
(3) tone-burst pulser–receiver excitation system.

E.3.2   Procedure and Setup

The consequence of skew angle in wave propagation is quite interesting. An ultrasonic wave may not go exactly where you try to send it. This effect is associated with (a) phase velocity variations that occur in anisotropic media as a function of direction and (b) the subsequent superposition of waves as they leave a transducer or excitation source. Because of the symmetry associated with many composite materials, normal beam incidence may not reveal any skew angle effect; there may be only a narrowing or broadening of the beam.

However, simply by cutting out a composite material at some nonsymmetric angle, a normal beam experiment can easily be conducted that will show the skewing effect; see Figure E.3-1. Another interesting experiment would be to study guided waves in a composite plate (see Figure E.3-2). In general, when waves are launched in a non-axisymmetric direction, skewing takes place. Analytical solutions can be more complex, and they call for a knowledge of slowness profiles in an anisotropic plate.

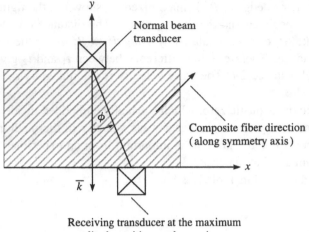

Figure E.3-1. Normal beam experiment showing skew angle ϕ on an angled cut-out thick block section of a composite material.

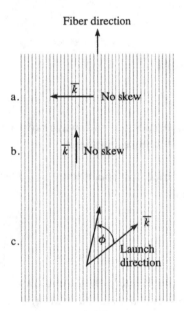

Figure E.3-2. Guided waves in a unidirectional graphite epoxy plate showing skew angle ϕ for angle beam excitation in the directions shown.

We now outline an experiment that is quite easy to conduct with just a thick composite material and a laboratory angle beam transducer.

(1) Place an angle beam transducer on a composite material, as shown in Figure E.3-3.
(2) Use this transducer to send a bulk wave with a frequency of approximately 0.8 MHz through the specimen.

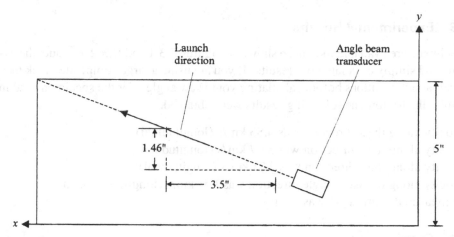

Figure E.3-3. Top view of a 3-inch-thick composite material test specimen with an angle beam transducer.

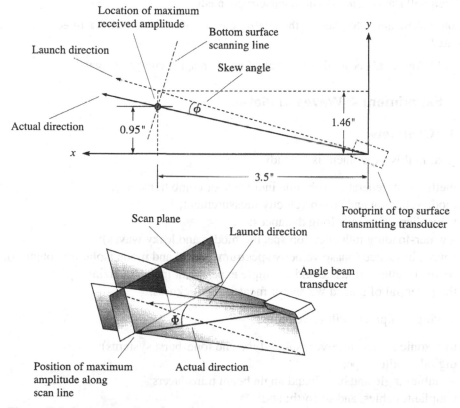

Figure E.3-4. Bottom view and three-dimensional rendering of the specimen in Figure E.3-3, showing the measurement of skew angle ϕ in the (x, y)-plane and the actual three-dimensional skew angle Φ between the launch and actual direction.

(3) Receive with a transducer on the bottom side of the composite, as shown in Figure E.3-4. Search for the signal in (x, y)-space. Try to peak the signal.

(4) Record the position of the maximum received amplitude. The skew angle can then be measured as shown in Figure E.3-4.

E.3.3 Experimental Results

If specimens are available like those shown in Figures E.3-1 and E.3-2, conduct the experiments and simply calculate your results. If you are using a large composite block then try several launch positions before calculating your skew angle. For the specimen used in this sample illustration, the following results were obtained:

- velocity along the x direction was 3.83 km/s (longitudinal);
- velocity along the y direction was 3.67 km/s (longitudinal);
- velocity along the z direction was 2.71 km/s (longitudinal);
- velocity along the skew angle direction was 1.51 km/s (longitudinal); and
- the measured skew angle was 7.74°.

E.3.4 Exercises

1. Why does skew angle occur?

2. When will skewing *not* occur in anisotropic media?

3. Could skew angle be used in the quality control of a newly manufactured composite plate?

4. Could skew angle be used in the evaluation of an aging composite plate?

E.4   Experiment 4: Waves in Plates

E.4.1 Overview

Our goal in this experiment is to study:

(1) methods of generation (oblique incidence or comb transducer);
(2) mode selection and group velocity measurement;
(3) wave dispersion over long distances;
(4) a water-loading influence (on specific modes and leaky waves);
(5) source influence (phase velocity spectrum control and mode isolation potential);
(6) wave structure influence and a sample experiment in defect sizing;
(7) the potential of guided waves for measuring thickness.

The following equipment will be required:

(1) ultrasonic pulser–receiver (shock-type and tone-burst systems);
(2) digital oscilloscope;
(3) variable-angle and broadband angle beam transducers;
(4) couplant, cables, and so forth; and
(5) an appropriate aluminum plate test specimen.

E.4.2 Generating Guided Waves with an Angle Beam Transducer

Before discussing any particular application of guided waves, it is first necessary to discuss the methods of generation and reception. There are, in fact, many methods for generating guided waves, such as normal beam acousto-ultrasound, electromagnetic acoustic transducers (EMATs), comb transducers, and angle beam longitudinal waves.

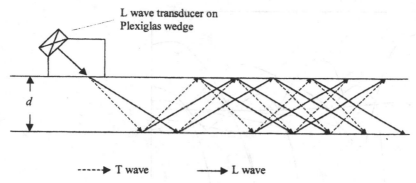

Figure E.4-1. Oblique incidence for the generation of guided waves.

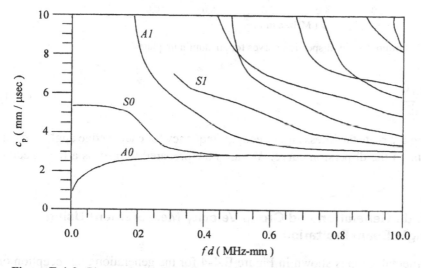

Figure E.4-2. Phase velocity dispersion curves for an aluminum plate.

Angle beam longitudinal waves offer an accurate and efficient method for – as well as a physical insight into – guided wave generation. Consider Figure E.4-1, which depicts the generation of a guided wave mode using a longitudinal transducer on a Plexiglas wedge. The longitudinal wave is incident at some incident angle and velocity. In accordance with Snell's law, the waves undergo mode conversion, reflection, and refraction at the interfaces. At some distance away from the transducer, the waves will no longer be individually identifiable but will have been superimposed into a wave packet. Now – for certain cases of incident angle, thickness d, and material properties – constructive interference will take place and a guided wave will propagate in the plate.

These conditions for constructive interference are actually met for many combinations of thickness, angle, and material properties. Thus, many modes are available. Theoretically, this problem can be formulated and solved for these wave resonances. If we set the material properties as constant, then guided wave dispersion curves can be generated as shown in Figures E.4-2 and E.4-3.

These curves are useful for developing guided wave–inspection techniques. For generation using a longitudinal wave transducer on a Plexiglas wedge, the phase velocity c_p bears a simple relation (via Snell's law) to the angle of incidence:

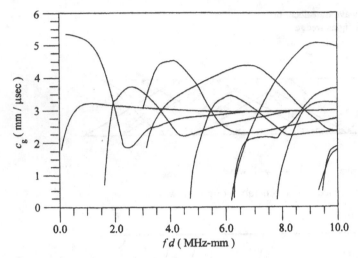

Figure E.4-3. Group velocity dispersion curves for an aluminum plate.

$$\frac{\sin \theta_i}{c_{\text{Plexi}}} = \frac{\sin 90}{c_p}. \qquad\qquad\qquad (\text{E.4.1})$$

Thus, by noting the plate thickness and sweeping frequency for each wedge angle, we can generate points on the dispersion curve. We are now prepared to discuss a few specific experiments.

E.4.3   Mode Generation and Group Velocity Measurement Using Angle Beam Excitation

The experimental setup is shown in Figure E.4-4 for the generation and reception of guided waves. The apparatus consists of two transducers mounted on Plexiglas wedges, an aluminum plate of known thickness, and a tone-burst source for signal generation. A point must be chosen on the phase velocity dispersion curve for generation – for example, the S0 mode at an fd of approximately 1.5 MHz-mm and a phase velocity of approximately 5.2 mm/μs. From equation (E.4.1), the angle of incidence for the chosen phase velocity is 32°. Therefore, a Plexiglas wedge with that particular incident angle should be used. Attach transducers with frequencies near 1.5 MHz to the wedges. A tone-burst signal is input to the transducers at 1.5 MHz; then, the frequency is adjusted to receive maximum signal amplitude. In this case, the frequency is 1.434 MHz. Note that, since the thickness is 1 mm, the frequency is numerically equal to the fd product. The resulting waveform is shown in Figure E.4-5.

In order to verify that the mode chosen is indeed the mode generated, one can measure the group velocity of the generated wave and compare it with the theoretical value. We may accomplish this as follows: Using the setup of Figure E.4-6, generate the mode and receive at point 2. Mark the location of the RF waveform on the scope with a cursor. Now, increment the receiving transducer to point 3. Mark the position of the waveform and measure the change in time between the original mark and the current mark. Calculate the group velocity ($\Delta t/x$).

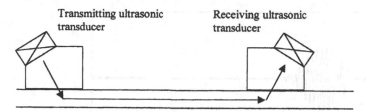

Figure E.4-4. Setup for guided wave generation and reception.

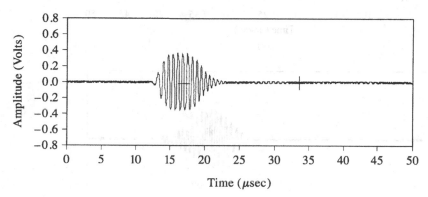

Figure E.4-5. RF waveform for the S0 mode at $fd = 1.434$ MHz-mm.

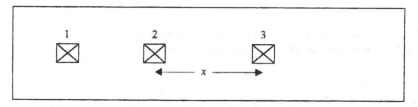

Figure E.4-6. Measurement of group velocity.

Two waveforms for this procedure are shown in Figure E.4-7 for the S0 mode at an fd of 1.434 MHz-mm. The distance x was three inches (76.2 mm), and the change in time was 15.875 μs, resulting in a group velocity of 4.8 mm/μs. From the group velocity curve, we note that this value is close to the theoretical value. Note the increase in pulse duration as the wave packet moves over three inches. This is because the point on the dispersion curve is somewhat dispersive in nature. Other points where phase velocity doesn't vary much with frequency could produce almost the same RF profile at the three-inch distance. Note also that the best way to measure group velocity is to extract time differences on the peak-to-peak (rather than the leading edge) position of the waveform. Table E.4-1 is provided for laboratory exercises to study other modes.

E.4.4   Comb Transducer Generation of Guided Waves

An alternative transducer array system for generating guided waves is illustrated in Figure E.4-8. The transducer element size and spacing dimensions (along with the excitation

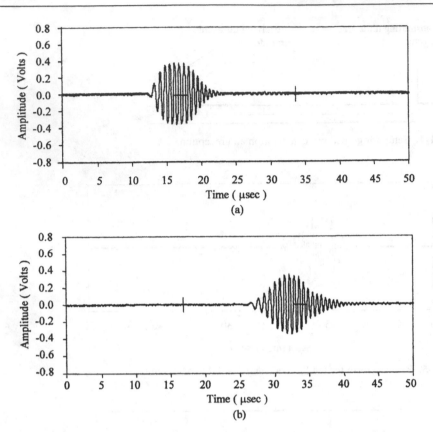

Figure E.4-7. Waveforms for the S0 mode at 1.434 MHz, where the receiving transducer for (b) has been incremented by 3 inches from its position for (a).

Table E.4-1. *Fill-in table for laboratory exercises*

Mode	c_p	θ_i (calculate)	f (fd)	c_g (theory)	c_g (exper.)
S0					
A1					
S1					

frequency variables) allow us to select modes and frequencies of choice on the guided wave dispersion diagrams. For a particular transducer design of fixed element and gap size and broadband frequency content, the excitation will be along a straight line drawn from the origin of the dispersion curve at a slope depending on the design parameters. See Rose et al. (1998).

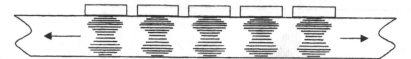

Figure E.4-8. Comb transducer technique for the generation of guided waves.

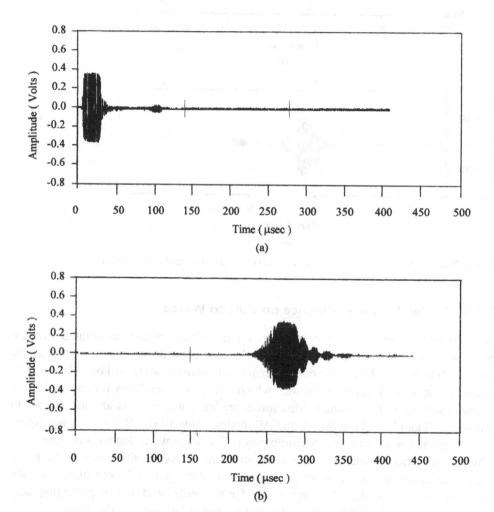

Figure E.4-9. The S0 mode propagating at (a) 4 inches and (b) 48 inches. Note that the shape of the waveform changes somewhat as a result of dispersion.

E.4.5 Distance and Dispersion

In the presence of a waveguide, propagating modes can travel over very long distances – up to 50 feet or more. To observe this behavior, we can generate and receive guided waves across an aluminum plate. Figure E.4-9 shows two waveforms, one propagating only 4 inches and the other 48 inches. Note the change in waveform shape due to dispersion.

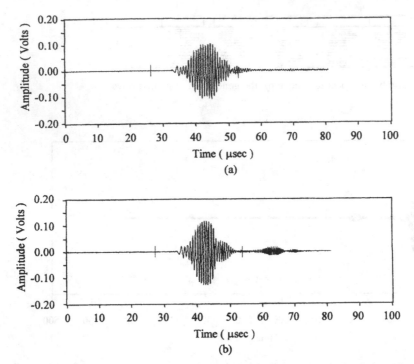

Figure E.4-10. S0 mode at 1.434 MHz-mm: (a) dry and (b) water-loaded specimen.

E.4.6   Water-Loading Influence on Guided Waves

Another interesting result for the propagation of guided waves is associated with the effect of water loading on an aluminum plate. Because of the difference in wave structure and power distributions for each mode, the reaction to water loading will be different. Often heard is the term "leaky Lamb wave," which refers to the tendency for energy to leak from the structure. In fact, some guided waves are leaky while others are not. Shown in Figures E.4-10 and E.4-11 are the S0 and A0 modes, respectively. Part (a) of each figure depicts propagation across a dry structure; part (b), across a water-loaded structure.

This phenomenon is related to the wave structure across the thickness of the plate. In-plane displacement loading on the outside surface of the plate will not leak energy into the fluid because of the shear loading effect. On the other hand, out-of-plane displacement on the outer surface of the plate will lead to energy leakage into the fluid.

E.4.7   Source Influence

The ability to select a specific point on the dispersion curve depends on both the frequency spectrum and the phase velocity spectrum associated with a given transducer source. For a specific angle of incidence, we do not generate a specific phase velocity only but rather a complete phase velocity spectrum. In general, for larger transducers the spectrum becomes more narrow. Theoretically, an infinite plane wave angle beam excitation source would produce just a single phase velocity value (see Rose, Ditri, and Pilarski 1994b).

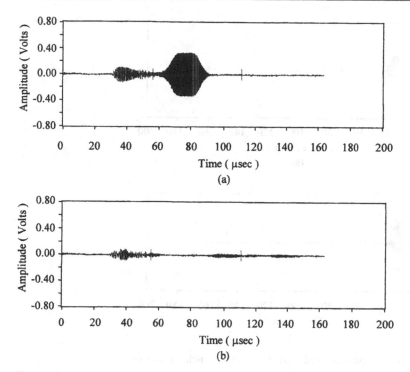

Figure E.4-11. A0 mode at 3.66 MHz-mm: (a) dry and (b) water-loaded specimen.

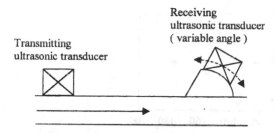

Figure E.4-12. Source influence experiment.

Consider now the experimental setup depicted in Figure E.4-12. According to Snell's law, a normal beam transducer will result in an infinite phase velocity – not too useful for generating specific guided wave modes! However, because of the source influence, lower phase velocities will also be possible. In fact, for a very small transducer we should be able to generate very low phase velocities. Receiving with an angle beam transducer allows the reception of a more narrow phase velocity spectrum. Therefore, we can isolate the modes being generated with the angle beam receiver and hence observe the amplitudes of the modes at their respective phase velocities.

We can send with two transducers of different sizes and observe the amplitudes of the specific modes generated. Figure E.4-13 shows waveforms received at 31° from a one-inch and a half-inch normal beam transducer at an fd of 0.845 MHz-mm (this would correspond to the S0 mode). Figure E.4-14 shows similar results for waveforms received at approximately 75° (this corresponds to the A0 mode). The smaller transducer more

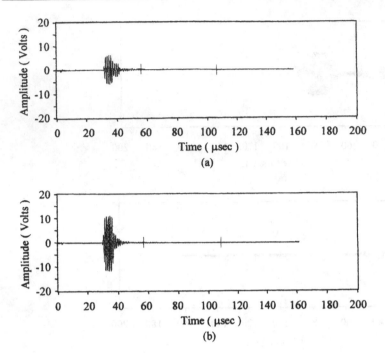

Figure E.4-13. Waveforms received at 31° from (a) one-inch transducer and (b) half-inch transducer.

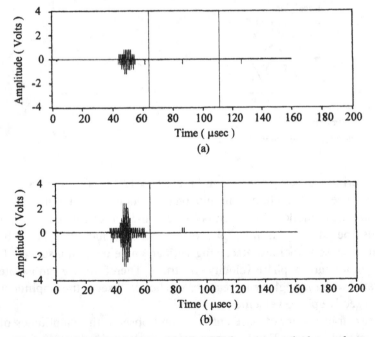

Figure E.4-14. Waveforms received at 75° from (a) one-inch transducer and (b) half-inch transducer.

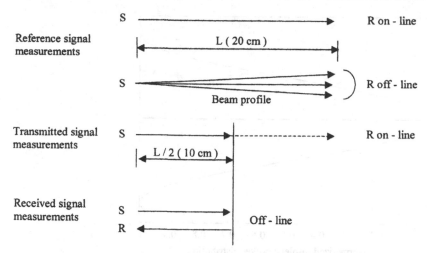

Figure E.4-15. Schematic showing the sending (S) and receiving (R) transducer arrangements used for measurement of the reference, transmitted, and reflected signal amplitudes, where the on-line setup corresponds to transmission factor measurements and the off-line setup corresponds to reflection factor measurements.

effectively generates both of these low–phase velocity modes because of the spectrum associated with the source influence. Many additional angle beam excitation experiments could be conducted to illustrate the source influence in mode selection and isolation.

E.4.8   Defect Sizing and Wave Structure Influence

Defect depth studies can be performed to study the potential of using the scattering characteristics of selected guided wave modes for defect depth classification. By examining the power distribution and wave structure, we will discover that power concentration at various depths in the material will provide different sensitivities and scattering characteristics. The experimental setup is shown in Figure E.4-15.

The wave structure for two different modes is shown in Figure E.4-16. When impinging on a crack of varying depth, we can expect that the different modes will generate different responses. For example, one would expect the A1 mode to be more sensitive to a surface crack and the S0 mode to be more sensitive to a flaw or inclusion located at the center of the material. A boundary element method (BEM) computation can be employed to predict the various sensitivities of different guided wave modes. An example of the BEM and corresponding experimental data are shown in Figure E.4-17; see Cho, Hongerholt, and Rose (1997).

E.4.9   Thickness Measurement Using a Tone-Burst Frequency Sweep Device

In this experiment we use a tone-burst function generator that performs a frequency sweep. The phase velocity being generated is fixed by the Plexiglas shoes, effectively producing a horizontal line across the phase velocity dispersion curves. When the frequency and plate thickness combination crosses the dispersion curve, a peak should occur. This

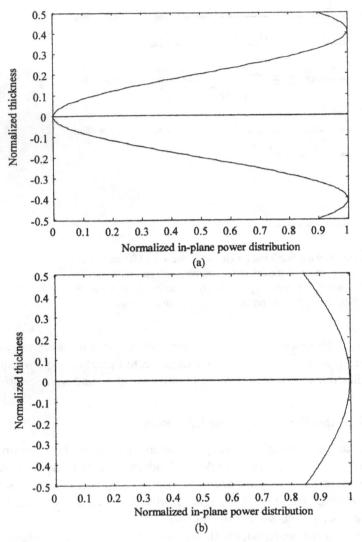

Figure E.4-16. Normalized in-plane power distributions for (a) A1 (2.5 MHz-mm) and (b) S0 (1.0 MHz-mm) as a function of defect depth.

is evident in the output of the tone-burst system for two different plate thicknesses (see Figure E.4-18). Since we know the phase velocity in the wedges, we can look at the dispersion curve to find the fd distance between two modes and then find the frequency distance from the tone-burst output. Dividing the fd distance by the frequency distance, we get the thickness of the plate, thus using guided waves for thickness measurement.

E.4.10   Exercises

1. How could a source influence and resulting phase velocity spectrum modify a theoretical phase velocity dispersion curve?

2. How are "leaky" Lamb waves produced?

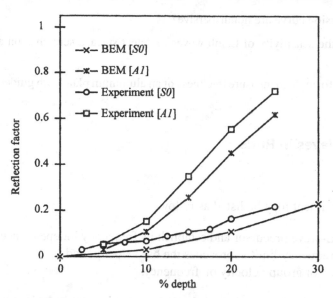

Figure E.4-17. Reflection factor variation for A1 (2.5 MHz-mm) and S0 (1.0 MHz-mm) as a function of percentage depth of the defect.

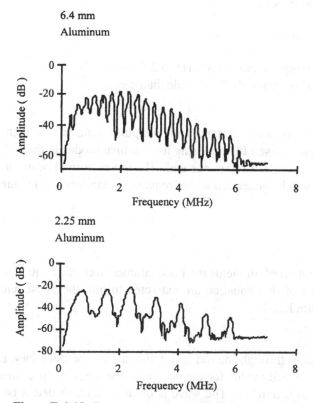

Figure E.4-18. Frequency sweep results for two aluminum plates of different thicknesses.

3. Which points on a dispersion curve are nondispersive?

4. How could you change the sensitivity of Lamb waves to detect surface scratches on a test specimen?

5. Develop an approximate formula to measure thickness of an aluminum plate with guided waves.

E.5 Experiment 5: Waves in Rods

E.5.1 Overview

Our objectives in this experiment may be listed as follows:

(1) to identify a possible L-wave precursor and establish its frequency independence;
(2) to show that the low-frequency limit approaches the bar velocity; and
(3) to show the dependence of group velocity on frequency.

The problem is to investigate the propagation of elastic harmonic waves in an infinite rod; our formulation is in cylindrical coordinates, assuming the z-axis is along the axis of the rod. Three different modes of vibration are possible: longitudinal, torsional, and flexural. For now, we deal with only the longitudinal modes in the solid cylindrical rod.

The following equipment will be needed:

(1) ultrasonic tone-burst pulser–receiver;
(2) digital oscilloscope;
(3) broadband transducers covering the range 100 kHz to 2.5 MHz;
(4) 1-ft-long solid steel rod test specimen (0.25" outside diameter);
(5) couplant, cables, etc.

The cylindrical specimen rod is fixed on a tripod stand, and both pulse-echo and through-transmission ultrasound techniques are used for exploring longitudinal modes in the rod. Normal beam incidence is used to excite the rod at one end. The use of normal beam incidence implies that many modes will be generated at the frequency–diameter (fd) value of excitation.

E.5.2 Results

The solution to our experimental problem yields the Pochhammer frequency equation for longitudinal modes. The roots of this equation are extracted to plot the phase and group velocity diagrams; see Figure E.5-1.

The L-Wave Precursor

Since a portion of the wave going through the rod is not affected by the boundary, it actually travels at the dilatational velocity value for the material – for the steel specimen used in these experiments, close to 6 mm/μs. The wave is often difficult to detect because of attenuation and its relative magnitude compared with that of the actual guided

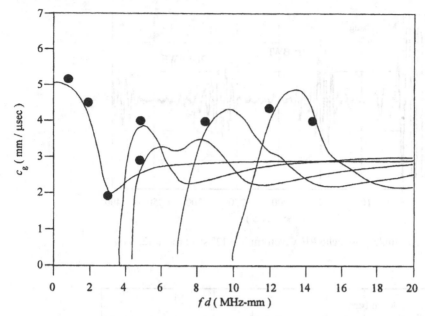

Figure E.5-1. Group velocity dispersion curve for a steel rod: ($c_L = 6.0$ mm/μs, $c_T = 3.1$ mm/μs); the heavy dots denote experimentally derived values.

wave modes in the rod. The group velocity of any guided wave modes in the rod is much less than 6 mm/μs. However, a large increase in gain will show the L-wave precursor. We can see that the precursor wave's position in the time domain doesn't change as the tone-burst frequency is changed, whereas the position of all other guided wave modes present *will* change with frequency. Observation can take place in either a pulse-echo or through-transmission mode. This precursor was observed in our laboratory for large gain near the 100-kHz frequency region and also at fd values near 12.

Low-Frequency Limit

The phase and group velocity diagrams show that, as the fd product approaches zero, the wave velocity approaches a limit value called the *bar wave velocity*. This is given by the formula $c_{bar} = \sqrt{E/\rho}$. For steel, $E = 2.1 \times 10^{11}$ N/m^2 and $\rho = 7{,}850$ kg/m^3, so the value for the bar velocity $c_{bar} = 5.17$ mm/μs. This may be verified by exciting the specimen (at as low a frequency as the transducer characteristics will permit) and then measuring the velocity. A typical result is shown in Figure E.5-2. The measured value for the group velocity is 5.16 mm/μs at an fd of 0.79, which is plotted on the group velocity curve in Figure E.5-1. The bar velocity is actually defined at the zero frequency limit shown in the curve for the S0 mode.

Frequency Dependence of Group Velocity

From theory and as shown in Figure E.5-1, the group velocity of the different modes are dependent on frequency. This can be observed by sweeping the frequency. The signal – which is a superposition of all of the modes at a particular fd value – changes position on

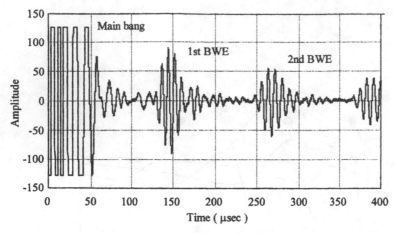

Figure E.5-2. Sample pulse echo RF waveform in a 12" steel rod at 125 kHz.

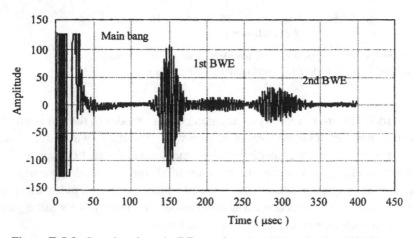

Figure E.5-3. Sample pulse echo RF waveform in a 12" steel rod at 300 kHz.

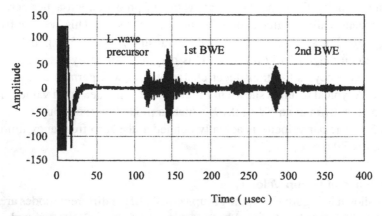

Figure E.5-4. Sample pulse echo RF waveform in a 12" steel rod at 1.9 MHz.

the oscilloscope, indicating that the group velocity is changing with frequency. The variation in group velocity can be clearly observed for small fd when there is only a single mode present and is often seen at higher fd values, even when multiple modes are present. Typical signals are presented for three different fd values in Figures E.5-2, E.5-3, and E.5-4 (note the L-wave precursor in Figure E.5-4). Additional experimental test points are plotted in Figure E.5-1.

E.5.3 Exercises

1. Provide an explanation for the fairly strong observation of an L-wave precursor for fd values near 0.5 and 12.

2. If multiple modes are present, how could you identify the particular modes? (This may call for changing the experimental apparatus or characteristics of the test specimen itself.)

3. Calculate the group velocity in Figures E.5-3 and E.5-4 and plot the values in Figure E.5-1. Use peak-to-peak measurements between the first and second backwall echoes (BWEs).

4. Estimate the acoustic zero points in Figure E.5-2, E.5-3, or E.5-4.

5. Explain a possible error source for the group velocity measurements shown in Figure E.5-1.

E.6 Experiment 6: Waves in Tubes

E.6.1 Overview

Our goal is to understand guided wave propagation possibilities in tubing. For axisymmetric waves, we shall cover:

(1) use of a specially designed conical element bore probe;
(2) observation of guided wave propagation in a 10-ft tube;
(3) observation of guided wave propagation in a U-bend tube;
(4) multiple defect detection in a 3-ft tube;
(5) observation of guided wave propagation in a liquid-loaded tube;
(6) resonance stage check via frequency sweeping; and
(7) circumferential location of a crack by using a segmented probe.

For non-axisymmetric waves, the emphasis will be on field distribution (a) along the distance between sender and receiver and (b) along the circumference of a tube.

Ultrasonic wave propagation in hollow cylinders has become a topic of great interest recently, primarily because of the millions of miles of tubing and piping that exist in our power generation and chemical processing facilities and the inspection requirements for safety and production efficiency. There are many experiments reported in the literature, and in this section we present a few examples of tubing inspection with guided waves. The required equipment is:

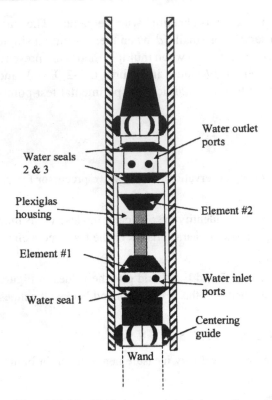

Figure E.6-1. Guided wave conical element bore probe.

(1) tone-burst system and digital oscilloscope;
(2) bore probe (conical shaped axisymmetric transducers designed for tubing inspec-
    tion – see Figure E.6-1); and
(3) partial circumferential loading transducers.

E.6.2   Results

Axisymmetric Waves

Figure E.6-2 displays the experimental results obtained by inserting the bore probe
shown in Figure E.6-1 into inconel tubing. A saw-cut defect (of about 10% through the
wall) is illustrated, along with the tube-end echoes. A study of successive backwall echoes
shows that the wave in this example easily travels a distance of 160 ft. Guided waves can
also travel around bent tubing, as shown in Figure E.6-3. Multiple defects can be found,
as illustrated in Figure E.6-4. In this case, all four defects were found with a single phase
velocity and frequency value. This may not always be possible, however; different phase
velocity and frequency values might be required to control mode selection and hence wave
structure. Tuning also might be necessary for each defect to produce its own wave resonance (see Shin and Rose 1998a).

Another interesting experiment is shown in Figure E.6-5. Note that the $L(0, 3)$ mode
disappears with water loading, owing to the out-of-plane displacement on the outside surface and energy leakage into the fluid. Dispersion curves as a reference for all of these

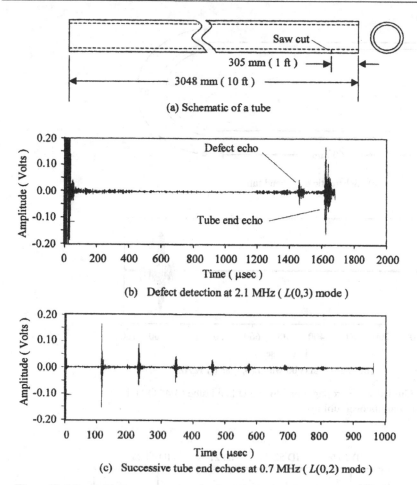

(a) Schematic of a tube

(b)   Defect detection at 2.1 MHz ($L(0,3)$ mode)

(c)   Successive tube end echoes at 0.7 MHz ($L(0,2)$ mode)

Figure E.6-2. Guided wave propagation in a 10-ft tube (3/4" O.D. by 0.0481" nominal wall inconel tubing).

experiments are illustrated in Figure E.6-6. For the bore probe of Figure E.6-1, the incident angle of 29° corresponds to a phase velocity of approximately 5.6 mm/μs. We therefore consider the frequency sweep shown in Figure E.6-7, where the mode crossings at positions a and b show up nicely. The bandwidth of excitation is associated with a source influence.

One additional experiment for axisymmetric loading is included here. Even though energy transmission is axisymmetric, the reflection from a defect is generally non-axisymmetric. As a result, a segmented receiver is required if one wishes to obtain an estimate of circumferential length. A sample result is shown in Figure E.6-8.

Non-Axisymmetric Waves

We can generate non-axisymmetric waves in tubing by using a partially loaded angle beam transducer, as illustrated in Figure E.6-9. Note the axial field variations along the tube in Figure E.6-10, as well as the circumferential distributions shown in Figure E.6-11. These waves seem to spiral down the tubing.

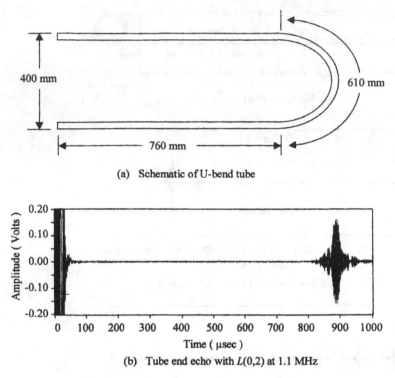

(a)   Schematic of U-bend tube

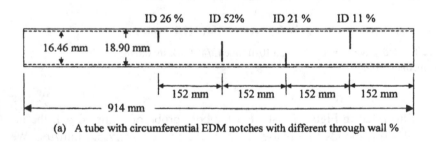

(b)   Tube end echo with $L(0,2)$ at 1.1 MHz

Figure E.6-3. Guided wave propagation along a U-bend tube (3/4" O.D. by 0.0481" nominal wall inconel tubing).

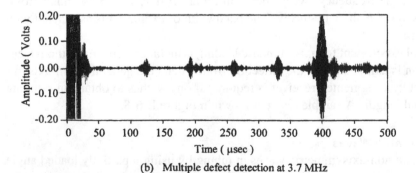

(a)   A tube with circumferential EDM notches with different through wall %

(b)   Multiple defect detection at 3.7 MHz

Figure E.6-4. Multiple crack detection in a 3-ft tube (3/4" O.D. by 0.0481" nominal wall inconel tubing).

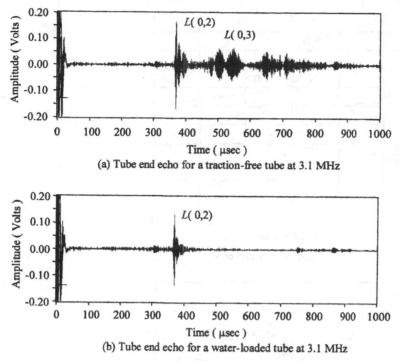

(a) Tube end echo for a traction-free tube at 3.1 MHz

(b) Tube end echo for a water-loaded tube at 3.1 MHz

Figure E.6-5. Traction-free and water-loaded test for ultrasonic guided wave propagation in an inconel tube (3/4" O.D. by 0.0481" nominal wall inconel tubing).

It is not always easy to generate the mode and frequency of choice. Since there are an infinite number of modes in hollow cylinders – each with its own propagating characteristics that might be useful in different applications – it is very important to use mode control and to understand the mode characteristics, particularly group velocities and wave structures. The parameters used to control mode generation are the frequencies, phase velocities, and applied loading conditions. Frequencies can be controlled by a tone-burst function generator; phase velocities can be controlled by using wedge techniques and by wedge angle. Element size and spacing must be controlled if we are using comb type transducers. Source design parameters (e.g., applied loading length and angle) can be used with the frequency and the phase velocity to control the mode excitation.

Figure E.6-12 shows theoretical and experimental results on the relationship between circumferential loading angle α and the amplitude of the generated guided wave modes. In this experiment we employed uniform pressure loading on an outer boundary of a sample hollow cylinder. For theoretical predictions of the amplitude factor as a function of the circumferential loading angle α, the normal mode expansion amplitude method was used (Ditri and Rose 1992). Figure E.6-12 shows results for α values of 360°, 180°, 90°, and 45°.

For 360° (axisymmetric) loading, all amplitude factors are zero except for the longitudinal modes, which means that none of the flexural modes (circumferential order > 0) can be generated except longitudinal modes (circumferential order = 0). Experimental and theoretical results indicate that 180° circumferential partial loading can generate an

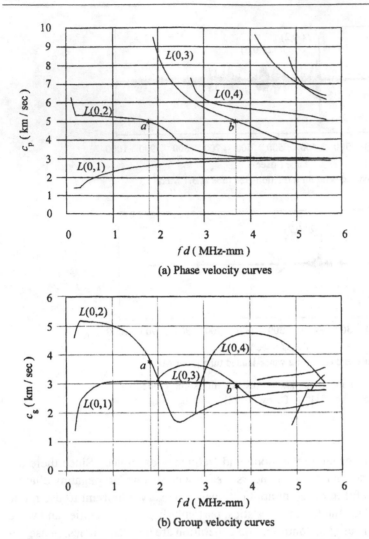

(a) Phase velocity curves

(b) Group velocity curves

Figure E.6-6. Theoretically calculated dispersion curves for (a) phase velocity and (b) group velocity, for type-304 stainless steel heat exchanger tubes (outer radius = 9.6 mm, tube wall thickness = 1.05 mm, longitudinal wave velocity = 5.7 mm/μs, shear wave velocity = 3.1 mm/μs).

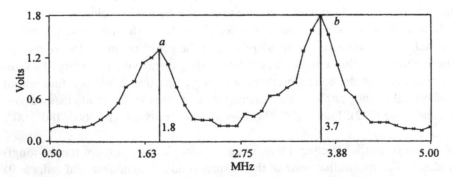

Figure E.6-7. Amplitude versus frequency curves showing resonance stage check for points a and b from theoretical dispersion curves.

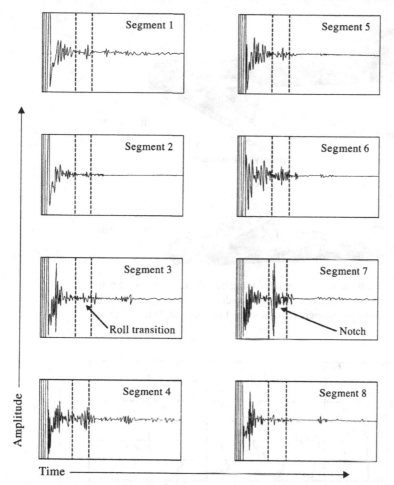

Figure E.6-8. Test results showing crack detection in roll transition area with a 20% through wall circular inside EDM (electro discharge machining) notch at 4.3 MHz, using a segmented probe from about 3 inches away.

axisymmetric mode well suited for tubing and piping inspection. If we decrease the loading angle to 90° then the amplitude factors of the flexural modes become larger.

In field applications, limited access to the inspected sample will almost always require non-axisymmetric loading conditions. Even for axisymmetric mode generation, the defects are usually non-axisymmetric and so the reflected signals consist of non-axisymmetric and axisymmetric modes both. Therefore, non-axisymmetric partial loading guided wave studies on hollow cylinders should be developed for advanced use; see Shin and Rose (1998b).

E.6.3   Exercises

1. For Figure E.6-5, wave structure figures are reported in Chapter 12. Explain the experimental results based on those curves.

2. Why does wave spiraling occur for partially loaded tubes around the circumference?

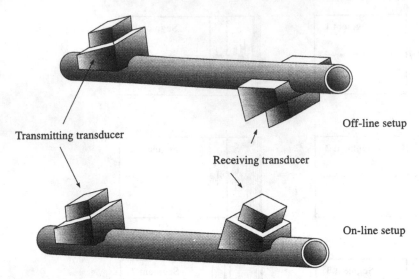

Transmitting transducer

Receiving transducer

Off-line setup

On-line setup

Figure E.6-9. Through-transmission partial loading transducer setups for the displacement field measurement, where a sending transducer is always placed on top of the tube. Specimen is a steam generator tube, and incident angle is 28° ($c_p = 5.73$ km/s).

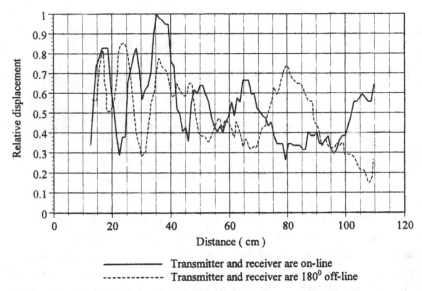

——————— Transmitter and receiver are on-line
- - - - - - - - - - - Transmitter and receiver are 180° off-line

Figure E.6-10. Displacement field variation with distance at 1.2 MHz and 28° incident angle ($c_p = 5.73$ km/s).

3. What minimum circumferential loading angle can be used to produce axisymmetric longitudinal waves in tubing?

E.7   Experiment 7: Horizontal Shear Waves

Our main goal in this experiment is to understand the generation, reception, and basic wave propagation characteristics of horizontal shear waves.

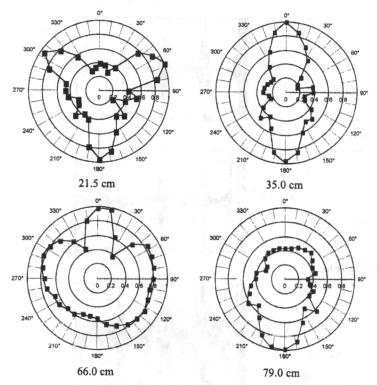

21.5 cm

35.0 cm

66.0 cm

79.0 cm

Figure E.6-11. Measured displacement field distributions along the circumference of the tube (3/4" O.D. by 0.0481" nominal wall inconel tubing).

E.7.1 Equipment

We shall use a tone-burst function generator system as well as an electromagnetic acoustic transducer (EMAT), which is capable of generating a variety of ultrasonic wave types. These include bulk longitudinal and shear waves, Lamb waves, and Rayleigh type surface waves. Specific transducer configurations and designs can be found in the literature for each particular wave type (see Ristic 1983). However, the basic underlying principles of ultrasonic wave generation by way of electromagnetic transduction remain the same. For ultrasonic wave reception, the wave in the metal produces an electric field that is sensed by the EMAT.

The operation of an EMAT relies on a current-carrying wire placed in a magnetic field **B**; this wire is then positioned near the surface of the metal in which the ultrasonic wave will propagate. Eddy currents **J** are induced in the metal and the magnetic field causes the material's electrons to deflect in the direction defined by the vector $\mathbf{J} \times \mathbf{B}$. The resultant body forces (Lorentz forces $\mathbf{F}_L$) generate ultrasonic energy. The ultrasonic wave type and direction are controlled by the magnetic field direction and the conducting wire configuration. Therefore, the EMAT design depends on the ultrasonic wave type desired.

As an example, consider the generation of both bulk longitudinal and bulk shear waves normal to the material surface. Longitudinal displacements are generated when the magnetic field is parallel to the surface, whereas shear displacements are generated when the magnetic field is perpendicular to the surface. The frequency of the displacements is

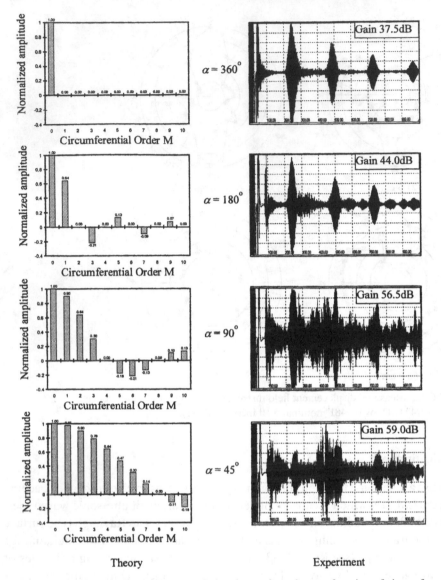

Figure E.6-12. Amplitude factors and experimental results as a function of circumferential loading angle α.

controlled by the applied voltage frequency used to generate current in the wire. The conducting wire is formed into planar elongated spiral coils positioned parallel to the specimen surface. Hence, a linearly polarized current distribution is induced in the specimen. The resulting Lorentz force vectors for longitudinal and shear waves are (respectively) perpendicular and parallel to the specimen surface. Figure E.7-1 is a schematic of the magnetic field and current configuration required for longitudinal and shear wave generation.

Because the coupling mechanism between the EMAT and the specimen are electromagnetic and not mechanical, the efficiency of EMATs is much lower than that of piezoelectric transducers. In fact, preamplifiers are required to increase the voltage levels obtained from the receivers so that the signal can be digitized using conventional ultrasonic equipment. This is the main disadvantage behind the use of EMATs. On the other hand, because the

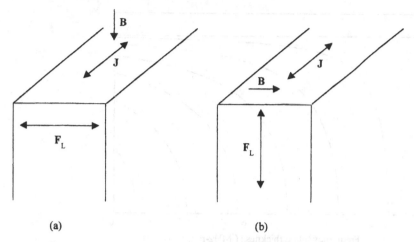

(a)   (b)

Figure E.7-1. Schematic describing the underlying principles behind the operation of EMAT transducers for generating (a) shear and (b) longitudinal waves. The direction of the eddy current **J** and the magnetic field **B** determines the Lorentz force (F_L) direction.

Shear Horizontal Dispersion Curves in Aluminum

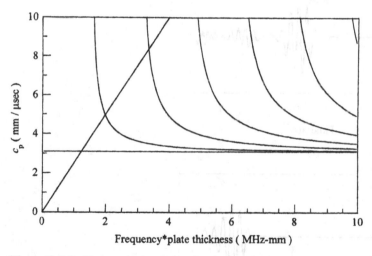

Figure E.7-2. Phase velocity curves for shear horizontal plate waves, showing the excitation line for a 6-mm wavelength EMAT.

coupling is electromagnetic, EMATs can be used as noncontact acoustic wave sources. This, in itself, promotes the use of EMATs for many applications where contact with the specimen is impossible. These applications usually involve specimens with rough or high-temperature surfaces.

E.7.2   Procedure

Using an aluminum plate and appropriate transducers, generate and measure horizontal shear wave velocities over various distances. Sample phase and group velocity curves are shown in Figures E.7-2 and E.7-3. Sample experimental results are shown in Figures E.7-4, E.7-5, and E.7-6.

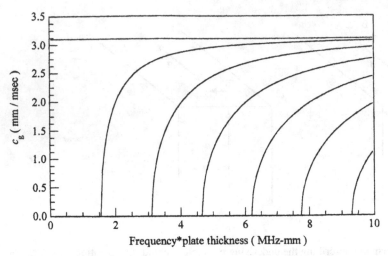

Figure E.7-3. Group velocity curves for shear horizontal plate waves.

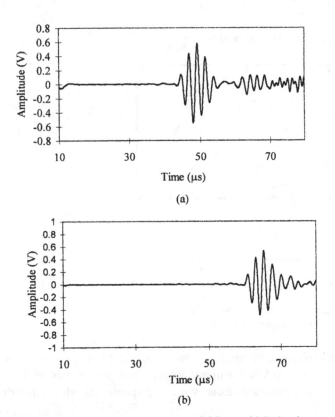

Figure E.7-4. SH_0 mode excited on 2.25-mm-thick aluminum plate with a 6-mm wavelength EMAT at 500 kHz ($fd = 1.12$ MHz-mm) and received by a shear wave transducer placed (a) 6 inches and (b) 8 inches away. The excitation point on the dispersion curves is shown in Figure E.7-2. The time of travel over the 2 inches (50 mm) is 16 μs, which yields a group velocity of 3.13 mm/μs; this value can be verified in Figure E.7-3.

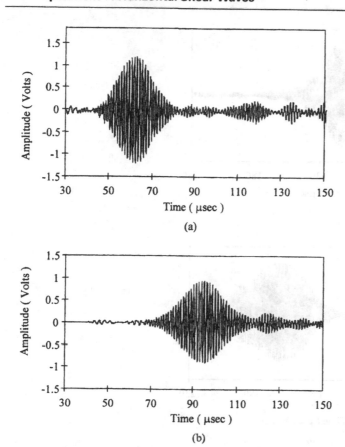

Figure E.7-5. SH_1 mode excited on 2.25-mm-thick aluminum plate with a 6-mm wavelength EMAT at 800 kHz ($fd = 1.8$ MHz-mm) and received by a shear wave transducer placed (a) 4 inches and (b) 6 inches away. The excitation point on the dispersion curves is shown in Figure E.7-2. The time of travel over the 2 inches (50 mm) is 32 μs, which yields a group velocity of 1.56 mm/μs; this value can be verified in Figure E.7-3.

E.7.3   Exercises

1. Verify the modes for the sample results shown in Figures E.7-4–E.7-6. Plot group velocities on Figure E.7-3.

2. Can shear horizontal plate waves travel at different velocities?

3. Is receiver angle important for proper reception of shear horizontal plate wave modes?

4. Is the polarity of the shear wave transducer important when receiving shear horizontal plate waves?

5. How can you assess the direction of shear wave polarization?

6. How can you generate both Lamb waves and horizontal shear plate waves with a wedge technique?

7. Are horizontal shear waves in a plate sensitive to water loading?

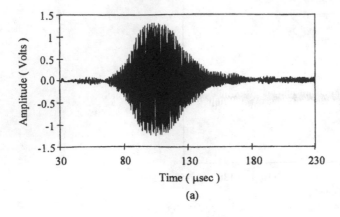

(a)

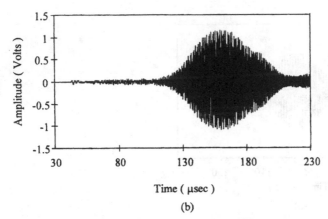

(b)

Figure E.7-6. SH$_2$ mode excited on 2.25-mm-thick aluminum plate with a 6-mm wavelength EMAT at 1.465 MHz ($fd = 3.3$ MHz-mm) and received by a shear wave transducer placed (a) 4 inches and (b) 6 inches away. The excitation point on the dispersion curves is shown in Figure E.7-2. The time of travel over the 2 inches (50 mm) is 60 μs, which yields a group velocity of 0.84 mm/μs; this value can be verified in Figure E.7-3.

Bibliography

Aboudi, J. (1981). Generalized effective stiffness theory for the modelling of fiber-reinforced composites, *Int. J. Eng. Sci.* 17(10): 1005–18.

Abo-Zena, A. (1979). Dispersion function computations for unlimited frequency values, *Geophys. J. Roy. Astr. Soc.* 58: 91–105.

Achenbach, J. D. (1976). Generalized continuum theories for directionally reinforced solids, *Arch. Mech.* 28(3): 257–78.

Achenbach, J. D. (1984). *Wave Propagation in Elastic Solids.* New York: North-Holland.

Achenbach, J. D. (1992). Mathematical modeling for quantitative ultrasonics, *Nondestr. Test. Eval.* 8/9: 363–77.

Achenbach, J. D., and Epstein, H. I. (1967). Dynamic interaction of a layer and a half space, *J. Eng. Mech. Division* 5: 27–42.

Achenbach, J. D., Gautesen, A. K., and McMaken, H. (1982). *Ray Methods for Waves in Elastic Solids.* Boston: Pitman.

Achenbach, J. D., and Keshava, S. P. (1967). Free waves in a plate supported by a semi-inifite continuum, *J. Appl. Mech.* 34: 397–404.

Adler, E. L. (1990). Matrix methods applied to acoustic waves in multilayers, *IEEE Trans. Ultrason. Ferroelec. Freq. Contr.* 37: 485–90.

Allen, D. R., and Cooper, W. H. B. (1983). A Fourier transform technique that measures phase delays between ultrasonic impulses with sufficient accuracy to determine residual stresses in metals, *NDT Int.* 16(4): 205–17.

Alleyne, D., and Cawley, P. (1995). The long range detection of corrosion in pipes using Lamb waves, in D. O. Thompson and D. E. Chimenti (Eds.), *Review of Progress in Quantitative Nondestructive Evaluation,* vol. 14, pp. 2073–80. New York: Plenum.

Alleyne, D., Lowe, M., and Cawley, P. (1996). The inspection of chemical plant pipework using Lamb waves: Defect sensitivity and field experience, in D. O. Thompson and D. E. Chimenti (Eds.), *Review of Progress in Quantitative Nondestructive Evaluation,* vol. 15, pp. 1859–66. New York: Plenum.

Al-Nassar, Y. N., Datta, S. K., and Shah, A. H. (1991). Scattering of Lamb waves by a normal rectangular strip weldment, *Ultrasonics* 29: 125–32.

Auld, B. A. (1990). *Acoustic Fields and Waves in Solids,* 2nd ed., vols. 1 and 2. Malabar, FL: Kreiger.

Auld, B. A., and Kino, G. S. (1971). Normal mode theory for acoustic waves and their application to the interdigital transducer, *IEEE Trans.* ED-18: 898–908.

Auld, B. A., and Tau, M. (1978). Symmetrical Lamb wave scattering at a symmetrical pair of thin slots, in *1977 IEEE Ultrasonic Sympos. Proc.,* vol. 61.

Baik, J. M., and Thompson, R. B. (1984). Ultrasonic scattering from imperfect interfaces: A quasi-static model, *J. Nondestr. Eval.* 4: 177–96.

Balasubramaniam, K. (1989). Guided plate waves and reflection factor for NDE of composite materials, Ph.D. thesis, Drexel University, Philadelphia.

Banerjee, P. K., Ahmad, S., and Manolis, G. D. (1987). Advanced elastodynamic analysis, in *Boundary Element Methods in Mechanics,* chap. 5, pp. 258–84. Amsterdam: North-Holland.

Barnes, R. B., and Burton, C. J. (1949). Visual methods for studying ultrasonic phenomena, *J. Appl. Phys.* 20: 286–94.

Basatskaya, L. V., and Ermolov, L. N. (1980). Theoretical study of ultrasonic longitudinal subsurface waves in solid media, *Defektoskopiya* 7: 58–65.

Bendec, F., Peretz, M., and Rokhlin, S. I. (1984). Ultrasonic Lamb wave method for sizing of spot welds, *Ultrasonics* 22(2): 78–84.

Beranek, L. L. (1990). *Acoustics.* New York: Acoustical Society of America, American Institute of Physics.

Bhatia, A. B. (1985). *Ultrasonic Absorption.* New York: Dover.

Birks, A. S., and Green, R. W. (Eds.) (1991). *Ultrasonic Testing* (Nondestructive Testing Handbook, vol. 7). Columbus, OH: American Society for Nondestructive Testing.

Bleistein, N., and Handelsman, R. A. (1986). *Asymptotic Expansion of Integrals.* New York: Dover.

Bond, L. J. (1990). Numerical techniques and their use to study wave propagation and scattering – A review, in *Elastic Waves and Ultrasonic Nondestructive Evaluation,* pp. 17–27. Amsterdam: North-Holland.

Born, M., and Wolf, E. (1965). *Principles of Optics,* 3rd ed. Oxford: Pergamon.

Bragar, A. M. B., and Herrmann, G. (1992). Floquet waves in anisotropic periodically layered composites, *J. Acoust. Soc. Am.* 91(3): 1211–27.

Bray, D. E. (1988). Application of the first higher-order (M_{21}) mode Rayleigh wave to the inspection of stainless steel overlays, in *New Directions in the Nondestructive Evaluation of Advanced Materials,* vol. 9, p. 73. Chicago: ASME.

Brebbia, C. A., Tells, J. C. F., and Wrobel, L. C. (1984). *Boundary Element Techniques.* Berlin: Springer-Verlag.

Brebbia, C. A., Umetani, S., and Trevelyan, J. (1985). Critical comparison of boundary element and finite element methods for stress analysis, in *BETECH 85* (Proceedings of the 1st Boundary Element Technology Conference), pp. 225–56. Adelaide: South Australian Institute of Technology.

Brekhovskikh, L. M. (1960). *Waves in Layered Media.* New York: Academic Press.

Cagle, C. V. (1972). *Handbook of Adhesive Bonding.* New York: McGraw-Hill.

Castagnede, B., Jenkins, J. T., Sachse, W., and Baste, S. (1990). Optimal determination of the elastic constants of composite materials from ultrasonic wave-speed measurements, *J. Appl. Phys.* 67(6): 2753–61.

Castaings, M., and Hosten, B. (1994). Delta operator technique to improve the Thomson-Haskell method stability for propagation in multilayered anisotropic absorbing plates, *J. Acoust. Soc. Am.* 95(4): 1931–41.

Cauchy, A. L. (1828). De la pression ou tension dans un systeme de points materiels, in *Exercices de mathematique,* vol. 3, p. 213.

Chadwick, P., and Currie, P. K. (1974). Stoneley wave at an interface between elastic crystals, *Q. J. Mech. Appl. Math.* 27: 497–503.

Chadwick, P., and Smith, G. D. (1977). Foundations of the theory of surface waves in anisotropic elastic materials, *Adv. Appl. Mech.* 17: 303–77.

Cheng, J. C., and Zhang, S. Y. (1966). Lamb wave modes propagating along arbitrary directions in an orthotropic plate, in D. O. Thompson and D. E. Chimenti (Eds.), *Review of Progress in Quantitative Nondestructive Evaluation,* vol. 15, pp. 253–9. New York: Plenum.

Chervinko, O. P., and I. K. Savchenkov (1986). Harmonic viscoelastic waves in a layer and in an infinite cylinder, *Sov. Appl. Mech.* 22: 1136.

Chimenti, D. E., and Nayfeh, A. H. (1986). Anomalous ultrasonic dispersion fluid-coupled, fibrous composite plates, *Appl. Phys. Lett.* 49: 492–3.

Chimenti, D. E., and Nayfeh, A. H. (1989). Ultrasonic leaky waves in a solid plate separating a fluid and vacuum, *J. Acoust. Soc. Am.* 85(2): 555–60.

Chimenti, D. E., Nayfeh, A. H., and Butler, D. L. (1982). Leaky waves on a layered half-space, *J. Appl. Phys.* 53(1): 170–6.

Chimenti, D. E., and Rokhlin, S. I. (1990). Relationship between leaky Lamb modes and reflection coefficient zeroes for a fluid-coupled elastic layer, *J. Acoust. Soc. Am.* 88(3): 1603–11.

Chinnery, P. A., Humphrey, V. F., and Beckett, C. (1997). The Schlieren image of two-dimensional ultrasonic fields and cavity resonances, *J. Acoust. Soc. Am.* 101(1): 250–6.

Cho Y. (1995). Defect detection and characterization potential with guided waves, Ph.D. thesis, Pennsylvania State University, University Park.

Cho, Y., Hongerholt, D. D., and Rose, J. L. (1997). Lamb wave scattering analysis for reflector characterization, *IEEE Trans. Ultrason. Ferroelec. Freq. Contr.* 44(1): 44–52.

Cho, Y., and Rose, J. L. (1996a). Guided waves in a water loaded hollow cylinder, *Nondestr. Test. Eval.* 12: 323–39.

Cho, Y., and Rose, J. L. (1996b). A boundary element solution for a mode conversion study on the edge reflection of Lamb waves, *J. Acoust. Soc. Am.* 99(4): 2097–2109.

Cho, Y., Rose, J. L., and Hongerholt, D. D. (1995). Defect characterization and sizing potential with guided waves: Theory, Paper presented at QNDE Conference (July 30 – August 4), Seattle.

Chou, P. C., Carleone, J., and Hsu, C. M. (1972). Elastic constants of layered media, *J. Comp. Mat.* 6: 80–93.

Chou, P. C., and Pagano, N. J. (1992). *Elasticity: Tensor, Dyadic, and Engineering Approaches.* New York: Dover.

Christensen, R. M. (1982). *Theory of Viscoelasticity. An Introduction,* 2nd ed. New York: Academic Press.

Chu L. L., Askar, A., and Cakmak, A. S. (1982). An approximate method for scattering in elastodynamics – The Born approximation II: SH waves in infinite and half-space, *Soil Dynamics and Earthquake Eng.* 1: 102–16.

Chu, Y. C., Degtyar, A. D., and Rokhlin, S. I. (1994). On determination of orthotropic material moduli from ultrasonic velocity data in non-symmetry planes, *J. Acoust. Soc. Am.* 95(6): 3191–3203.

Claeys, J. M., and Leroy, O. (1982). Reflection and transmission of bounded sound beam on half-spaces and through plates, *J. Acoust. Soc. Am.* 72: 585–90.

Cooper, H. F. (1967). Reflection and transmission of oblique plane waves at a plane interface between viscoelastic media, *J. Acoust. Soc. Am.* 42: 1065–9.

Cooper, R. M., and Naghdi, P. M. (1957). Propagation of nonaxially symmetric waves in elastic cylindrical shells, *J. Acoust. Soc. Am.* 29: 1365–73.

Coquin, G. A. (1964). Attenuation of guided waves in isotropic viscoelastic materials, *J. Acoust. Soc. Am.* 36: 1074.

Couchman, J. C., and Bell, J. R. (1978). Prediction, detection and characterization of a fast surface wave produced near the first critical angle, *Ultrasonics* 16: 272–4.

Cruse, T. A., and Rizzo, F. J. (1968). A direct formulation and numerical solution of the general transient elastodynamic problem: I, *J. Math. Anal. Appl.* 22: 244–59.

Datta, S. K., Al-Nassar, Y., and Shah, A. K. (1991). Lamb wave scattering by a surface-breaking crack in a plate, in D. O. Thompson and D. E. Chimenti (Eds.), *Review of Progress in Quantitative Nondestructive Evaluation,* vol. 10, pp. 97–104. New York: Plenum.

Datta, S. K., Ledbetter, K. M., Shindo, Y., and Shah, A. K. (1998a). Phase velocity and attenuation of plane elastic waves in a particle-reinforced composite medium, *Wave Motion* 10(2): 171–82.

Datta, S. K., Shah, A. K., Bratton, R. L., and Chakraborty, T. (1988b). Wave propagation in laminated composite plates, *J. Acoust. Soc. Am.* 83(6): 2020–6.

Datta, S. K., Shah, A. K., Chakraborty, T., and Bratton, R. L. (1988c). Wave propagation in laminated composite plates; anisotropy and interface effects, in A. K. Mal and T. C. T. Ting (Eds.), *Wave Propagation in Structural Composites* (AMD, vol. 90), pp. 39–52. New York: ASME.

Davies, B. (1985). *Integral Transforms and Their Applications,* 2nd ed. New York: Springer-Verlag.

Dayal, V., and Kinra, V. K. (1989). Leaky Lamb waves in an anisotropic plate. I: An exact solution and experiments, *J. Acoust. Soc. Am.* 85: 2268–76.

De Billy, M., and Quentin, G. (1984). Experimental investigation of reflection coefficients for lossy liquid–solid–liquid systems, *Ultrasonics* 22: 249–52.

Deputat, J. (1990). Application of the acoustoelastic effect in measurements of residual stresses, *Arch. Acoust.* 15(1/2): 69–92.

Deschamps, M., and Hosten, B. (1992). The effects of viscoelasticity on the reflection and transmission of ultrasonic waves by an orthotropic plate, *J. Acoust. Soc. Am.* 91: 2007–15.

Ditri, J. J. (1992). Some theoretical and experimental aspects of the generation of guided elastic waves using finite sources, Ph.D. thesis, Drexel University, Philadelphia.

Ditri, J. J. (1994a). On the determination of the elastic moduli of anisotropic media from limited acoustical data, *J. Acoust. Soc. Am.* 95(4): 1761–7.

Ditri, J. J. (1994b). On the propagation of pure plane waves in anisotropic media, *Appl. Phys. Lett.* 64(6): 701–3.

Ditri, J. J. (1994c). Utilization of guided elastic waves for the characterization of circumferential cracks in hollow cylinders, *J. Acoust. Soc. Am.* 96: 3769–75.

Ditri, J. J. (1997). Determination of nonuniform stresses in an isotropic elastic half space from measurements of the dispersion of surface waves, *J. Mech. Phys. Solids* 45(1): 51–66.

Ditri, J. J., and Hongerholt, D. (1996). Stress distribution determination in isotropic materials via inversion of ultrasonic Rayleigh wave dispersion data, *Int. J. Solids Structures* 33(17): 2437–51.

Ditri, J. J., and Rose, J. L. (1992). Excitation of guided elastic wave modes in hollow cylinders by applied surface tractions, *J. Appl. Phys.* 72(7): 2589–97.

Ditri, J. J., and Rose, J. L. (1993). An experimental study of the use of static effective modulus theories in dynamics problems, *J. Comp. Mat.* 27(9): 934–43.

Ditri, J. J., and Rose, J. L. (1994). Excitation of guided waves in generally anisotropic layers using finite sources, *ASME J. Appl. Mech.* 61: 330–8.

Ditri, J., Rose, J. L., and Chen, G. (1991). Mode selection guidelines for defect detection optimization using Lamb waves, in *Proceedings of the 18th Annual Review of Progress in Quantitative NDE,* vol. 11, pp. 2109–15. New York: Plenum.

Dominguez, J. (1985). Applications of boundary element methods in elastodynamics, in *BETECH 85* (Proceedings of the 1st Boundary Element Technology Conference), pp. 105–27. Adelaide: South Australian Institute of Technology.

Eder, J. E., and Rose, J. L. (1995). Composite cure evaluation using obliquely incident ultrasonic waves, in D. O. Thompson and D. E. Chimenti (Eds.), *Review of Progress in Quantitative Nondestructive Evaluation,* vol. 14, p. 1279. New York: Plenum.

Eringen, A. C., and Suhubi, E. S. (1975). *Linear Theory* (Elastodynamics, vol. 2). New York: Academic Press.

Every, A. G., and Sachse, W. (1990). Determination of the elastic constants of anisotropic solids from acoustic-wave group-velocity measurements, *Phys. Rev. B* 42(13): 8196–8205.

Every, A. G., Sachse, W., Kim, K. Y., and Niu, L. (1991). Determination of elastic constants of anisotropic solids from group velocity data, in D. O. Thompson and D. E. Chimenti (Eds.), *Review of Progress in Quantitative Nondestructive Evaluation,* vol. 10B, pp. 1663–7. New York: Plenum.

Ewing, W. M., Jardetsky, W. S., and Press, F. (1957). *Elastic Waves in Layered Media.* New York: McGraw-Hill.

Farnell, G. W. (1970). Properties of elastic surface waves, in W. P. Mason and R. N. Thurston (Eds.), *Physical Acoustics,* vol. 6, pp. 109–66. New York: Academic Press.

Fedorov, F. I. (1968). *Theory of Elastic Waves in Crystals.* New York: Plenum.

Fiorito, R., Madigosky, W., and Uberall, H. (1979). Resonance theory of acoustic waves interacting with an elastic plate, *J. Acoust. Soc. Am.* 66: 1857–66.

Flugge, W. (1975). *Viscoelasticity,* 2nd ed. Berlin: Springer-Verlag.

Folk, R., and Herczynski, A. (1986). Solutions of elastodynamic slab problems using a new orthogonality condition, *J. Acoust. Soc. Am.* 80(4): 1103–10.

Friedman, B. (1990). *Principles and Techniques of Applied Mathematics.* New York: Dover.

Fu, C. Y. (1946). Studies on seismic waves: II. Rayleigh waves in a superficial layer, *Geophys.* 11(1/4): 10–23.

Fung, Y. C. (1965). *Foundations of Solid Mechanics.* Englewood Cliffs, NJ: Prentice-Hall.

Gazis, D. C. (1959a). Three-dimensional investigation of the propagation of waves in hollow circular cylinders, I. Analytical foundation, *J. Acoust. Soc. Am.* 31(5): 568–73.

Gazis, D. C. (1959b). Three-dimensional investigation of the propagation of waves in hollow circular cylinders, II. Numerical results, *J. Acoust. Soc. Am.* 31(5): 573–8.

Good, M. S., and Van Fleet, L. G. (1986). Ultrasonic beam profiles in coarse-grained materials, Paper presented at the 8th International Conference on NDE in Nuclear Industry (December), Orlando.

Graff, K. F. (1991). *Wave Motion in Elastic Solids.* New York: Dover.

Green, A. E., and Zerna, W. (1992). *Theoretical Elasticity.* New York: Dover.

Green, W. A. (1982). Bending waves in strongly anisotropic elastic plates, *Q. J. Mech. Appl. Math.* 35: 485–507.

Grewal, D. S. (1996). Improved ultrasonic testing of railroad rail for transverse discontinuities in the rail head using higher-order Rayleigh (M_{21}) waves, *Mat. Eval.* 54(9): 983–6.

Gubernatis, J. E., Domany, E., Krumhansl, J. A., and Huberman, M. (1977). The Born approximation in the theory of the scattering of elastic waves by flaws, *J. Appl. Phys.* 48(7): 2812–19.

Gurtin, M. E. (1981). *Topics in Finite Elasticity.* Philadelphia: Society for Industrial and Applied Mathematics.

Hashin, Z. (1962). The elastic moduli of heterogeneous materials, *J. Appl. Mech.* 29(2): 143–50.

Haskell, N. A. (1953). The dispersion of surface waves on multilayered media, in G. D. Lauderback, H. Benioff, and J. B. Macelwane (Eds.), *Bulletin of the Seismological Society of America,* vol. 43, pp. 17–34. Berkeley: University of California Press.

Hayes, M. (1969). A simple statical approach to the measurement of the elastic constants in anisotropic media, *J. Mat. Sci.* 4: 10–14.

Hayes, M., and Rivlin, R. S. (1961a). Propagation of a plane wave in an isotropic elastic material subjected to pure homogeneous deformation, *Arch. Rational Mech. Anal.* 8(15): 15–22.

Hayes, M., and Rivlin, R. S. (1961b). Surface waves in deformed elastic materials, *Arch. Rational Mech. Anal.* 8: 358–80.

Heelan, P. A. (1953). On the theory of head waves, *Geophys.* 18: 871–6.

Helbig, K. (1984). Anisotropy and dispersion in periodically layered media, *Geophys.* 49(4): 364–73.

Henneke, E. G. II (1972). Reflection–refraction of a stress wave at a plane boundary between anisotropic media, *J. Acoust. Soc. Am.* 51: 210.

Herczynski, A., and Folk, R. (1989). Orthogonality conditions for the Pochammer–Chree modes, *Q. J. Mech. Appl. Math.* 42 (part 4): 523–36.

Hildebrand, F. B. (1976). *Advanced Calculus for Applications,* 2nd ed. Englewood Cliffs, NJ: Prentice-Hall.

Hirao, M., Fukuoka, H., and Hori, K. (1981). Acoustoelastic effect of Rayleigh surface wave in isotropic material, *ASME J. Appl. Mech.* 48: 119–24.

Hirao, M., Fukuoka, H., and Murakami, Y. (1992). Resonance acoustoelasticity measurement of stress in thin plates, *Res. Nondestr. Eval.* 4: 127–38.

Hirose, S., and Yamano, M. (1996). Scattering analysis and simulation for Lamb wave ultrasonic testing, in D. O. Thompson and D. E. Chimenti (Eds.), *Review of Progress in Quantitative Nondestructive Evaluation,* vol. 15, pp. 201–8. New York: Plenum.

Hochstadt, H. (1989). *Integral Equations.* New York: Wiley.

Hosten, B. (1991). Reflection and transmission of acoustic plane waves on an immersed orthotropic and viscoelastic solid layer, *J. Acoust. Soc. Am.* 89: 2745–52.

Hosten, B., Deschamps, M., and Tittmann, B. R. (1987). Inhomogeneous wave generation and propagation in lossy anisotropic solids. Application to the characterization of viscoelastic composite materials, *J. Acoust. Soc. Am.* 82(5): 1763–70.

Hueter, T. F., and Bolt, R. H. (1955). *Sonics Techniques for the Use of Sound and Ultrasound in Engineering and Science.* New York: Wiley.

Hughes, D. S., and Kelley, J. L. (1951). Second-order elastic deformation of solids, *Phys. Rev.* 92(5): 1145–9.

Husson, D. (1985). A perturbation theory for the acoustoelastic effect of surface waves, *J. Appl. Phys.* 57(5): 1562–8.

Husson, D., and Kino, G. S. (1982). A perturbation theory for acoustoelastic effects, *J. Appl. Phys.* 53(11): 7250–8.

Ingard, K. U. (1988). *Fundamentals of Waves and Oscillations.* Cambridge University Press.

Jackins, P. D., and Gaunaurd, G. C. (1986). Resonance acoustic scattering from stacks of bonded elastic plates, *J. Acoust. Soc. Am.* 80: 1762–75.

Jeong, P. Y. (1987). Ultrasonic characterization of centrifugally cast stainless steel, EPRI document no. NP-5246, Electric Power Research Institute, Palo Alto, CA.

Jeong, P. Y., and Rose, J. L. (1986). Ultrasonic wave propagation consideration for centrifugally cast stainless steel pipe inspection, Paper presented at the 8th International Conference on NDE in Nuclear Industry (December), Orlando.

Jiao, D., and Rose, J. L. (1991). An ultrasonic interface layer model for bond evaluation, *J. Adhesion Sci. Tech.* 5(8): 631–46.

John, F. (1955). *Plane Waves and Spherical Means Applied to Partial Differential Equations.* New York: Wiley.

Karim, M. R., Mal, A. K., and Bar-Cohen, Y. (1990). Determination of the elastic constants of composites through inversion of leaky Lamb wave data, in D. O. Thompson and D. E. Chimenti (Eds.), *Review of Progress in Quantitative Nondestructive Evaluation,* vol. 9. New York: Plenum.

Kawashima, K., Fujii, I., Yamamoto, A., Shimizu, Y., and Fukushima, A. (1997). Measurement of anisotropic elastic constants of sandwiched SiC/Ti composite by double transmission critical angle and leaky surface wave techniques, in D. O. Thompson and D. E. Chimenti (Eds.), *Review of Progress in Quantitative Nondestructive Evaluation,* vol. 16, pp. 1135–42. New York: Plenum.

Kennett, B. L. N. (1983). *Seismic Wave Propagation in Stratified Media.* Cambridge University Press.

Kino, C. S. (1987). *Acoustic Waves: Devices, Imaging and Digital Signal Processing.* Englewood Cliffs, NJ: Prentice-Hall.

Kinra, V. K., Jaminet, P. T., Zhu, C., and Iyer, V. R. (1994). Simultaneous measurement of the acoustical properties of a thin-layered medium: The inverse problem, *J. Acoust. Soc. Am.* 95: 3059.

Kinsler, L. E., Frey, A. R., Coppens, A. B., and Sanders, J. V. (1982). *Fundamentals of Acoustics.* New York: Wiley.

Kline, R. A. (1992). *Nondestructive Characterization of Composite Media.* Lancaster, PA: Technomic Publishing.

Kline, R., and Jiang, L. (1996). Using Rayleigh wave dispersion to characterize residual stresses, in D. O. Thompson and D. E. Chimenti (Eds.), *Review of Progress in Quantitative Nondestructive Evaluation,* vol. 15, pp. 1629–36. New York: Plenum.

Knopoff, L. (1956). Diffraction of elastic waves, *J. Acoust. Soc. Am.* 28(2): 217–29.

Knopoff, L. (1964). A matrix method for elastic wave problems, *Bull. Seism. Soc. Am.* 54: 431–8.

Kobayashi, S. (1987). Elastodynamics, in *Boundary Element Methods in Mechanics,* chap. 4, pp. 192–255. Amsterdam: North-Holland.

Kobayashi, S. (1988). Recent progress in BEM for elastodynamics, in *Boundary Elements* (Geomechanics, Wave Propagation and Vibrations, Computational Mechanics Publications, vol. 4). New York: Springer-Verlag.

Kolodner, I. I. (1966). Existence of longitudinal waves in anisotropic media, *J. Acoust. Soc. Am.* 40: 730–1.

Kolsky, H. (1963). *Stress Waves in Solids.* New York: Dover.

Koshiba, M., Karakida, S., and Suzuki, M. (1984). Finite element analysis of Lamb waves scattering in an elastic plate waveguide, *IEEE Trans. Sonics and Ultrason.* 31(1): 18–25.

Krautkramer, J., and Krautkramer, H. (1966). *Wekstoffprufung mit Ultraschall.* Berlin: Springer-Verlag.

Krautkramer, J., and Krautkramer, H. (1990). *Ultrasonic Testing of Materials,* 4th ed. New York: Springer-Verlag.

Kreyszig, E. (1962). *Advanced Engineering Mathematics.* New York: Wiley.

Kundu, T. (1988). On the nonspecular reflection of the bounded acoustic beams, *J. Acoust. Soc. Am.* 83: 18–24.

Kundu, T., and Mal, A. K. (1985). Elastic waves in multilayered solid due to a dislocation source, *Wave Motion* 7: 459–71.

Kupperman, D. S., Reimann, K. J., and Abrego-Lopez, J. (1987). Ultrasonic NDE of cast stainless steel, *NDT Int.* 20(3): 145–52.

Kwun, H., and Bartels, K. A. (1997). Magnetostrictive sensor (MsS) technology and its application, Paper presented at the Ultrasonic International Conference (June), Delft, Netherlands.

Lancaster, P., and Tismenetsky, M. (1985). *The Theory of Matrices,* 2nd ed. Orlando: Academic Press.

Lavrentyev, A. I., Degtyar, A. D., and Rokhlin, S. I. (1996). Absolute ultrasonic measurements of residual stresses, in D. O. Thompson and D. E. Chimenti (Eds.), *Review of Progress in Quantitative Nondestructive Evaluation,* vol. 15, pp. 1653–60. New York: Plenum.

Lee, Y. C., Kim, J. O., and Achenbach, J. D. (1995). Acoustic microscopy measurement of elastic constants and mass density, *IEEE Trans. Ultrason. Ferroelec. Freq. Contr.* 42: 253–64.

Lekhnitskii, S. G. (1981). *Theory of Elasticity of an Anisotropic Body.* Moscow: Mir.

Lewis, P. A., Temple, J. A. G., and Wickham, G. R. (1996). Elastic waves defraction at cracks in anisotropic materials, in D. O. Thompson and D. E. Chimenti (Eds.), *Review of Progress in Quantitative Nondestructive Evaluation,* vol. 15, pp. 41–8. New York: Plenum.

Li, H. U., and Negishi, K. (1994). Visualization of Lamb mode patterns in a glass plate, *Ultrasonics* 32(4): 243.

Li, Y., and Thompson, R. B. (1990). Influence of anisotropy on the dispersion characteristics of guided ultrasonic plate modes, *J. Acoust. Soc. Am.* 87(5): 1911–31.

Lindsay, R. B. (1968). *Mechanical Radiation.* New York: McGraw-Hill.

Littles, J. W., Jacobs, L. J., and Zureick, A. K. (1997). The ultrasonic measurement of elastic constants of structural FRP composites, in D. O. Thompson and D. E. Chimenti (Eds.), *Review of Progress in Quantitative Nondestructive Evaluation,* vol. 16, pp. 1807–14. New York: Plenum.

Love, A. E. H. (1926). *Some Problems of Geodynamics.* Cambridge University Press.

Love, A. E. H. (1944a). *Mathematical Theory of Elasticity,* 4th ed. New York: Dover.

Love, A. E. H. (1944b). *A Treatise on the Mathematical Theory of Elasticity.* New York: Dover.

Lowe, M. J. S. (1995). Matrix techniques for modeling ultrasonic waves in multilayered media, *IEEE Trans. Ultrason. Ferroelec. Freq. Contr.* 42(4): 525–42.

Mal, A. K. (1988a). Guided waves in layered solids with interface zones, *Int. J. Eng. Sci.* 26: 873–81.

Mal, A. K. (1988b). Wave propagation in layered composite laminates under periodical surface loads, *Wave Motion* 10: 257–66.

Mal, A. K., Gorman, M., and Prosser, W. (1992). Material characterization of composite laminates using low frequency plate wave dispersion data, in D. O. Thompson and D. E. Chimenti (Eds.), *Review of Progress in Quantitative Nondestructive Evaluation,* vol. 11B, pp. 1451–8. New York: Plenum.

Mal, A. K., and Lee, J. (1995). Scattering of elastic waves by multiple inclusions and cracks, in D. O. Thompson and D. E. Chimenti (Eds.), *Review of Progress in Quantitative Nondestructive Evaluation,* vol. 14, pp. 1303–10. New York: Plenum.

Mal, A. K., Lih, S. S., and Bar-Cohen, Y. (1994). Characterization of the elastic constants of unidirectional laminates using oblique-incidence pulsed data, in D. O. Thompson and D. E. Chimenti (Eds.), *Review of Progress in Quantitative Nondestructive Evaluation,* vol. 13, pp. 1149–56. New York: Plenum.

Mal, A. K., and Rajapakse, Y. O. S. (Eds.) (1990). *Impact Response and Elastodynamics of Composites* (AMD, vol. 116). New York: ASME.

Mal, A. K., and Ting, C. T. (Eds.) (1988). *Wave Propagation in Structural Composites* (AMD, vol. 90). New York: ASME.

Mal, A. K., Xu, P. C., and Bar-Cohen, Y. (1990). Leaky Lamb waves for the ultrasonic nondestructive evaluation of adhesive bonds, *J. Eng. Mat. Tech.* 112: 255–9.

Malvern, L. E. (1969). *Introduction to the Mechanics of a Continuous Medium.* Englewood Cliffs, NJ: Prentice-Hall.

Mason, W. P. (Ed.) (1973). *Physical Acoustics – Principles and Methods,* vols. 6 and 7 (1970), vol. 10 (1973). New York: Academic Press.

Meeker, T. R., and Meitzler, A. K. (1964). Guided wave propagation in elongated cylinders and plates, *Phys. Acoust.* 1 (part A): 111–67.

Meitzler, A. H. (1961). Mode coupling occurring in the propagation of elastic pulses in wires, *J. Acoust. Soc. Am.* 33: 435–45.

Meitzler, A. H. (1965). Backward-wave transmission on stress pulses in elastic cylinders and plates, *J. Acoust. Soc. Am.* 38: 835–42.

Merkulov, L. G. (1963). Ultrasonic waves in crystals, *Appl. Mat. Res.* 2: 231–40.

Miklowitz, J. (1978). *The Theory of Elastic Waves and Waveguides.* New York: North-Holland.

Mindlin, R. D. (1955). *An Introduction to the Mathematical Theory of Vibrations of Elastic Plates.* Fort Monmouth, NJ: U.S. Army Signal Corps Engineering Laboratories.

Mindlin, R. D. (1958). Vibrations of an infinite elastic plate at its cut-off frequencies, *Proc. 3rd U.S. Nat. Congr. Appl. Mech.* p. 225.

Mindlin, R. D., and McNiven, H. D. (1960). Axially symmetric waves in elastic rods, *J. Appl. Mech.* 27: 145–51.

Mindlin, R. D., and Medick, M. A. (1959). Extensional vibrations of elastic plates, *J. Appl. Mech.* 26: 561–9.

Morse, P. M., and Feshbach, H. (1953). *Methods of Theoretical Physics.* New York: McGraw-Hill.

Müller, D. E. (1956). A method for solving algebraic equations using an automatic computer, *Mathematical Tables and Other Aids to Computation* 10: 208–15.

Murnaghan, T. D. (1951). *Finite Deformation of an Elastic Solid.* New York: Wiley.

Musgrave, M. J. P. (1954). *On the Propagation of Elastic Waves in Aeolotropic Media.* Teddington, Middlesex: National Physical Laboratory.

Musgrave, M. J. P. (1959). *The Propagation of Elastic Waves in Crystals and Other Anisotropic Media.* Teddington, Middlesex: National Physical Laboratory.

Musgrave, M. J. P. (1970). *Crystal Acoustics.* San Francisco: Holden-Day.

Nagy, P. B., and Adler, L. (1989). Nondestructive evaluation of adhesive joints by guided waves, *J. Appl. Phys.* 66: 4658–63.

Nagy, P. B., and Rose, J. H. (1993). Surface roughness and the ultrasonic detection of subsurface scatterers, *J. Appl. Phys.* 73: 556.

Nayfeh, A. H. (1995). *Wave Propagation in Layered Anisotropic Media with Applications to Composites.* Amsterdam: North-Holland.

Nayfeh, A. H., and Chimenti, D. E. (1988). Propagation of guided waves in fluid-coupled plates of fiber-reinforced composite, *J. Acoust. Soc. Am.* 83: 1736–43.

Nayfeh, A. H., Taylor, T. W., and Chimenti, D. E. (1988). Theoretical wave propagation in multilayered orthotropic media, in A. K. Mal and T. C. T. Ting (Eds.), *Wave Propagation in Structural Composites* (AMD, vol. 90), pp. 17–27. New York: ASME.

Nemat-Nasser, S., and Minagawa, S. (1977). On harmonic waves in layered composites, *J. Appl. Mech.* 44: 689–95.

Nemat-Nasser, S., and Yamada, M. (1981). Harmonic waves in layered transversely isotropic composites, *J. Sound and Vib.* 79(2): 161–70.

Neubauer, W. G. (1973). Observation of acoustic radiation from plane and curved surfaces, in W. P. Mason and R. N. Thurston (Eds.), *Physical Acoustics,* vol. 10, pp. 61–126. New York: Academic Press.

Ngoc, T. D. K., and Mayer, W. G. (1980). A general description of ultrasonic nonspecular reflection and transmission effects for layered media, *IEEE Trans. Sonics and Ultrason.* 27(5): 229–36.

Nikiforov, L. A., and Kharitonov, A. V. (1981). Parameters of longitudinal subsurface waves excited by angle-beam transducers, *Defektoskopiya* 6: 80–5.

Norris, A. (1988). On the acoustic determination of the elastic moduli of anisotropic solids and acoustic conditions for the existence of symmetry planes, *Q. J. Mech. Appl. Math.* 42 (part 3): 413–26.

Nowacki, W. (1963). *Dynamics of Elastic Systems.* New York: Wiley.

Ogden, R. W., and Sotiropoulos, D. A. (1995). Interfacial waves in pre-stressed layered incompressible elastic solids, *Proc. R. Soc. London A* 450: 319–41.

Ogilvy, J. A. (1986). Ultrasonic beam profiles and beam propagation in an austenitic weld using a theoretical ray tracing model, *Ultrasonics* 24(6): 337–47.

Onoe, M. A. (1955). *A Study of the Branches of the Velocity–Dispersion Equations of Elastic Plates and Rods* (Report of the Joint Committee on Ultrasonics of the Institute of Electrical Communication Engineers and the Acoustical Society of Japan).

Onoe, M. A., McNiven, H. D., and Mindlin, R. D. (1962). Dispersion of axially symmetric waves in elastic rods, *J. Appl. Mech.* 29: 729–34.

Pain, H. J. (1993). *The Physics of Vibrations and Waves.* New York: Wiley.

Pao, Y. H., and Mindlin, R. D. (1960). Dispersion of flexural waves in an elastic circular cylinder, *J. Appl. Mech.* 27: 513–20.

Pao, Y. H., and Mow, C. C. (1973). *Diffraction of Elastic Waves and Dynamic Stress Concentrations.* New York: Crane-Russak.

Papadakis, E. P., T. Patton, Y. Tsai, D. O. Thompson, and R. B. Thompson (1991). The elastic moduli of a thick composite as measured by ultrasonic bulk wave pulse velocity, *J. Acoust. Soc. Am.* 89(6): 2753–7.

Pelts, S. P., Menon, S. M., and Rose, J. L. (1996). Guided wave mode propagation influence for a shear horizontal wave source in an anisotropic viscoelastic layer, in D. O. Thompson and D. E. Chimenti (Eds.), *Review of Progress in Quantitative Nondestructive Evaluation,* vol. 15, pp. 223–30. New York: Plenum.

Pelts, S. P., and Rose, J. L. (1996). Source influence parameters on elastic guided waves in an orthotropic plate, *J. Acoust. Soc. Am.* 99(4): 2124–9.

Perlis, S. (1991). *Theory of Matrices.* New York: Dover.

Pierce, A. D. (1989). *Acoustics: An Introduction to its Physical Principles and Applications.* Woodbury, NY: Acoustical Society of America.

Pilarski, A. (1983). The coefficient of reflection of ultrasonic waves from an adhesive bond interface, *Arch. Acoust.* 8: 41–54.

Pilarski, A. (1995). Ultrasonic evaluation of the degree in layered joints, *Mat. Eval.* 43: 765–70.

Pilarski, A., Ditri, J. J., and Rose, J. L. (1993). Remarks on symmetric Lamb waves with dominant longitudinal displacements, *J. Acoust. Soc. Am.* 93(4) (part 1): 2228–30.

Pilarski, A., and Rose, J. L. (1988a). A transverse-wave ultrasonic oblique-incidence technique for interfacial weakness detection in adhesive bonds, *J. Appl. Phys.* 63(2): 300–7.

Pilarski, A., and Rose, J. L. (1988b). Ultrasonic oblique incidence for improved sensitivity interface weakness determination, *NDT Int.* 21(4): 241–6.

Pilarski, A., and Rose, J. L. (1989). Utility of subsurface longitudinal waves in composite material characterization, *Ultrasonics* 27: 226–33.

Pilarski, A., Rose, J. L., and Balasubramaniam, K. (1990). The angular and frequency characteristics of reflectivity from a solid layer embedded between two solids with imperfect boundary conditions, *J. Acoust. Soc. Am.* 87(2): 532–42.

Pilarski, A., Szelazek, J., Deputat, J., Ditri, J., and Rose, J. L. (1992). High-frequency Lamb modes for ultrasonic tensometry, in C. Hallai and P. Kulcsar (Eds.) *Non-Destructive Testing 92* (Proceedings of 13th World Conference on NDT), pp. 1044–8. Amsterdam: Elsevier.

Pitts, L. E., Plona, T. J., and Mayer, W. G. (1977). Theory of nonspecular reflection effects for an ultrasonic beam incident on a solid plate in a liquid, *IEEE Trans. Sonics and Ultrason.* 24: 101–9.

Plona, T. J., Baharavesh, M., and Mayer, W. G. (1975). Rayleigh and Lamb waves at liquid–solid boundaries, *Ultrasonics* 13(4): 171–4.

Pollard, H. F. (1977). *Sound Waves in Solids.* London: Pion Ltd.

Qu, J., Berthelot, Y., and Li, Z. (1996). Dispersion of guided circumferential waves in a circular annulus, in D. O. Thompson and D. E. Chimenti (Eds.), *Review of Progress in Quantitative Nondestructive Evaluation,* vol. 15, pp. 169–76. New York: Plenum.

Rajana, K. M., Cho, Y., and Rose, J. L. (1996). Utility of Lamb waves for near-surface crack detection, in D. O. Thompson and D. E. Chimenti (Eds.), *Review of Progress in Quantitative Nondestructive Evaluation,* vol. 15, pp. 247–52. New York: Plenum.

Rayleigh, J. W. S. (1945). *The Theory of Sound.* New York: Dover.

Redwood, M. (1960). *Mechanical Waveguides.* New York: Pergamon.

Ristic, V. M. (1983). *Principles of Acoustic Devices.* New York: Wiley.

Roberts, R. A., and Kupperman, D. S. (1987). Ultrasonic beam distortion in transversely isotropic media, in D. O. Thompson and D. E. Chimenti (Eds.), *Review of Progress in Quantitative Nondestructive Evaluation,* vol. 7A, pp. 49–56. New York: Plenum.

Rokhlin, S. I. (1980). Diffraction of Lamb waves by a finite crack in an elastic layer, *J. Acoust. Soc. Am.* 67(4): 1157–65.

Rokhlin, S. I. (1981). Resonance phenomena of Lamb waves scattering by a finite crack in a solid layer, *J. Acoust. Soc. Am.* 69(4): 922–8.

Rokhlin, S. I., Bolland, T. K., and Adler, L. (1986a). Reflection and refraction of elastic waves on a plane interface between two generally anisotropic media, *J. Acoust. Soc. Am.* 79(4): 906–18.

Rokhlin, S. I., and Chimenti, D. E. (1990). Reconstruction of elastic constants from ultrasonic reflectivity data in a fluid coupled composite plate, in D. O. Thompson and D. E. Chimenti (Eds.), *Review of Progress in Quantitative Nondestructive Evaluation,* vol. 9, pp. 1411–18. New York: Plenum.

Rokhlin, S. I., Hefets, M., and Rosen, M. (1981). An ultrasonic interface wave method for predicting the strength of adhesive bonds, *J. Appl. Phys.* 52: 2847–51.

Rokhlin, S. I., Lewis, D. K., Graff, K. F., and Adler, L. (1986b). Real-time study of frequency dependence of attenuation and velocity of ultrasonic waves during the curing reaction of epoxy resin, *J.
Acoust. Soc. Am.* 79: 1786–93.

Rokhlin, S. I., and Marom, D. (1986). Study of adhesive bonds using low-frequency obliquely incident
ultrasonic waves, *J. Acoust. Soc. Am.* 80: 245–58.

Rokhlin, S. I., and Wang, W. (1992). Double through-transmission bulk wave method for ultrasonic phase
velocity measurement and determination of elastic constants of composite materials, *J. Acoust. Soc.
Am.* 91(6): 3303–12.

Rokhlin, S. I., and Wang, Y. J. (1991a). Analysis of boundary conditions for elastic wave interaction with
an interface between two solids, *J. Acoust. Soc. Am.* 89: 503–15.

Rokhlin, S. I., and Wang, Y. J. (1991b). Equivalent boundary conditions for thin orthotropic layer between two solids, reflection, refraction and interface waves, *J. Acoust. Soc. Am.* 89: 503–15.

Rokhlin, S. I., Wu, C. Y., and Wang, L. (1990). Application of coupled ultrasonic plate modes for elastic
constant reconstruction of anisotropic materials, in D. O. Thompson and D. E. Chimenti (Eds.), *Review of Progress in Quantitative Nondestructive Evaluation,* vol. 9, pp. 1403–10. New York: Plenum.

Rose, J. H., and Richardson, J. M. (1982). Time domain Born approximation, *J. Nondestr. Eval.* 3(1):
45–53.

Rose, J. L. (1975). Ultrasonic field analysis and approximation parameters, *Brit. J. Nondestr. Test.* 14(4):
109–13.

Rose, J. L., Balasubramaniam, K., and Tverdokhlebov, A. (1989a). A numerical integration Green's function model for ultrasonic field profiles in mildly anisotropic media, *J. Nondestr. Eval.* 8(3): 165–79.

Rose, J. L., Cho, Y., and Ditri, J. J. (1994a). Cylindrical guided wave leakage due to liquid loading, in
D. O. Thompson and D. E. Chimenti (Eds.), *Review of Progress in Quantitative Nondestructive
Evaluation,* vol. 13, pp. 259–66. New York: Plenum.

Rose, J. L., Ditri, J. J., Huang, Y., Dandekar, D., and Chou, S.-C. (1991). One sided ultrasonic inspection
technique for elastic constant determination, *J. Nondestr. Eval.* 10(4): 159–66.

Rose, J. L., Ditri, J., and Pilarski, A. (1994b). Wave mechanics in acousto-ultrasonic nondestructive evaluation, *J. Acoust. Emission* 12(1/2): 23–6.

Rose, J. L., Ditri, J. J., Pilarski, A., Zhang, J., Carr, F. T., and McNight, A. (1992). A guided wave inspection technique for nuclear steam generator tubing, in C. Hallai and P. Kulcsar (Eds.) *Non-Destructive
Testing 92* (Proceedings of 13th World Conference on NDT), pp. 191–5. Amsterdam: Elsevier.

Rose, J. L., and Goldberg, B. B. (1979). *Basic Physics in Diagnostic Ultrasound.* New York: Wiley.

Rose, J. L., Jiao, D., Pelts, S. P., Barshinger, J. N., and Quarry, M. J. (1997a). Hidden corrosion detection
with guided waves, Paper no. 292, NACE International Corrosion 97 (March 10–14), New Orleans.

Rose, J. L., Jiao, D., and Spanner, J., Jr. (1996). Ultrasonic guided wave NDE for piping, *Mat. Eval.*
54(11): 1310–13.

Rose, J. L., and Meyer, P. A. (1975). Model for ultrasonic field analysis in solids, *J. Acoust. Soc. Am.*
57: 598–605.

Rose, J. L., Mortimer, R. W., and Chou, P. C. (1972). Applications of dynamic photoelasticity in flaw
detection analysis, *Mat. Eval.* 30(11): 242.

Rose, J. L., Nayfeh, A., and Pilarski, A. (1989b). Surface waves for material characterization, *J. Appl.
Mech.* 57: 7–11.

Rose, J. L., Pelts, S. P., and Quarry, M. (1997b). A comb transducer model for guided wave NDE, *Ultrasonics* 36(1/5): 163–8.

Rose, J. L., Pilarski, A., Balasubramanian, K., Dale, J., and Diprimeo, D. (1988). Wave scattering and
guided wave considerations in anisotropic media, in D. O. Thompson and D. E. Chimenti (Eds.),
Review of Progress in Quantitative Nondestructive Evaluation, vol. 7A, p. 88. New York: Plenum.

Rose, J. L., Pilarski, A., and Huang, Y. (1990). Surface wave utility in composite material characterization, in *Research in Nondestructive Evaluation,* vol. 1, pp. 247–65. New York: Springer-Verlag.

Rose, J. L., and Shin, H. J. (1998). Ultrasonic guided waves for finned tubing evaluation, Paper presented
at the 7th Annual Research Symposium, ASNT Spring Conference (March 23–27), Anaheim, CA.

Rose, J. L., Zhu, W., and Zaidi, M. (1998). Ultrasonic NDE of titanium diffusion bonding with guided
waves, *Mat. Eval.* 56(4): 535–9.

Rousseau, M., and Gatignol, Ph. (1986). Asymptotic analysis of nonspecular effects for the reflection and the transmission of a Gaussian acoustic beam incident on a solid plate, *J. Acoust. Soc. Am.* 80: 325–33.

Sachse, W. (1974). Measurement of the elastic moduli of continuous-filament and eutectic composite materials, *J. Comp. Mat.* 8: 378–90.

Scholte, J. G. (1942). On the Stoneley wave equation, *Proc. Kon. Nederl. Akad. Wetensch.* 45: 20–5, 159–64.

Schwab, F. A. (1970). Surface-wave dispersion computations: Knopoff's method, *Bull. Seism. Soc. Am.* 60: 1491–1520.

Scott, R. A., and Miklowitz, J. (1967). Transient elastic waves in anisotropic plates, *J. Appl. Mech.* 34: 104–10.

Segal, E., and Rose, J. L. (1980). Nondestructive testing of adhesive bond joints, in R. S. Sharp (Ed.), *Research Techniques in Nondestructive Testing,* chap. 8. London: Academic Press.

Serway, R. A. (1990). *Physics for Scientists and Engineers with Modern Physics,* 3rd ed., chap. 41. Philadelphia: Saunders.

Sezawa, K. (1927). Dispersion of elastic waves propagated on surface of stratified bodies and curved surfaces, *Bull. Earthquake Res. Inst., Univ. Tokyo* 3: 1–8.

Shah, A. H., and Datta, S. K. (1982). Harmonic waves in a periodically laminated medium, *Int. J. Solids Structures* 18: 397.

Shin, H. J., and Rose, J. L. (1998a). Guided wave tuning principles for defect detection in tubing, *J. Nondestr. Eval.* 17(1): 27–36.

Shin, H. J., and Rose, J. L. (1998b). Non-axisymmetric ultrasonic guided waves in pipes, Paper presented at the 7th Annual Research Symposium, ASNT Spring Conference (March 23–27), Anaheim, CA.

Silk, M. G., and Bainton, K. P. (1979). The propagation in metal tubing of ultrasonic wave modes equivalent to Lamb waves, *Ultrasonics* 17: 11–19.

Simmons, J. A., Dreswcher-Krasicka, E., and Wadley, H. N. G. (1992). Leaky axisymmetric modes in infinite clad rods, *J. Acoust. Soc. Am.* 92(2): 1061–90.

Singh, G. P., and Rose, J. L. (1982). A simple model for computing ultrasonic beam behavior of broadband transducers, *Mat. Eval.* 40(7): 880–5.

Sneddon, I. N. (1995). *Fourier Transforms.* New York: Dover.

Sokolnikoff, I. S. (1956). *Mathematical Theory of Elasticity.* New York: McGraw-Hill.

Solie, L. P., and Auld, B. A. (1973). Elastic waves in free anisotropic plates, *J. Acoust. Soc. Am.* 54: 1.

Sommerfeld, A. (1964). *Mechanics of Deformable Bodies.* New York: Academic Press.

Sotiropoulos, D. A., and Sifniotopoulos, C. G. (1995). Interfacial waves in pre-stressed incompressible elastic interlayers, *J. Mech. Phys. Solids* 43(3): 365–87.

Spetzler, H., and Datta, S. (1990). Experimental flaw detection by scattering of plate waves, in S. K. Datta, J. D. Achenbach, and Y. S. Rajapakse (Eds.), *Elastic Waves and Ultrasonic Nondestructive Evaluation,* pp. 135–41. Amsterdam: Elsevier.

Stokes, G. G. (1876). Smith's prize examination, Cambridge. [Reprinted 1905 in *Mathematics and Physics Papers,* vol. 5, p. 362, Cambridge University Press.]

Stoneley, R. (1924). Elastic waves at the surface of separation of two solids, *Proc. Roy. Soc. London* 106: 416–28.

Talbot, R. J., and Przemieniecki, J. S., Finite element analysis of frequency spectra for elastic waveguides, *Int. J. Solids Structures* 11: 115–38.

Tamm, K., and Weiss, O. (1961). Wellenausbreitung in unbegrenzten Scheiben und in Scheibenstreifen, *Acustica* 11: 8–17.

Tattersall, H. G. (1973). The ultrasonic pulse-echo technique as applied to adhesion testing, *J. Phys. D* 6: 819–32.

Tauchert, T. R., and Guzelsu, A. N. (1972). An experimental study of dispersion of stress waves in a fiber reinforced composite, *J. Appl. Mech.* 39: 98–102.

Thompson, R. B., and Gray, T. A. (1983). A model relating ultrasonic scattering measurements through liquid–solid interfaces to unbounded medium scattering amplitudes, *J. Acoust. Soc. Am.* 74(4): 1279–90.

Thompson, R. B., Lee, S. C., and Smith, J. F. (1987). Relative anisotropies of plane waves and guided modes in thin orthotropic plates: Implication for texture characterization, *Ultrasonics* 25: 133–7.

Thompson, R. B., and Newberry, B. P. (1987). A model for the propagation of Gaussian beams in anisotropic media, in D. O. Thompson and D. E. Chimenti (Eds.), *Review of Progress in Quantitative Nondestructive Evaluation,* vol. 7A, pp. 31–9. New York: Plenum.

Thomson, W. T. (1950). Transmission of elastic waves through a stratified solid medium, *J. Appl. Phys.* 21: 89–93.

Timoshenko, S. P., and Goodier, J. N. (1987). *Theory of Elasticity,* 3rd. ed. New York: McGraw-Hill.

Tokuoka, T., and Iwashimizu, Y. (1968). Acoustical birefringence of ultrasonic waves in deformed isotropic elastic media, *Int. J. Solids Structures* 4: 383–9.

Toupin, R. A., and Bernstein, B. (1961). Sound waves in deformed perfectly elastic materials. Acoustoelastic effect, *J. Acoust. Soc. Am.* 33(2): 216–25.

Tverdokhlebov, A. (1983). On the acoustoelastic effect, *J. Acoust. Soc. Am.* 73(6): 2006–12.

Tverdokhlebov, A., and Rose, J. (1988). On Green's functions for elastic waves in anisotropic media, *J. Acoust. Soc. Am.* 83(1): 118–21.

Tverdokhlebov, A., and Rose, J. L. (1989). On application domain of Green's function approximation for the mild anisotropic media, *J. Acoust. Soc. Am.* 86(4): 1606–7.

Uberall, H. (1973). Surface waves in acoustics, in P. Mason and R. N. Thurston (Eds.), *Physical Acoustics,* vol. 10, pp. 1–60. New York: Academic Press.

Van Buskirk, W. C., Cowin, S. C., and Carter, R., Jr. (1986). A theory of acoustic measurement of the elastic constants of a general anisotropic solid, *J. Mat. Sci.* 21: 2759–62.

Varadan, V. K., and Varadan, V. V. (Eds.) (1980). *Acoustic, Electromagnetic and Elastic Wave Scattering – Focus on T0 Matrix Approach.* New York: Pergamon.

Verdict, G. S., Gien, P. H., and Burger, C. P. (1992). Finite element study of Lamb wave interactions with holes and through thickness defects in thin metal plates, in D. O. Thompson and D. E. Chimenti (Eds.), *Review of Progress in Quantitative Nondestructive Evaluation,* vol. 11, pp. 97–104. New York: Plenum.

Viktorov, I. A. (1967). *Rayleigh and Lamb Waves – Physical Theory and Applications.* New York: Plenum.

Weiss, O. (1959). Uber die Schallausbreitung in verlustbehafteten Median mit komplexem Schub und kompressions Modul, *Acustica* 9: 387–99.

Wu, R. S., and Aki, K. (Eds.) (1990). *Scattering and Attentuation of Seismic Waves.* Basel: Birkhäuser.

Yapura, C. L., and Kinra, V. (1995). Guided waves in a fluid–solid bilayer, *Wave Motion* 21: 35–46.

You, Z., Lord, W., and Ludwig, R. (1987). Numerical modeling of elastic wave propagation in anisotropic materials, in D. O. Thompson and D. E. Chimenti (Eds.), *Review of Progress in Quantitative Nondestructive Evaluation,* vol. 7A, pp. 23–30. New York: Plenum.

ZANLY (1992). In *FORTRAN Subroutines for Mathematical Applications,* vol. 2. Houston: IMSL.

Zemanek, J., Jr. (1972). An experimental and theoretical investigation of elastic wave propagation in a cylinder, *J. Acoust. Soc. Am.* 51(1) (part 2): 265–83.

Zhang, S. Y., Shen, J. Z., and Ying, C. F. (1988). The reflection of the Lamb wave by a free plate edge: Visualization and theory, *Mat. Eval.* 46: 638–41.

Zimmer, J. E., and Cost, J. R. (1970). Determination of the elastic constants of a unidirectional fiber composite using ultrasonic velocity measurements, *J. Acoust. Soc. Am.* 47(3): 795–803.

Index